DIAKONIA

DIAKONIA

Re-interpreting the Ancient Sources

JOHN N. COLLINS

New York   Oxford
OXFORD UNIVERSITY PRESS
1990

Oxford University Press
Oxford New York Toronto
Delhi Bombay Calcutta Madras Karachi
Petaling Jaya Singapore Hong Kong Tokyo
Nairobi Dar es Salaam Cape Town
Melbourne Auckland

and associated companies in
Berlin Ibadan

Copyright © 1990 by John N. Collins

Published by Oxford University Press, Inc.,
200 Madison Avenue, New York, NY 10016

Oxford is a registered trademark of Oxford University Press

All rights reserved. No part of this publication may be reproduced,
stored in a retrieval system, or transmitted, in any form or by any means,
electronic, mechanical, photocopying, recording or otherwise,
without the prior permission of the publisher.

Library of Congress Cataloging-in-Publication Data
Collins, John N. (John Neil), 1931–
Diakonia : Re-interpreting the Ancient Sources / John N. Collins.
p. cm.
Includes bibliographical references.
ISBN 0-19-506067-9
1. Service (Theology) 2. Clergy—Office.
3. Diakonia (The Greek word) I. Title.
BT738.4.C65 1990
262'.14—dc20   89-36612   CIP

1 3 5 7 9 8 6 4 2

Printed in the United States of America
on acid-free paper

To my wife Carolyn
with love

and in memory especially of
Richmond, Muswell Hill, Mecklenburgh Square
1971–1976

Preface

There was a time in 1970s when I would look at newly published theological theses and take good note of the length of years between presentation of the thesis and its publication. On one such book I recall a loving dedication to the scholar's children "without whom"—or some such words—"the book would have been written sooner." This too is the book of a thesis of many years ago. The research and writing were originally done in the Department of New Testament Studies in the University of London King's College under the astute and benign guidance of Christopher F. Evans, then head of the department, during the period October 1971 to October 1975. The thesis, "Διακονεῖν and Associated Vocabulary in Early Christian Tradition," was successfully examined in June 1976. The same month the thesis was accepted for publication with the request that the Greek be made more accessible to the nonspecialist—a word about this later—and that the sections of the thesis devoted to separate examinations of the Greek verb and the Greek nouns be combined around common themes for the purpose of eliminating repetition. The latter task has occupied me in subsequent years and proved so complex that in the main I could undertake it only in term holidays from teaching when I could enjoy some opportunity for extended periods of concentration.

The common themes I refer to are now in the form of chapters 5, 6, and 7 under the titles "Word," "Deed," and "House and Table." In the writing of these chapters the material has been enriched at different points by sources that either had not been available to me when I was working on the thesis or I had not been able to fit into the thesis for reasons of space. I might instance additional matter arising from the Greek novel, relating to the ancient banquet, to *The Testament of Job,* and to "the ministry of love." I have reworked chapter 8, which deals with an interpretation of Epictetus disputed between Dr. Dieter Georgi and me. Ironically, the day after I had completed the manuscript of the book I entered my theological bookshop in Melbourne's Mitcham to find on display the English translation of his *Die Gegner,* in the epilogue to which he takes issue with my

1974 article; I have entered a note or two to take up his points. In discussing the papyri in chapter 9—a difficult task once I was no longer within reach of a specialist library—I had to take account of Zbigniew Borkowski's *Une description topographique des immeubles à Panopolis,* which the author most kindly sent to me from New York in April 1976.

As well as these rearrangements and additions, a major piece of new writing is chapter 1. The thesis contained only fifteen pages sketching the issues of "diakonia" and church order discussed at length here; much more had indeed been written but, at the last minute and again for reasons of space, had to be discarded. This chapter developed in its present form in the course of a fellowship which the thesis attracted from the British Trust for Tantur at the Ecumenical Institute of Tantur near Bethlehem. The chapter reviews ideas that went into the making of contemporary views of ministry and ministries, and this is meant to give relevance to the breadth of research undertaken in part II of the book where a new enquiry begins, on an unprecedented scale, into the meaning of the ancient Greek words underlying early Christian perceptions about ministry. The pages may also be seen as an invitation to the reader to keep pace with the long enquiry—in effect, to sidestep the urge to jump ahead to summaries that are to be found in an appendix or to consult only the new interpretations of early Christian sources that are presented in part III. Read only in such ways, the study could give the impression here and there of seeking mainly to be provocative or of being idiosyncratic in places, whereas its claims are rooted in the linguistic and cultural matters teased out in part II.

A word about the use of the ancient Greek language. Since the matters at issue revolve around the meaning of a small set of Greek words, both writer and reader have no option but to tangle with the original language in ancient sources of various kinds. If readers are to feel confident that the writer is not leading them astray, they will want to enter as closely as possible into the process of interpreting these sources. For those who read Greek this does not present a problem and could indeed be a stimulating exercise, even if I seem to spell things out in more detail than is necessary. The spelling out in such detail is of course not for them but for the larger number of readers who do not read Greek but who may be directly affected by issues raised. These readers, I suggest, would benefit from familiarising themselves with at least the main Greek words that recur over and over: διακονία (transliterated and pronounced *diakoni'a*), διάκονος *(dia'konos),* and διακονεῖν *(diakonein');* if readers go to this trouble, they will find themselves not just gliding over meaningless symbols but actually reading words and entering that much more closely into the ancient Greek mind. (A note on endings to Greek words: the three words listed are, in order, an abstract noun ending in *-ia,* a common noun ending in *-os,* and a verb ending in *-ein;* these are the endings of the words in what we might call their pure state—nominative case for the nouns and the infinitive form for the verb—whereas in the living language the endings vary enormously.) An appendix provides a key to the Greek alphabet, and some readers may wish to take advantage of that for the purpose of reading other words and phrases. In any case, with very few exceptions—and reasons for the exceptions are given in such instances—everything in Greek, and indeed almost everything from other non-English sources, is translated. Unless otherwise attributed, the translation is my own.

One other way of facilitating closer study of the work is to make use of the index of sources. Very often in the run of discourse or argument, references to various ancient sources are provided in brackets in the main text or in footnotes; such sources are cited in support or in illustration almost only when they are themselves discussed elsewhere in the book. To confirm their relevance and weight, the reader will need to consult the index for referral to pages where each is examined.

The reader may be interested to know that the research was not undertaken with questions of church, ministry, or ecumenism in mind, nor for the writer are such questions, in spite of the prominence given to them in the opening chapter and in the afterword, necessarily the major import or interest of the finished work. The research into the meaning of the ancient Greek became necessary, rather, in the pursuit of an answer to a much narrower question about the connection between two ideas in a totally significant saying attributed to Jesus in the Gospel according to Mark: the first noted that the Son of man came to serve, and the second that he would give his life as a ransom (Mark 10:45). What kind of relationship did the author intend us to see between service ($\delta\iota\alpha\kappa\text{o}\nu\epsilon\tilde{\iota}\nu$ in his language) and the giving of a life for others? Once having begun to answer this question, however, I was very soon to realise that the enquiry along this line could not proceed until, as chapter 2 illustrates, there was agreement as to what a writer of Greek in the first Christian generations would have had in mind with talk of $\delta\iota\alpha\kappa\text{o}\nu\epsilon\tilde{\iota}\nu$. Accordingly, my reflections on what I first set out to come to grips with conclude the book.

The rest of what I want to say here consists of expressions of thanks to people who have helped me at various stages of the work. First, to Peter R. Ackroyd, then professor of Old Testament at King's College, who, late at night in September 1970, after a lecture in Melbourne on his Australian tour, was kind enough to give me his time and then set about, while he was still on tour, arranging an introduction to his colleague at King's in New Testament studies, Christopher F. Evans. Second, to Christopher Evans himself, who directed my thesis ever firmly but was tolerant of enthusiasms, requiring only accountability and clarity. I was conscious of the privilege of working with someone as highly respected and widely loved as he. He was always kind in the midst of busy-ness, and his own lively sense of church as "a continuing, developing community" encourages both backward and forward vision.

To other members of King's College at the time I express thanks for assistance in particular matters: to Dr. Gossage for advice on an early draft of the chapter on Plato; to M. A. Knibb for considering and advising on a proposal concerning the Aramaic background of Mark 10:45; and to C. J. A. Hickling for introducing me to Ernest Best's Society for New Testament Studies Seminar on 2 Corinthians. A fellow visitor to King's College, Günter Unger of Erlangen, was of the greatest assistance to me in 1974 in providing me at that late stage with a copy of Brandt's *Dienst und Dienen;* two years later he was also a solicitous host on my visit to the Evangelical Faculty of Theology in Munich. That same visit to Germany allowed me to meet Hannes Kramer, his wife, and colleagues at the International Centre for the Diaconate in Freiburg im Breisgau, where I was a guest of the ever-kind Alex Gondan, for many years editor of the centre's journal *Diakonia Christi*. I am indebted to two scholars who have contributed to questions of early church order: Manuel Guerra y Gomez—whose contributions have not received the attention they deserve—for the stimulation of a meeting with him during a pleasant interlude in Burgos, and André Lemaire for helpful correspondence. I am indebted too to Dr. Walter Wegner, rector of the Ecumenical Institute of Tantur during my fellowship 1976–1977, for supporting my candidature for the Israeli government's Rose Zweitel Memorial Scholarship, as I am to John Todd, then of the publishers Darton, Longman and Todd of London, and to Bishop Alan C. Clark of East Anglia for supporting my earlier candidature for the fellowship by the British Trust for Tantur. John Todd initiated the idea of this book at a time when I was asking him to support a simpler project. I understood him as saying that scholarship has something to offer churches that are struggling to express themselves in a new age; everything he does, from his work on Luther to his reviewing of religious literature, seems to be coloured by this vision of newness. To both of these men I am grateful also for the warmth of friendship, expressed in Bishop Clark's case in special ways to myself, my wife, and our children both in Norfolk and happily also on the bayside of Melbourne. To Miss J. S. M. Dannatt and to Miss J. C. Morrogh, then warden and deputy warden respectively of William Goodenough House in Mecklenburgh Square, London, my wife and I would like to record a particular word of thanks for their care.

In Melbourne, which has been home again since late in 1977, I have appreciated the encouragement of Dr. Francis J. Moloney, S.D.B., and indeed the way he has urged me towards the finish; for support of a similar kind, expressed over the past two years in several particular and time-consuming ways, I am grateful also to Dr. William J. Dalton, S.J., presbyter extraordinary and unpretentious patron. Lastly, in Australia, I would like to thank Dr. G. H. R. Horsley, then of The Ancient History Documentary Research Centre of Macquarie University, Sydney, for providing me with items of value to this study in the late stages of its completion.

To unnamed staff of several libraries I am also indebted. In order, these have been the invariably helpful staff of the then British Museum, where, in the Venice edition 1486 of Rufinus's Latin translation of Josephus, the most beautiful book of my experience came into my hands, the libraries of King's College, University College, and of the Senate House in the University of London; then of Dr. Williams's Library—so intriguingly rich—of Gordon Square, and on the other side of the square the Classical Institute, where most of the Greek was eventually pursued. In the Institute I had the good fortune of encountering and of being advised in some papyrological details (these are noted in chapter 9) by Dr. Borkowski, whose other kindness has already been recorded. Finally, to the Benedictine monks who were librarians at Tantur at the time—in temporary exile from Montserrat—a very warm word of thanks not only for their care of the library and of its clients but for showing in their life and liturgy the heart of ecumenism.

As noted earlier, in Australia the rewriting has been advanced slowly in the midst of the work and rewards of teaching. Teaching the William Goldings and the Ken Keseys of the curriculum does not always sit well with putting together some observations of the usage of Philo. A post in a theological institute instead of in a secondary school would, of course, have created a more suitable climate for writing, and I can only express disappointment that my status in the Roman Catholic church as a "laicised" priest disbarred me from any such appointment in the region where we have chosen to live.

There remain my dearest and deepest thanks to my wife and indeed now to our children, Catherine and Johnny, who both, since the years when their chief interest in *diakonia* was to create ripple effects on bookshelves, have shown consideration, interest in my progress, and much good humour. The book is dedicated to my wife with love, as the inscription shows. It was the reason we both left Melbourne for London, and has accompanied our marriage all the years since we first began our life together in the church of St. Elizabeth of Portugal on the hill above the Thames in Richmond, Surrey. Her practical support really began thereafter, as she kept us both happily housed and well fed in our years in London through jobs that were not entirely congenial; she also did the major part of the final rushed typing of the thesis. To one other person, our friend Helen Luxton of Melbourne, I give my last word of thanks; her profound loyalty led her to support this undertaking at the critical moments of its beginning.

Seaford, Melbourne   J. N. C.
February 1990

Contents

I ASSUMPTIONS, 3

1. The Latter-Day Servant Church, 5
   *Word Studies, 6   "Diakonia" and the German Evangelical Diaconate, 8
   A Theological Conception, 11   The Second Vatican Council, 14
   The World Council of Churches, 20   Counterpart to Witness, 24
   The Doctrine of Ministry, 26   The Language of Office, 35   Deacons, 41*

2. The Servant Son of Man, 46
   *Service as a Saving Action, 48   Jesus' Service at the Supper, 52
   The Community's Eucharist, 54   The Isaian Servant, 55   The New Ethic, 56
   Rule in the Community, 58   Functions in the Community, 58
   Ecclesiastical Office, 59   Attendance on the Rabbi, 59   In Fealty to God, 59
   In Summary, 60*

3. The Early Servant Church, 63
   Works of Mercy, 63   "Diaconiae" at Rome, 66   Deacons of Old, 69

   Conclusion to Part I, 71

II NON-CHRISTIAN SOURCES, 73

4. The Go-Between, 77
   *A Self-Sufficient Community, 78   Trade, 78   A Colourless Term, 79
   Functions of the Subordinate, 81   Functions of the Go-Between, 84
   The Functions as Menial, 86   Functions of the Attendant, 87   Plato's Usage, 88
   Speed, 89   Hermes, 90   Greek Slaves, 92   A Crossroads in Lexicography, 93*

5. Word, 96

   *A Servant Girl and a Scholar, 96   To Heaven, 98
   From Heaven with Hermes and Iris, 100   Messengers Less than Gods, 104
   Constantine, 105   A Saviour and a Trickster, 107   The Making of Aesop, 109
   The Brother of a Reluctant Prophet, 110   Jeremiah, 111   An Unlikely Prophet, 111
   "How Sweetly You Do Minister to Love", 115   Errands, 124
   A Letter from Chaerea, 126   The Twins' Petition, 127   In High Places, 128
   Communicating, 131*

6. Deed, 133

   *A Virgin's Ransom from Hermes, 133   A Little Philosophy, 136
   Double-Crossing Agents and Others, 138   Doing Caligula in, 140
   The Piety of Petronius, 142   The Loving Cup, 144   Civic Duties. 145
   A Little Flattery, 146   At the Behest of a Deity, 147*

7. House and Table, 150

   *Housebound, 150   Staff, 151   In the Abstract, 153   Waiters at Work, 154
   A Sense of Occasion, 156   "The God Took Joy", 157   Decorum, 159
   Making Friends, 160   A Ritual Fit for a King, 161   A Guild of Worshippers, 163
   Temple Liturgies, 164   The Charity of Job, 165   More Lasting than Bronze, 166*

8. A Question of Diplomacy, 169

   *"Authorised Representative", 170   Stoic Attitudes, 171
   Civil Servants and Agents in God's Commonwealth, 173   The Cynic's Mission, 175*

9. In the Language of the Papyri, 177

   *A Personal Name, 179   A Coptic Nickname, 181   Odd Jobs, 182
   Mistaken Identity, 184   Panopolis, 185*

III   FIRST CHRISTIAN WRITINGS, 193

10. Spokesmen and Emissaries of Heaven, 195

    *The Apostle as Spokesman, 195   The Spokesman as Medium, 203
    Paul's Sacred Mission to the Gentiles, 211   Paul and the Twleve in Acts, 212
    The Later Christian Mission, 215   Prophets and Angels, 215*

11. Emissaries in the Church, 217

    *The Collection for Jerusalem, 217   The Delegation from Antioch to Jerusalem, 221
    Emissaries of the Apostle, 222   Emissaries of the Community, 224*

12. Commissions under God, Church, and Spirit, 227

    *Christ and Rome, 227   Local Office, 228   Extending Ministry, 230
    Moral Duties, 231   Gifts, 232*

13. Deacons, 235

    *Philippi, 235   The Church of Timothy, 237   Didache, 238   Clement of Rome, 238
    Ignatius and Polycarp, 239   Hermas and Justin, 242   Origins of the Title, 243*

14. The Gospels, 245
   Jesus as Waiter, 245   Servant of All, 247   The Son of Man, 248

Afterword, 253
Notes, 265
Appendix I. Meanings of the Greek Words for Ministry, 335
Appendix II. Uses of the διακον- Words in the New Testament, 338
Appendix III. Key to the Greek Alphabet, 340
Abbreviations, 341
Index of Sources of διακον- Words, 345
   Non-Christian and Post-Biblical Writings, 345
   Papyri, 357
   Inscriptions, 358
   Biblical and Early Christian Writings, 358
   Papyri Relating to Early Deacons, 361
Index of Other Greek Terms, 362
Index of Authors, 363

DIAKONIA

DIAKONIA

I

ASSUMPTIONS

Boswell records a conversation about language between the Corsican patriot General Paoli and Dr. Johnson in which the general had this to say: "We may know the direct signification of single words; but by these no beauty of expression, no sally of genius, no wit is conveyed to the mind. All this must be by allusion to other ideas."[1] This observation about our ways with words was much admired by Dr. Johnson and says much about what this book is about: the book is about the meaning of the ancient Greek word διακονία *(diakonia)* and its cognates. A small number of English words has traditionally been used to provide us with what the Corsican general called its direct signification—words like "ministry" and "service"—but to come closer to what the Greek word said to the Greek mind we need to reach out into the range of ideas it is associatted with. To know a word, it helps to know the company it keeps.

The Greek word and two of its cognates—the verb διακονεῖν *(diakonein)* and the common noun διάκονος *(diakonos)*—occur about a hundred times in the New Testament, and they have generally been translated into English by words of the "ministry" group, although in today's Bibles it is very often by words of the "service" group. In one set of passages, instead of being translated in one or other of these ways, the Greek common noun is merely reproduced in English letters more or less corresponding to the Greek letters, and so we have come by the English word "deacon"; by the same process we have also received a few more words relating to "diaconate" from other early Christian literature. A similar phenomenon, embodying this distinction between "ministry" and "diaconate," is to be observed in other European languages, beginning with Latin itself.

In recent decades what has been happening in the course of the modern shift from "ministry" to "service" is a profound change in perceptions not only of the ordained ministry called diaconate but of ordained ministry itself. Whatever may have been the intention of the churchmen at the dawn of Europe in establishing their distinction between "ministry" and "diaconate," many of today's churches are seemingly intent on understanding both terms as synonymous with "service." They have even enshrined its value in a word that has become part of theological and pastoral terminology and that is a transliteration of the Greek abstract noun, "diakonia," and this is conveniently stated by one writer to be "service in the church and of a more general nature besides."[2] A detailed exposition of what "diakonia" entails is the matter of the first chapter. This traces the comparatively recent provenance of "diakonia" and illustrates what that concept has been taken to imply for the structure of the church, for the relation of church with church, and of church with society; but the chapter will also be suggesting that "diakonia" is not an entirely happy word.

One reason for this may be put at once. The word "service" with which "diakonia" is frequently alternated has many uses in English and usually we have no trouble understanding what people are talking about. We recognise immediately, for instance, the particular kinds of service referred to in such everyday statements as "his service to the nation was duly rewarded"; "her service to the sick was an inspiration to many"; "you get wonderful service at that supermarket"; "it's getting expensive to service your car"; and "it's all part of the service, madam." Take the word into a theological setting and other meanings appear, but here replace the English word with "diakonia" and the statements at once become less precise. This is because the native wit by which we distinguish between the kinds of service in the phrases just cited does not go to work on a loan word like "diakonia." Sentences neither look nor sound the same as they do when speaking of "service"; they change their shape and weight. And because "diakonia" would be occurring by reason of its weight in theology, all types of service of a Christian kind tend to be forced into the same mould. As a consequence we have no way of reckoning, for example, whether an ordained minister's "diakonia" within his congregation is different from the "diakonia" that the congregation itself might be asked to extend to the poor. We could suspect we are being told that they are one and the same thing. Some find this an exhilarating possibility. Others find it confusing, perhaps even distressing.

Where did such talk of "diakonia" come from? A principal source is always given as what Jesus says at Mark 10:45—that he came not to be served but to serve—but from a survey of what interpreters make of this we shall see in chapter 2 that they are by no means of one mind as to what this service or "diakonia" consisted in. A number of other statements also refers to a "diakonia" of Christians, and a review of these in chapter 3 will suggest that "diakonia" could indeed be a misnomer and that one is likely to come to an understanding of the cognate Greek words in Christian sources only after becoming more broadly informed about uses of the words in Greek at large. The second part of the study, accordingly, will undertake that kind of investigation, while in the third part we will return for a review of usage in Christian writings.

1

The Latter-Day Servant Church

Interest in "diakonia" is largely attributable to an earlier emphasis on linguistic research in the study of the Bible. Since the 1930s, when Kittel's *Theological Dictionary of the New Testament* began to appear in Germany, the great mass of lexical information suddenly within arm's reach of any student of early Christian literature tended to encourage the delineation of themes and theological conceptualisations among a variety of the ancient authors. Theology has been enriched in the process but at certain points also it has certainly been distorted. To establish the existence of a theological conception in the works of many writers of different backgrounds, literary skills, and theological interests is difficult, but the narrower the base from which one hopes to do this the more exacting the task becomes. The mere presence of a word can easily be taken as evidence of the line of thought being pursued, whereas reflection might show that the author is using the word for a different theological purpose or for no theological purpose at all.

In any such enquiry one will move with caution and remain aware of a network of influences. Ideas may be Christian in origin or part of a theological deposit shared by Christians and Jews. Even as Christian they may stem from a Palestinian or a Hellenistic milieu. In the latter case the formulation of the ideas may owe less to Jewish than to pagan religion and culture. Whatever their provenance, the ideas come to us in Greek that may be the native language of a Hellenist or the second language of a Jew or the literary offering of a translator. In these conditions there is going to be room both for differences of opinion among interpreters and for error, and when the enquiry centres on the meaning of a small group of words error is sometimes invited. In addition to factors like these, however, which relate to the patterns of thought and expression of an era now past, there are the patterns of thought that the modern reader brings to the task. In the case of the conceptualisation called "diakonia" such thought patterns played a large part in the initial word study, a fact that those unfamiliar with the history of German theology at this point could not suspect. Some background will thus be useful here. This will instruct us about what "diakonia" stands for at the same time as it will provide us with a picture of theologians coming up with results that were exactly the kind they were hoping for.

Word Studies

The standard word study is that by H. W. Beyer in Kittel's dictionary in 1935.[1] Beyer was a young professor in the faculty of evangelical theology at Greifswald in what is now the German Democratic Republic. He had been a student of Hans Lietzmann's, had wide interests, and was unfortunately to die in World War II before he was forty-five years of age. He was perhaps already beginning to work on his material for the dictionary when an attractive theological study appeared called—to translate its title—*Service and Serving in the New Testament*. This was the doctoral dissertation of Wilhelm Brandt, Beyer's senior by a few years and a man whose career, both academic and pastoral, revolved around the Inner Mission, an organisation within German Lutheranism that owed its existence to certain conceptions about Christian diaconate or, in its German form, "Diakonie."[2] The book is different from Beyer's essay in covering several other word groups apart from the one related to "diakonia" and also in the more extensive examination it undertakes of the notion of service in the ancient Greek world. It is not strictly a philological treatise, however, or an essay in semantics but rather an evaluation of service in early Christian circles as contrasted with non-Christian attitudes.

Brandt records in his foreword that the book owed its inspiration to the fact that he had shared in the work of the Kaiserswerth community of deaconesses and to his experience of the loving care expended on their patients by members of the Bethel community. Graciously professing that such service was a living exposition of the New Testament, he states that the book's main purpose is to present Jesus the Servant in lowly, exemplary bondage with all that implies for service in the community. These personal attitudes are recalled here to set the stamp more quickly than an exposition of his ideas would do on the kind of service Brandt had set himself to examine. It was a service of love. This too is what Beyer concluded in what was professedly an essay in lexicography. His indebtedness to Brandt is expressed and is obvious in almost all that he writes. Like him, Beyer writes of service ($\delta\iota\alpha\kappa\text{ο}\nu\epsilon\tilde{\iota}\nu$) as "active Christian love for the neighbour." In the case of Jesus such service is the quality by which he saves.

We will look more closely at the ideas of both of these writers in chapter 2 where we will be able to set their views in the context of other scholarly treatments of the words, but the few judgements about the words which we have just recorded lie at the root of all subsequent talk of "diakonia." In fact all advocacy of "diakonia" is ultimately founded on the work of one or other of these writers. In the English-speaking world, since Brandt's book was never translated and is rare even in its German form, this is usually with reference to Beyer, whose opinions carry in addition the authority attaching to Kittel's dictionary.

In the few decades since then, but particularly over the last twenty years, an extensive literature has accumulated around "diakonia." Some of this will shortly be reviewed but only rarely in any of it has a reexamination of the meaning of the words underlying "diakonia" been undertaken on a basis other than of occurrences in the New Testament. That is to say that those who write expressly of "diakonia" or who have occasion to draw upon some aspect of the notion are generally satisfied that the uses of the words in non-Christian literature have been adequately dealt with by Brandt and Beyer and have nothing further to offer for an understanding of Christian usage. Thus one normally finds that occurrences of the words are assembled in a way that permits the emergence of an aspect of service suitable for the discussion in hand, whether it concerns ethics, ecclesiastical office or authority, the diaconate, ministry more generally conceived, or the ministry of Jesus, and the author's effect is not usually achieved without the free association of sometimes dubiously related passages—as, to put the situation broadly, between passages in the gospels which happen to use words of the group and passages in the letters where the same words occur.

To emphasise the limitations of this approach it will be useful to mention briefly those few other writers who have taken the enquiry some way back into non-Christian sources and whose work will be considered in greater detail at various later stages. These are Eduard Schweizer, Jean Colson, Manuel Guerra y Gomez, Dieter Georgi, André Lemaire, and Jürgen Roloff. With the exception of Georgi none of these has attempted to modify Beyer's conclusion that within Christian writings the words took on meanings distinctively Christian. Roloff has reservations about Beyer's method in respect of the synoptic gospels and accordingly seeks to account for the special character of Christian usage in the light of modern tradition criticism.[3] Colson has maintained an individual approach in respect of the title "deacon," which he judges to be the Greek equivalent of a Jewish title from Qumran rather than a title of Hellenistic origin, but in most respects his understanding of the word group is barely distinguishable from that widely received.[4] Schweizer examined the word group in relation to the question of church order in general and is now cited almost as frequently as Beyer in support of "diakonia."[5] Lemaire and Guerra, like Colson, have related their studies chiefly to an understanding of the ancient diaconate. Lemaire provides a useful compendium of ancient sources but has offered no new judgement of a linguistic kind on that basis and, in otherwise treating of "diakonia" in relation to the general doctrine of ministry, shows himself to be in accord with the current trend.[6] Guerra is more useful again in that he provides the largest collection to date of ancient non-Christian material, and he is also original in the detail and direction of his interpretation. This points mainly at the religious character of the non-Christian usage; it is only surprising that in the treatment of Christian usage he makes little attempt at the correlation of the two usages that his survey invites. In fact, like Lemaire, Guerra has subsequently shown that he accepts "diakonia" as an expression for distinctively Christian values.[7]

Among these writers Georgi is alone in asserting that Beyer was wrong in principle when he depicted usage in the New Testament as distinctively Christian. Further, he does not accept that in the New Testament the words operate mainly within the field of service to one's neighbour or fellow man and woman. Instead they should be viewed in the entirely different field of service under God in the delivery of his revelation, and here their uses would coincide with those in the religious language of the Stoics and Cynics, whose proselytising activities paralleled those of the Christians. These judgements are contained in a few pages of the book in which he investigated the background of the opposition to Paul's apostolate in Corinth,[8] and in opening up a new kind of relationship between Christian and non-Christian usage they cut directly across the linguistic arguments for the existence of a theological conception known as "diakonia." More will be said of this in chapter 3. Not all of those with an interest in the question of Paul at Corinth have been convinced that Georgi's picture of the linguistic relationships is accurate—and again chapter 8 will be given to exploring these relationships—but all have had to keep the picture in view. The point for the moment, however, is that although a keen discussion has turned for many years on the question that Georgi raised, those with an interest in "diakonia" as a theological term have not been drawn to review the grounds on which they see it as referring necessarily to what Beyer called "active Christian love for the neighbour." One might say that Guerra's observations on the religious nature of non-Christian usage have also been neglected, but Guerra did not go on to contest Beyer's interpretation of the New Testament. The neglect of Georgi's work, on the other hand, even by the few theologians who are not satisfied that "diakonia" is a useful conceptualisation, is an indication of how entrenched "diakonia" has become in certain areas of theology.[9] It rests on linguistic studies, certainly, but when one looks beyond the studies themselves to the purposes for which they were initially undertaken one is able to appreciate that theology is sometimes made less by taking meaning from words than by attempting to put meaning into Christian life.

"Diakonia" and the German Evangelical Diaconate

Those who first spoke of "diakonia" were not the linguists or theologians of our day but Lutheran churchmen of nineteenth-century Germany. They were seeking to establish a form of Christian ministry among the displaced, the delinquent, the sick, the illiterate, and all those adversely affected in the aftermath of the Napoleonic wars and by the onset of industrialisation. They were aware of the approach taken to these problems within Roman Catholicism, where religious orders of men and women were versatile instruments in many fields of charitable activity, but within Protestantism any overt approach by way of monasticism was unacceptable. There were also obstacles to a purely lay ministry, for a movement in this direction first would have had to overcome an inertia induced by generations of an inward-looking pietism and then would have needed to alter attitudes in regard to pastoral activity, which was considered to be the preserve of the ordained ministry. Men were expected to be fathers of families and guardians of the word of God within the household, and the women to be wives and mothers or dutiful daughters. And yet in the judgement of those who saw the need for a new pastoral initiative it was largely the skills and temperament of women that were called for. To this view there was undoubtedly a romantic angle—women being held by nature to be, in the words of the Strassburg pastor Franz Heinrich Härter, "not domineering but given to helping others . . . in a gentle and undemonstrative spirit"[1]—but it also reveals the character of the ministry for which a need was felt and which led several to work energetically towards the establishment of an order of Lutheran deaconesses.

In 1835 Count Adalbert von der Recke-Volmarstein published his pamphlet on deaconesses[2] and in the following year both he and Pastor Theodor Fliedner, who had been impressed by what he had seen of deaconesses among Mennonite communities in Holland and who owed much in his undertakings to the assistance of his wife Friederike, founded separate houses near Düsseldorf for the reception and training of women in this ministry. Fliedner's work, situated at Kaiserswerth, is perhaps the best known outside of Germany; with him in 1851 Florence Nightingale was able to experience for the first time nursing for some months in a hospital, an interlude that was of the highest inspiration to her.[3] Parallel with vocational training was the religious ideal Fliedner put before his candidates. He envisaged "a band of German missionary women who will take the mercy of Christ to the forsaken sick, the neglected children, the disheartened poor and the wayward prisoners throughout the nation";[4] they would live a life in common of prayer and work in order to ensure, as he preached at the induction ceremony of 1839, that they would truly be "servants of the Lord Jesus, servants of the sick for the sake of Jesus, and servants of one another."[5] Challenged by his co-religionists on the grounds that he was introducing orders of nuns into Protestantism, he replied that he was merely restoring to the church what it had once possessed in the deaconesses of apostolic times.[6]

This conviction about the nature of the ancient diaconate lay at the root of the movement. Luther himself had seen the diaconate as the ministry of distributing the wealth of the church to the poor, although he had not succeeded in reforming the office along these lines. In an even poorer condition than the Roman Catholic diaconate, which for centuries had been but a step towards the higher rank of priesthood within the hierarchy and which Luther had curtly dismissed as the ministry of reading the epistle and gospel at Mass,[7] the nineteenth-century Lutheran diaconate was for all practical purposes defunct. Thus the canonical diaconate ("Diakonat" in German) had little to offer the newer reformers. Conceived of as "diakonia" ("Diakonie" in German), however, it offered a diaconate in action, the "care of the weak, sick, needy and suffering members of the Body of Christ," as Fliedner defined it.[8]

There was much sympathy with this position when it was argued by Fliedner and others before the High Consistory of the Prussian church at the Monbijou conference of 1856[9] but, as was intimated to Recke-Volmarstein as early as 1835 by the then Crown Prince and later King Frederick Wilhelm IV of Prussia, the communities of women could not rightly be considered to constitute an order of deaconesses until they were so recognised by ecclesiastical authority,[10] a recognition that the authorities were unwilling to give. Johann Hinrich Wichern was the most forceful advocate of the necessity of such recognition. He resisted the use of the formal title "deacon" for the brothers of his community in Hamburg and at the same time had long been the most celebrated advocate of a more general "diakonia" exercised by the whole church. "Diakonia" was the manifestation of love and the necessary counterpart of the preaching of the word, but it would not truly be manifested in the church until it was embodied in an independent ecclesiastical office. "The true and full awakening of the church's diakonia," he wrote, "is dependent upon the renewal of the apostolic diaconate."[11]

Lack of canonical recognition, however, did not prevent the spread of houses of deacons and deaconesses. Operating outside of the church's official ministry, these people nonetheless saw themselves as the inheritors of the early church's diaconate. "Diakonia is care of the poor," wrote Wilhelm Löhe, founder of the Neuendettelsau deaconesses in Bavaria, "and as such is an office of the church; that is a conclusion one must take from the New Testament."[12] In a more exact account of the relationship existing in the early church between "diakonia" and diaconate, Löhe wrote:

> In sacred scripture all offices of the Holy Spirit are called "diakonia," that is, office or service. In the same way all those who hold offices or have dealings with men under a mandate from God—from Christ himself down to the least significant office-holder—are at times called "diakonos," that is, servant. In the first days of the Jerusalem community, nevertheless, the same words "diakonia" and "diakonos" were used as official designations of the office of corporal works of mercy and of those engaged therein. Thus we have come to understand that the word "diakonia" means nothing other than the sacred work of caring for the poor and that the term "deacons" designates none other than the Seven, who first held this office, and their successors.[13]

In this passage we see Löhe coming to terms with the fact that in the New Testament "diakonia" does not necessarily designate care of the poor; further, the conviction which he shared with most theologians that the Seven were deacons and that their diaconate was essentially an office for works of mercy would not long stand up to critical examination from either an exegetical or a historical point of view. Before the end of the century the historian of the new diaconal movement, Gerhard Uhlhorn, who was entirely sympathetic to the idea enshrined in "diakonia," found it necessary to point out that the Seven were more likely to have been presbyters than deacons and that history showed the diaconate not to have been an "office of mercy" but one that in the first place was auxiliary to the office of presbyter and bishop.[14] If this suggests that a certain ambiguity could underlie the understanding of the diaconate as essentially "diakonia," the ambiguity was not such as to obscure the objectives or hold back the development of the German movement. In 1957 an important part of the plan outlined by Wichern was realised with the formal inauguration of "Das Diakonische Werk" (diakonic organisation) by which the evangelical churches summoned all their members to the ministry of charity.[15] "Diakonia," in the words of its modern historian Herbert Krimm, was and remains "that organic, inalienable and unchangeable function of the body of Christ which has to do with the church's fight against poverty, sickness and every kind of earthly need."[16]

Such clarity of definition and public recognition of an idea mark a great advance on the day in 1836 when the Fliedners joyfully received at Kaiserswerth a middle-aged woman named Gertrud Reichardt, whom the pastor did not feel free to register at the time as anything other than "nurse" but who is affectionately remembered as "the first deaconess of modern times."[17] One hundred fifty years after that event more than 10,000 women bore her title in the Federal Republic of Germany, where there were also more than 7,000 deacons, and the Diakonische Werk had a staff of 260,000 in some 18,000 institutions operating across a wide range of social services.[18]

To the non-German, undertakings of such a nature fall conveniently under the heading of charity or social work, and a description of them is so entered, for example, in *The Encyclopedia of the Lutheran Church*.[19] Accordingly it is not surprising that a degree of confusion occurs when the Lutheran conception of "diakonia" is assimilated into circles where the same basic Christian activities and organisations had previously only been spoken of in terms of charity. As one observer noted in regard to an attempt by Roman Catholic theologians in Germany to distinguish between charity and "diakonia," the question would seem to be one largely of terminology. "Where the catholic church," he pointed out, "designates works undertaken for love of the neighbour by 'charity,' for the same kind of undertakings the evangelical church uses the term 'diakonia.' "[20] Instead of distinguishing between the two, therefore, we should probably be asking whether the newer term "diakonia" has more to offer theologically than the term "charity." We recall that the newer term took root initially as an alternative designation for the diaconate and that then, because the diaconate of the early church was conceived of as an office for works of mercy and love, "diakonia" became a general term for merciful and loving assistance. After the century of pastoral endeavour initiated by men like Fliedner and by force of which this acceptation was consolidated within Lutheran circles in Germany, a theologian of those circles, Wilhelm Brandt, wrote the learned monograph on "service and serving in the New Testament" which proposed that the original Greek word $\delta\iota\alpha\kappa o\nu i\alpha$ was the most succinct and distinctive expression of the early Christian conception of service and that it did indeed represent the ideals the evangelical diaconate sought to embody. Given Beyer's scholarly support in Kittel's dictionary, this conclusion became an accepted fact of Christian lexicography, and neologisms like "diakonia" began to appear in languages other than German in reference not only to a diaconate exercised by certain members of the church but to a diaconate to which each member of the church and the church as a body were called. This change could hardly have been effected without the philological work of Brandt and Beyer. Conversely, however, one might suspect that the philological approach of these theologians may have been different had the experiences of a nineteenth-century church not created a climate of opinion in which "diakonia" was already equated with works of loving service. This in turn would suggest that the foundations of "diakonia" as a theological conception may be neither broad nor sound.

A Theological Conception

The shift from talk about diaconate to a notion called "diakonia" has been clearly registered in dictionaries of theology and encyclopaedias of church life. In books of this kind around the turn of the century articles were devoted to the diaconate itself, to the nineteenth-century forms of "Diakonie" in German evangelical churches, and to the institutions that were known in Rome and elsewhere from early medieval times as "diaconiae" and which will form a topic in a later chapter. In these articles the approach was directly historical in the main, and linguistic comment when offered was not aimed at inculcating any teaching about service but would in fact leave little scope for developing such teaching. Thus, writing on "Deacon," James Strong emphasised the broad range of meaning of the Greek term in the New Testament, recalled that in the opinion of the philologist Buttmann its meaning must be related to *"runner,* i.e. messenger," and advised against trying to relate the title "deacon" to mere service at table.[1] J. Armitage Robinson was just as little interested in presenting ideas about serving people when he observed that with the Greek common noun the emphasis is on the performance of a task in hand.[2] These judgements will find their place in the semantic question to be treated in chapter 4 and are recorded here because they afford a clear contrast with the customary approach to these words in the decades since the appearance of the article on them in Kittel's dictionary.

In these more recent articles, linguistic observations have tended to form the basis for statements about the theological and ethical content of "diakonia" itself. Here, because the linguistic premise is that the Greek words speak of service at table and other similar forms of lowly service, the title "deacon" emerges as a technical instance of "servant" among other instances in Christian writings where the notion of service is expressed less technically in respect of Christ, apostle, community leader, or believing Christian. In respect of the diaconate this approach has been judged helpful because the origins of the office within Christian communities have always been obscure and attempts to find them outside of Christian circles in either a Jewish or a Hellenistic institution have generally been taken as inconclusive.[3] More to the point here, however, is the result for the notion of service itself. This is far reaching because service becomes the common ground of Christ, Christians, and their institutions at the level of their respective missions. As such it becomes a category within theology, and has been so recorded in the bibliographies,[4] and takes its place in the encyclopaedias as an individual topic or as part of topics like ministry, office, or authority.

The treatment of "diakonia" along these lines in one such handbook will serve as a convenient summary of the leading ideas and philological principles that are encountered in these works of reference as they are also in a wide range of theological literature and religious journalism.[5] According to the author of this handbook, in non-Christian Greek the various uses of the Greek verb διακονεῖν developed from a basic sense of "to serve at table." In Christian Greek the conception behind its application to discipleship (John 12:26) "smashes through" this previous field of meaning and becomes specifically Christian. In Mark 10:45, then, the term has developed into "the characteristic word for loving activity, stemming from love of God, in respect of one's brothers and neighbours." Correspondingly the abstract noun διακονία expresses the idea of a "service of love" and is used either generally, in regard, for example, to what the household of Stephanas did for the community at Corinth (1 Cor. 16:15), or for the specific kindness of taking gifts to the poor in Jerusalem (Rom. 15:31). At the same time the daily service recorded at Acts 6:1 shows that table fellowship is also basic to the meaning of the word group in the New Testament. Paul is then said to extend the range of meaning within this general context by using the words to designate apostolic activity (Rom. 11:13) and the whole work of salvation (2 Cor. 3:7–9), while the statement in Eph. 4:12 shows how "diakonia" becomes "a force which conditions the organism of the body of Christ in its entirety" with the result that the Christian community is to be understood as "a living instrument of service [Dienstorganismus] in the world."

In a complementary note to this essay the lexicon's editor, L. Coenen, comments after the manner of several other writers on the fact that the early Christians showed a marked preference for words like διακονεῖν over words of the λειτουργεῖν group which are prominent in the cultic terminology of the Septuagint translation of the Old Testament (and which produced words like "liturgy" in English). For Christians, he observes, "the service which counts is not that which is performed at the altar but the service which reaches out from altar to human kind; the true liturgy of the Christian community is its 'diakonia.' "

In the course of the main essay the author draws attention to a few passages where the ideas expressed do not conform to the pattern of "diakonia" he has outlined. The most significant of these are Rom. 13:4 and Gal. 2:17, in the first of which the common noun is applied to the pagan state and in the second to Christ in a context where the specifically Christian meaning cannot be upheld for the person who is the source and model of "diakonia." In a similar way W. Jannasch has noted the difficulty of accounting semantically for the connection between the general run of the words' uses in the New Testament and the particular use in respect of deacons. He further points out that words that were used in respect of such basic theological matters as the mission of Jesus and the apostleship of Paul could have been expected to produce more derivatives in modern languages than narrowly ecclesiastical terms like "deacon."[6] Observing the same linguistic fact, the members of a working party established by the World Council of Churches for the purposes of examining the biblical notion of "diakonia" called it "a great dilemma."[7]

Despite such anomalies, however, writers have generally proceeded to present a bold and coherent outline of the development of "diakonia" as a concept specifically Christian and theologically significant. As a result "diakonia" is now widely accepted as a finished product of modern reflection on the linguistic data of the New Testament representing what Jesus was and did, how disciples were related to him and to each other, and both the scope and style of the Christian community's responsibilities. Restored to the language of theology and church life, "diakonia" is understood as enabling Christians to view the church from a perspective that relates it closely with the Jesus whom early tradition recognised as the man who "went about doing good" (Acts 10:38), the "man for others" in the modern phrase, and enabling both ordained and lay Christians to view themselves as co-workers in a servant church. This vision has contributed to the development of some new attitudes and initiatives in those churches especially where the relationship of church to society had previously been problematical or where by force of a strongly institutionalised character a church tended to be identified with its ordained ministers. The vision has also served to bring into a clearer focus the ground on which the various Christian denominations might be able to achieve unity. In the following sections then we will look at the way "diakonia" has been an influential idea in the social ethic of the churches and in ecclesiology, the latter especially in its ecumenical dimension.

The Second Vatican Council

The core of the ecclesiology traditionally espoused in the Roman Catholic church was the authority and power exclusive to ordained ministers by which they might teach, govern, and sanctify the faithful. This doctrine remains. There were, however, many reasons both theological and pastoral, even plainly historical and sociological, that led the Second Vatican Council to replace the one-dimensional view of the church that the doctrine projected with a view that was multifaceted. This occurred when the church was presented in its documents as being the people of God before being a hierarchically structured institution. The shift is a significant one in the history of the doctrine of ministry because these documents are the most thoroughgoing attempt ever undertaken by a church to declare what church is and because in this church authenticity of ministry is held to be the foremost mark of the church of God. Accordingly, when the council proclaimed the common priesthood of the faithful—the latter word a typically preconciliar term for members of the Church who do not belong to the ordained priestly state—and accorded to them a role in the ministerial function of the church, it is not surprising that the Protestant observer spoke of the development as "unprecedented"[1] and that the Roman Catholic theologian used words like "sensational" and "remarkable."[2]

This development has released great energies in the Roman Catholic community. At the same time it has left Roman Catholics with problems that are far from resolved. In the twenty years that the community has been attuning itself to the new emphasis, the lay person has been asking what kind of minister the local priest might be if all his congregation share Christ's priesthood in baptism, and the priest himself has often been confused at seeing the sudden erosion of his ministerial status. The questions are not elucidated for either party by the council's pronouncement that the ministerial and common preisthood differ essentially and not merely in degree[3] because this formula cannot be explained satisfactorily on the basis of the council's other teaching. That this is so is more effectively illustrated than argued. In one of the earliest assessments of the council's achievements, G. A. Lindbeck, the Protestant observer just cited, accurately pointed to fundamental "ambiguities" and "equivocations" in the council's major document, the Constitution on the Church, whereby at such a point as this "the Constitution can indeed be understood in radically different ways by honest and competent scholars."[4] Thus on the one hand a Dominican theologian will reassure priests of "the sacramental fact" that according to the council there is "an 'objective' fact of participation in the priesthood of Christ . . . which takes place through the conferring of charisms and powers through hierarchical channels along a line of uninterrupted 'tradition' which goes back to the Apostles and to Christ."[5] On the other hand a theologian engaged on the Roman Catholic side of dialogue with Lutherans emphasises the "sustained parallelism in the council's description of the common priesthood of the laity and the special priesthood of the ordained"; he concludes that "what they have in common is far more fundamental, far more extensive and far more decisive than the distinction between them," suggesting then that "it would be faithful to Vatican II to say that there is no unbridgeable gap between the ordained priesthood and the priesthood of all the faithful."[6]

The latter view undoubtedly represents the predominant trend in unofficial thinking, and the church authorities themselves tacitly acknowledged what many see as "the recognised inadequacy"[7] of the council's statements on the special priesthood in briefing the International Theological Commission to examine the subject anew[8] and in then taking the subject up in the third synod of 1971.[9] Of course the problem of who the minister is is not confined to Roman Catholic circles but is there a particularly sensitive one, and attention is focused on it in that tradition because the problem arises in large part from the very novelty of the council's language in dealing with the ministerial responsibility of the people of God as a whole, and because intimately related to the comprehensive use of terms like "ministry" are assumptions about ministry as "diakonia." In looking, then, at this aspect of the council's teaching the purpose is not to undertake a new search for its express doctrine of ministry but to illustrate that a notion of ministry pervades the teaching without ever being clearly delineated.

The word "diaconia" (the Latin form of our Greek) occurs only twice, both times in the Constitution on the Church. One of these instances refers to the diaconate not unnaturally as "the *diaconia* of liturgy, word and charity"[10] and need not detain us. Of greater interest is the instance relating to the office of bishops. After a traditional dogmatic statement about the mission and authority of bishops as successors to the apostles, the paragraph adds the statement (its pertinent Latin terms are given here in parenthesis), "That office [munus] . . . is, in the strict sense of the term, a service [servitium], which is called very expressively in sacred scripture a *diakonia* [diaconia] or ministry [ministerium]."[11] Here, the Latin words "servitium," "diaconia," and "ministerium" are synonymous, and they are used to counterbalance the high idea of "munus"—which is the traditional conceptualisation of office previously described in the constitution—with the idea of office as service, indeed as lowly service. The same contrast is introduced in one of the constitution's paragraphs on the foundation of the church. Identified with the kingdom of God announced by Jesus, the church is said to be manifested above all in the person of Christ who came "to minister and give his life as a ransom for many." The citation here is from Mark 10:45 and it suggests, beyond the saving power of Jesus' death, that the church's mission must be carried out in the spirit of Jesus' own ministry, namely by "observing his precepts of charity, humility and abnegation" (art. 5). The same scriptural passage is used to much the same effect in another leading document of the council, the Pastoral Constitution on the Church in the Modern World. Here the introductory paragraph on "An offer of service to mankind" (art. 3 of the translation we have been using) concludes with words of an obvious intent: "The Church is not motivated by an earthly ambition but is interested in one thing only—to carry on the work of Christ under the guidance of the Holy Spirit, for he came into the world to bear witness to the truth, to save and not to judge, to serve and not to be served."

These few statements suffice to illustrate a main emphasis in the council's statements about ministry. While it is clear that the church must continue the preaching and saving work of Jesus and that this is a work of the church as a whole within which the bishops act by reason of office, ministry itself is invoked to convey the idea that the work is a service that is lowly and in the broadest terms beneficent. This is by no means to force the meaning of "ministry" but it is to introduce into doctrine about ministry an element that had not previously been present. Traditionally ministry was the work of word and sacrament and was exercised through the office of ordained ministers. The normal usage is exactly represented in the phrase about men entering "the ministry." On the council's broader presentation, by contrast, ministry is something within the reach of all church members and is a responsibility of all church members, a fact that probably makes it impossible to state in terms of ministry what is office and what is not. Certainly emphasis upon ministry as the context of all Christian activity obscures the demarcation between the ministry of the ordained and the ministry of the baptised. In *We Who Serve* Cardinal Bea recorded his surprise that ministry or service ("Dienen" in his original German) had emerged as a major theme of the council[12] and has presented an interpretation of the theme in the light of biblical statements about service. Documenting the theme in the council itself, he pointed out that independently of passages that treat of ministry in a technical sense there are 190 passages where the broader notion is used to explain "the attitude of the hierarchy towards the People of God, reciprocal relations between members of the Church and finally the position of the Church and all its members in relation to the whole human family."[13] The cumulation of the terminology within this wide field has accordingly encouraged the assumption that even the ordained ministry is but a form of the service expressed in "attitudes" and in "reciprocal relations," and in a later section of this chapter we will refer to attempts made in postconciliar theology to give theological depth to this broad form of service. For the moment we can measure the influence of the approach in just two informed but less academic sources. In the first of these, a book on Roman Catholicism since Vatican II, Peter Hebblethwaite argues that the conciliar doctrine, which he calls "a much more demanding view of the priest's role" than the one prevailing since the council of Trent, has contributed to the personal crisis within members of the ordained priesthood. He writes:

> Many priests were simply incapable of understanding it, and fell back into doing what they knew best. Others worked out the full logic of the new emphases. Ministry is for service. This was the key principle. It represented a return to New Testament thinking and was of vital importance in ecumenical discussion. It was no longer possible to view the priest simply as a privileged person in the Church endowed with special powers. Now, on the contrary, he was seen essentially in his relationship to the community he existed to serve.[14]

A reviewer, writing again for the general public, seized on these pages of the book as pointing to a development that could destroy the identity of the Roman Catholic church:

> What . . . is the importance of the liturgical upheaval compared with the whole shift of emphasis concerning the nature of ministerial priesthood and the participation of the "laity"? The age-long hierarchical structure of the Roman Church could easily survive the loss of the Tridentine Mass but not the threat of drastic declericalisation.[15]

In contrast to this alarmist attitude to a perceived new line of thinking is the welcome accorded the new emphasis by those who, as Hebblethwaite put it, work out the full logic of ministry as service. The logic enables them to present the church in a way that is challenging, and in their view also refreshing, for all members of the church whether they hold office by ordination or not. A typical presentation of the new emphasis is the following on "The Church of Christ" where the Catholic principle of order is brought into line with the principle of service. The passage is taken from an English catechetical booklet[16] and is thus a measure of the importance which educators attach to the new doctrinal approach:

> There is no doubt at all that Peter and his fellow apostles were appointed to special positions of authority within the new people of God. We should be very careful to notice, however, just what Christ had to say about the way they should use their authority. [Here the Son of man's words about service are cited from Matt. 20:25–28, and reference is made to Jesus washing the disciples' feet.] The true follower of Christ—no matter what his position within the people—must always be the servant of others. And as the apostles set about organising the Church, it was a command they were to keep very much in mind. . . .
>
> The key to a true understanding of the organisation of the Catholic Church is the command of Christ to serve. St Paul mentions in his epistles various offices such as elders, presbyters, bishops, priests, and deacons. These men and women were given positions of authority in the new people of God, but their authority was to be Christ-like. They were there to serve the people and not to lord it over them. In fact, the usual word in the New Testament to refer to their authority is "diakonia," a word which means service. The organisation of the Catholic Church is still based on the command of Christ to serve. . . . Every member of the new people of God is called to a life of service. We must serve God by serving each other. But if our service is to be truly Christ-like we need leaders to guide and organise our efforts. They, like the apostles before them, are our servants.

The passage proceeds to outline within this context the role of bishops and other ordained leaders, but even the young people for whom this instruction was prepared would surely sense that there is something futile in attempting to transform the understanding of office in the church by tagging followers of Christ, officebearers and nonofficebearers alike, as "the servants of others." Is the instruction trying to take into account a truth of another order altogether, namely, that at the roots office is alien to the church? If so, would this not be more approximately expressed in the words of Brother Roger Schutz of Taizé to the World Federation of Catholic Youth, "The Christian vocation is a vocation of service, but only a church stripped of power will be able to change the course of history"?[17] Or is the truth here the more homely one voiced by an Irish bishop at his installation, "Of course there are times when the exercise of authority demands the use of discipline and restraint and the making of difficult and even unpopular decisions. However, even here authority is a service"?[18] On the former view we have a democratised church; on the latter we have the schoolmasterly attitude of "I know this is not nice but I am doing it for your own good."

The tensions produced by an undifferentiated use of the notion of service were keenly felt by fathers of the council themselves and found expression for example in the preparation of the Decree on the Apostolate of the Laity. In their attempt to attribute "apostolate" to "laity," members of the preparatory commission were "again and again," as F. Klostermann has recorded, "in danger of being forced into extreme positions which were diametrically opposed to each other",[19] and in the final text of the decree the opposition of "hierarchy and non-hierarchy" becomes problematic in his view because of the overlap in "very many callings and ministries."[20] At these points obscurity remains, and the International Theological Commission, in the chapter entitled "Priest or Servant?" from the document on priestly ministry that it prepared for the third synod, rightly pointed out the part that a misuse of the notion of service plays in this. The commission itself did not elucidate the matter, however, when it insisted that the Son of man's word on service (Mark 10:45) exposed "the radical newness" of Christ's priesthood because it took this as "necessary for an understanding in depth of the Christian condition as well as for a rediscovery of priestly ministry in its theological dimension, Vatican II having defined the work of the church and the priestly ministry by 'service' ";[21] here service is that in which the ordained and the nonordained have a common Christian identity and thus cannot also be the ground on which they perform essentially different roles.

In that document the commission understood service as the obedient and lowly service of the Suffering Servant of the book of Isaiah and of early Christian tradition, a conceptualisation that is authentically theological. When instead of service of this kind, which is a form of sacred servitude under God, the service is taken as performed to man, and is thus a beneficent activity, the thinking would seem to lose touch entirely with previous conceptualisations of ministry. For example, when Charles Moeller, an authoritative commentator on the council, spoke of "the ministerial priesthood" as "a 'ministry' in the exact sense of the term, that is, as a service to the People of God," he placed neither priesthood, which he recognises as pertaining to "the threefold sacerdotal, royal and prophetic dignity in which *all* Christians, lay and clerical alike, share," nor ministry in a context that allows "ministerial priesthood" to convey the idea that this ministry is hierarchical and different from that of the universal priesthood.[22]

In the twenty years since the Second Vatican Council the theme of ministry as service has done much in some quarters to redress the imbalance between the roles of clergy and laity, but in the process the word "ministry" would seem to have been rendered useless for designating that role for which certain Christians have always been ordained. The preceding discussion has illustrated this by concentrating on the question that immediately arose[23] of the distinction between ordained and nonordained ministry. The problem would be just as acute, however, were one to concentrate on the nature of the ordained ministry itself. Here it is not enough to speak of service in an almost emotive way as "the vital principle" by which the church reproduces "the mode of existence" of its head and founder,[24] because by that principle the whole church and not just the hierarchy would be ministerially activated. The obscurity that the council's emphasis on service introduces to discussion at this point has been analysed in terms similar to the preceding by L. J. Cameli.[25] Cameli sees very well that no adequate doctrine can rest on ministry if that is open to being interpreted as merely a "style of noncoercive and, indeed, service-oriented activity bringing men to faith,"[26] and accordingly undertakes to discover what can legitimately be made of this conceptualisation. Because his exegetical work on the scriptural passages most often invoked by the council in its teaching on ministry (Mark 10:35–45; Matt. 20:20–28; Luke 22:24–27; John 13:1–20) is thoroughly informed, his judgement as to where the conceptualisation fits is to be noted, and that is within the domain of spirituality, not of systematic theology. As he puts it, service awareness or ministerial consciousness is not "a 'given' with ordination" like the ministerial role at the Eucharist[27] but is a spiritual condition to be cultivated so that "the subjective experience and impact of service"[28] will infuse all ministry, especially that of the ordained priest.

If Cameli's is reliably the best case that can be put for ministry as service, and if into the bargain it effectively discredits service as a principle for the theology of church order, there might appear to be little point in pursuing the present survey of "diakonia" in today's church. The conclusion to which the survey is moving, however, is not that we should make some other use of the notion of service if it is doing no good in theology, but that if extended discussion has not succeeded in finding an assured place for it then the notion itself might be suspect. Cameli was led to reexamine what he called "the root issues of service in the gospel tradition."[29] That may have been to beg the question.

We have referred to Cardinal Bea's surprise that service should have emerged as so strong a theme in the council. In spite of placing it, as the International Theological Commission would also do, within the context of the biblical Suffering Servant, he described it in terms of reciprocal relations between members of the church and between church and world. This is undoubtedly the aspect of service accentuated in the council's statements, and an English document on postconciliar pastoral strategy went so far as to say, "Vatican II really found its direction when it saw clearly that the nature of the Church was to be the servant of the world."[30] Cameli, whose survey of "Service" in the council is divided into sections on the church's "Diaconal Nature," "Diaconal Structure," and "Diaconal Style,"[31] correctly attributed the emphasis on beneficent service to the modern conceptualisation of "diakonia." This conceptualisation came to the fore, as we have seen, within nineteenth-century German Protestantism, and its sudden emergence at an important stage in the history of Catholicism is truly noteworthy.

Prior to the council Roman Catholic writers paid little attention to "diakonia" as Protestants conceived of it. It is not even prominent in the large collection of studies entitled *Diaconia in Christo,* which was published in the first year of the council and which explored the possibilities of orienting the ancient order of the diaconate to the church and world of today.[32] Indeed from the purely linguistic point of view Jean Colson's contribution on the New Testament in this volume may be said to have worked against placing the "diaconia" of the diaconate within the context of the contemporary notion of "diakonia,"[33] and it is perhaps significant that by the end of the council when numbers of the essays in the volume were reedited for a French collection Colson's was replaced by one that accurately reflected the modern thinking.[34] In 1962 an influential article by Congar on "hierarchy as service" did invoke "diakonia" in the sense advocated by Brandt and Beyer but the conception was by no means the leitmotiv and to a degree was accommodated within a study that ranged through and beyond the New Testament (to patristic and medieval notions about "praeesse" and "prodesse" and "utilitas" as "caritas") in order to make the case that of itself hierarchy is not a deviation from but is an authentic form of Christian ministry.[35] The article cites K. H. Schelkle, and it would seem to be this German Catholic theologian who wrote most explicitly in the preconciliar period about "diakonia" in relation to the church's ministry.[36] Fairly well known as Schelkle's book became, however, the council would not have been indebted to it to any significant degree.

The real factors favouring the emergence of service as a basic theme were the spirit of penitence in which the council set out to eschew an uneasy triumphalism, the intention to reform ecclesiastical structures which were felt by many to have rendered much of the church's ministry remote and inept, the sociological tensions that strict clericalism was fostering in a democratic age, and the impact upon systematic theology of the newer biblical sciences. To churchmen meeting in these circumstances, "diakonia," as it already existed in Protestant theology, was an attractive motif. It spoke of humility, it spoke of service as ministry in action, it spoke of brotherly give and take at all levels of church life, and it had a biblical pedigree. Many of the council fathers themselves would have been aware that a year before their own first assembly "diakonia" had been a major theme of the Third Assembly of the World Council of Churches at New Delhi, and several commentators have in fact alluded to the influence on the council of ecumenical thinking in this regard.[37] In particular Charles Moeller has recorded that the emergence of the idea of service in the Pastoral Constitution on the Church in the Modern World was "no accident, but the result of collaboration with the separated brethren." The collaboration was initiated by a letter in April 1963 from Lukas Vischer of the Faith and Order section of the World Council of Churches which spoke among other things of "the trilogy: communion *(koinonia), diaconia* and witness," a theme that, as Moeller notes, had been running through the World Council and its commission on Faith and Order.[38] This being the case, it will be of interest to extend the present survey of "diakonia" in today's church to the area of official ecumenism.

The World Council of Churches

Vischer's formula had come to its most prominent expression in the New Delhi Assembly of 1961. The assembly's sections were "Witness," "Service," and

"Unity," where unity was understood strictly as the Greek "koinonia" and where witness and service formed within it, in Vischer's later phrase, "one single coherent responsibility."[1] The assembly's theme was the plainly christological "Jesus Christ—the Light of the World," to which the three topics would seem to have a clear relevance, and the reports, which had been prepared after a wide and novel form of consultation with the churches,[2] were presented to the assembly in the favourable atmosphere created by the integration of the International Missionary Council with the World Council and by the important new membership of the Orthodox Church of Russia. The report on service in particular is also to be seen in the light of the strong tendency within the World Council to orient theology towards man's engagement with the world.[3] One practical manifestation of this tendency at New Delhi was the extension of the mandate of the Division of Inter-Church Aid and Service to Refugees to include World Service, a development that was influenced by what the division's report to the central committee prior to the assembly had called "the many theological discussions about the meaning of *Diakonia*."[4] More expressly, the director of the division stated the belief at the conclusion of his preliminary review of the section report to the assembly that "through the reflection which has been taking place in recent years among theologians on the significance of *diakonia* God has been preparing them to move into and to live in new political and social situations,"[5] and he grounded the belief most closely on reflection in Germany[6] where, as we have seen, service to human need was first presented as a biblical doctrine of "diakonia." Against this background we may note from the assembly's diary the "rich stimulus" of the day on which the topic of service was introduced. The day began with a sermon on "Bear ye one another's burdens and so fulfil the law of Christ" and concluded with the voicing of "much appreciation" for and "an unusually wide and alert discussion" of Masao Takenaka's address "Called to Service."[7] The sympathetic reception extended to the address is significant because Takenaka was proposing "humble service to the world" not merely as a duty in the line of ethics but, after the manner of an earlier World Council study on "The Lordship of Christ over the World and the Church," as "the true basis of Christian humanism."[8] This was to elevate service to the level of theological principle in the problematical area of church and world. The original study had put the principle in the following words:

> The Church manifests the Lordship of Christ over the world also by its humble service *(diakonia)* to the world. The Church's service is the making real in human life of Christ's own love for the world for the salvation of which he died, and thus it effects the restoration of man's true humanity as it was intended by God in the creation, according to the pattern of Christ who is himself the new Adam, the firstfruits of the new order. This is the true basis of Christian humanism.[9]

Thus "diakonia" is the transference to man in his humanity of the love that Christ manifested in effecting the world's salvation, and is a mandate for the churches in political and social situations, as logically inferred by L. E. Cooke.[10] The World Council had not always seen the relationship of church to world in so clear a light, however, nor did the vision remain for long unobscured.

The two preceding assemblies at Amsterdam in 1948 and at Evanston in 1954 had both sought to express the basis of this relationship in concepts drawn from natural theology: man as a social creature within "a responsible society."[11] An attempt in a scholarly symposium before the Amsterdam assembly to find the basis closer to hand in biblical theology had proved singularly unsuccessful.[12] It brought to light early Christianity's detachment from society,[13] described the problem of a biblical social ethic as acute,[14] and even raised a question around Karl Barth's principle that Christian man is "made for the service of his fellow men."[15] At Evanston there was indeed an emphasis on "ministry to the world" but the phrase did not carry the connotation of "diakonia." It was used in an evangelical sense,[16] and "works of service, compassion and identification" were seen merely as confirming the Christian's testimony.[17] Similarly, in acknowledging that "the ministry of the laity" is a share in "Christ's ministry to the world" the assembly was basically envisaging new attitudes to "daily living and work,"[18] and even the statement "The Church is sent into the world as a ministering community," which in later years could only point to "diakonia," was inculcating responsible involvement in one's job and social environment.[19] The difference between this conception of ministry and that of "diakonia" is apparent from the attitude to it that carried over to the assembly from the Advisory Commission. Weighing Christian responsibility in the face of "vast multitudes of hungry, homeless, and hopeless people," they urged the Christian to undertake his "due share" of social action and his duties in the "common life" he shared with other men—a type of service, in other words, in which governmental agencies and private institutions were already engaged.[20] The preparatory survey on evangelism had perhaps suggested the possibility of a more distinctively Christian service in the question it posed at the end of its deliberations, "What are the practical implications of the identification in service with others of which Paul speaks in 1 Cor. 9?"[21] However, the section of the assembly that dealt with social questions and was best equipped to provide an answer took the pragmatic line of "We are not called upon to shoulder the burden of this world, but to seek justice, freedom and peace to the best of our ability in the social order."[22] In this they reflected the conviction of their own preparatory survey that contemporary "social theology" was at best inadequate.[23]

At New Delhi an attempt was made to supply for this inadequacy by a concentration on service. Although reflection in this area was oriented chiefly to the kind of work undertaken by the Division of Inter-Church Aid, Refugee and World Service, service also constituted a section of the assembly as a whole and was as basic as witness to the view the delegates held of the church's role. The idea was accordingly set in a substantial theological context, as the opening paragraphs of the report indicate:

> Christian service, as distinct from the world's concept of philanthropy, springs from and is nourished by God's costly love as revealed by Jesus Christ. Any Christian ethic of service must have its roots there. The measure of God's love for men is to be seen in the fact that his Son was willing to die for them.
>
> Such is the God we worship and whose creatures we are called to love and serve for Christ's sake. All our service is a response to the God who first loved us. Justice is the expression of this love in the structures of society. In serving him and them, we follow the Christ who deliberately refused the way of force and chose the role of a servant. As the Father sent him, so he sends us to sacrifice

ourselves in his service. As Christ took the form of a servant and gave himself for the redemption and reconciliation of the whole man and the whole world, Christians are called to take their part in his suffering and victorious ministry as servants of the Servant-Lord. The power for service is given by the Holy Spirit who used the Church as his instrument in manifesting the Kingdom of God and Lordship of Jesus Christ in all human relations and all social structures. Service thus is a part of adoration of God and witnesses to his love for us and all men.[24]

The aim of this passage is to establish in service the church's mandate for involving itself in "all human relations and all social structures," where the term "structures" makes of service much more than an ethical value. The notion itself, however, is presented largely in terms of Christ the Suffering Servant and of the Christian in the service of God, and at no point is it clear why service of this nature, which is worshipful service of God and in theological terms has to do with the strictly religious matter of salvation, should necessarily manifest itself in a type of service that reaches to the structures of society. Had the emphasis been instead on service as "diakonia," which in its modern acceptation is truly service to man, an opening of the church to the world could have been argued more closely. Later paragraphs of the report do indicate that Christian service is "diakonia"[25] and in preliminary considerations this had been taken as "humble service to the world," but in the leading theological statement just cited the perspective and argument is different.

If in retrospect the New Delhi statement is deficient on the theological issue, the delegates themselves were aware of limitations in the practical sphere. Gratified as they were to see a spirit of service emphasised, they realised that this said nothing about ways and means, a deficiency frankly admitted in the report itself.[26] Moreover, although the report assumed the existence of a "distinctive Christian meaning of Service,"[27] it did not succeed in stating this in a way to convince delegates that Christian service was different, for example, from service in a welfare state.[28] Corporate Christian service tends to appear as a stopgap[29] and the service of the individual Christian as a religiously motivated participation in existing social services,[30] with the result that the theology of service appears to be competing on uneven terms with contemporary socialist theory.

It is not surprising therefore that a theology of service was only briefly to the fore in the World Council's social theory. The World Conference on Church and Society in 1966 did profess that the church can no longer seek to be "the governing, dominating institution" and that it had now "a chance to restore one of the essential marks of Christ's Church, namely to be a serving community in the world,"[31] but this formed no more than a minor motif at a conference where complex questions of world development took precedence over theology.[32] The question evoked similar passing references to service at the assemblies in Uppsala and Nairobi, but the social, economic, and political problems these assemblies set out to confront were of such a scale as to reveal the total unpreparedness of theology to cope with them.[33] The potential of theology to contribute to their solution was one of the first questions the joint Roman Catholic–World Council Committee on Society, Development and Peace (Sodepax) set itself to examine, and the theme of service does not seem to have been relevant to their purpose unless we except a reference to "Christ's relationship to his neighbours, his service in society as recorded in the Gospels and reflected in the theological and ethical statements of the Epistles," which is buried in a list of more than a score of themes and topics judged worthy of further investigation.[34]

Counterpart to Witness

There are many indications that although "diakonia" yielded ground to notions like freedom and justice in theological statements about the relationship of church and society it yet remains entrenched both as the expression of an essential characteristic of the church within any social system and as the particular designation of the Christian duty towards all those in any kind of need. Christ is "the *Diakonos,* the One True Servant," according to the Sixth Assembly of the Conference of European Churches, and as such is the pattern of what the church is to be in regard to men and women in their worldly needs, just as Christ "the *Leitourgos,* the One True Worshipper" and "the *Apostolos,* the One True Proclaimer" establishes the pattern in the religious area.[1] The pattern is established in the Son of man's word about the service "performed for us" (Matt. 20:28) from which "we know what helps man to be truly human."[2]

The human dimension has been appreciated especially in the socialist countries of Eastern Europe because it opens up the prospect of joint effort between Marxists and Christians "in the service of the people," which yet remains a form of Christian testimony within society at a time when the church has lost much of its position as an institution of witness over against society.[3] Within the Orthodox churches also Bishop Antonie Plămadeălă was led to do for Orthodox theology what he saw that "the rediscovery of diakonia" had done for Protestant and Roman Catholic theology in the West. From scriptures, Orthodox tradition, and modern writing, he disengaged a picture of "the Church of good works," a church that has "a clear theology of service" and "cannot justify itself on earth except through service."[4] This is a strong position to adopt. In 1978 Orthodox member churches of the World Council received a formal opportunity to further clarify and amplify their view of "diakonia" in a consultation on church and service sponsored by the World Council in Crete. The main paper was "Liturgical Diaconia" by Alexandros Papaderos. In this, where the basic notion of "diakonia" as "service to mankind" is derived from Beyer's article in Kittel's *Theological Dictionary,* the scope of "diakonia" is grandly summarized.

> Christian diaconia is understood as:
> a) *proclamation* of mercy . . . ;
> b) service of the *neighbour,* especially the 'least' . . . ;
> c) service of the *creation* . . . ;
> d) service of the *whole* human being . . . ;
> e) the service of the *whole* community . . . ;
> f) *liturgical diaconia* . . . a diaconia which equips us with the required "spirituality for combat" . . . , and finally
> g) *eschatological diaconia* (protecting diaconal aims from chiliastic-messianic

expectations . . .); diaconia as "sign" and foretaste of the kingdom of God and as criterion for acceptance into fellowship with God.[5]

This approach provides a clear delineation of "diakonia" as the service to human need which must accompany the church's witness and life of worship. The conception thus comes back to—even as it enhances—that held within the evangelical churches of Germany where, as the president of the Diakonische Werk stated, witness is not reduced to service or service substituted for witness and neither is separated in practice from the other.[6] This relationship is clearly maintained for example in the Leuenberg Agreement presented for consideration by some sixty Lutheran and Reformed Churches of Europe,[7] but during the 1960s it was often obscured by the admixture within the notion of "diakonia" of ideas about the service that effected salvation—that of the Suffering Servant—with the result that service appeared to envelop so much of the church's mission as to displace witness[8] and to occasion a conflict about priorities in the church's mission, which an observer once saw as "the most significant principle of division within Christianity."[9] Since the assembly at New Delhi, those most directly involved in service within the ecumenical sphere have themselves clarified their understanding of it. In 1965 the Division of Inter-Church Aid, Refugee and World Service published a study in preparation for the World Conference on Church and Society where the following definition of "diakonia" occurs: "the service which the individual Christian as well as the Church as a corporate unit is called to render to every needy person in all kinds of suffering and alienation."[10] comprehensive as the definition is, it does not overlap witness and, as this study and the later note on "The Biblical and Theological Bases of Inter-Church Aid"[11] make plain, is rooted with witness in the one mandate from Jesus Christ.

Any problem that "diakonia" now presents to the commission (as the Division of Inter-Church Aid since became) is not one of definition or of priorities but of method and scope. Reporting to the Nairobi assembly, the Programme Unit on Justice and Service had only this to say: "The Unit is still struggling to give better expression to the way it does theology and draws out today's theological understanding of the biblical words 'justice' and *diakonia.*"[12] Among the messages here is that the meaning of "diakonia" has been satisfactorily wrested from the scriptures and that the term has an accepted place within this sector of ecumenical work. The following definition appeared at that time in the World Council publication *Ecumenical Terminology:* "the practical service of the Church and the individual following Christ's example and in obedience to him, and in confirmation of the Gospel of God's love in Christ."[13] The German definition printed in parallel speaks of "the biblically based service of helping love."

Ten years later, in July 1984, on the fortieth anniversary of work of the commission, the central committee of the World Council issued the statement "The Diaconal Task of the Churches Today," which republicised the broad vision of "diakonia" projected at the Vancouver Assembly:

> Diakonia as the church's ministry of sharing, healing and reconciliation is of the very nature of the church. . . . diakonia cannot be confined within the institutional framework. It should transcend the established structures and boundaries

of the institutional church and become the sharing and healing action of the Holy Spirit through the community of God's people in and for the world.[14]

A reading of the essays accompanying the statement shows how what began as a venture to give expression to *koinonia* or fellowship is now presented as an embodiment of "diakonia." In this spirit the Larnaca Declaration made at the commission's world consultation in Crete in 1986 stated: "As the third millennium A.D. approaches, we dedicate ourselves, from this day forward, to work for justice and peace through our diakonia."[15]

In such an ecumenical setting, then, "diakonia" clearly has an assured place. One writer speaks of "diakonia" as "a well-worn word," even "rather old-fashioned . . . perhaps a bit retrograde."[16] Outside these circles, by contrast, not all are confident that the word, which in the parlance of an ecumenical technocrat might pass as old-fashioned, is not designating a newfangled idea difficult to find room for in the traditional perception of the Christian calling. An anecdote from Alexandros Papaderos about the reaction of an Orthodox priest in Crete to "diakonia" in practice nicely illustrates this.

> One of our priests . . . had great difficulty in fitting all this [the development of a farmers' cooperative] into his understanding of the mission and ministry of the Church. After the first seminar therefore he circulated a memorandum. . . . And heresy was precisely what the good priest thought we were engaged in. His argument was this: Christ came into the world to sow the seed of God's Word (Matt. 13) and not . . . tomatoes and cucumbers.[17]

The Doctrine of Ministry

The formula "witness and service" carries over at times from discussion of the church's ministry to the world to discussion of ministry within the church. It serves to emphasise that the church is built up not only through preaching and worship but also through complementarity within the fellowship of believers. Charitable and social "diakonia" is thus first exercised within the community before being extended to the world community. The emphasis is especially marked in statements like those of the Second Vatican Council about the pattern or style to which the ordained ministry should conform and is so widespread that in an extensive study of the history and modern content of the theology of ministry B. J. Cooke has been almost apologetic about introducing a separate chapter on "The New Testament Ideal of Service." He takes the opportunity nonetheless of reflecting on modern ministry as "taking care of the community" in the light of the modern truism that ministry in the early church was seen as "diakonia" and interpreted as "serving others."[1] This spirit is embodied in a particular way in modern forms of the order of deacons but beyond that is presented as the context for the exercise of authority, administrative procedures, academic theology, and in fact for life at all levels within the church.[2] As well, however, as setting a "diakonic" style for ministerial and nonministerial activities, modern theology has also and more frequently invoked "diakonia" for a very different purpose. This is for the defining of ministry, from which it emerges that ministry pertains to the body of Christ in its totality and makes ministers of all its members. In the phrase adopted at the national conference of French bishops in 1973, the church is "tout entière 'ministérielle' " or ministerial through and through.[3]

A measure of how widely this view is now represented is provided by the many agreed statements on ministry issued from consultations between representatives of different churches. Here there is an insistence that ministries exist over and above ordained ministries and that all ministries pertain to the one ministry of the church or are forms of what the Canterbury statement intriguingly calls "ministerial service." Thus that statement goes on to say, "The ordained ministry can only be rightly understood within this broader context of various ministries, all of which are the work of one and the same Spirit."[4] Putting forward the same principle, the St. Louis statement says, "We are convinced that the special Ministry must not be discussed in isolation but in the context of the ministry of the whole people of God"[5] and points in this to a procedural advantage for churchmen of different ministerial traditions. As explained by L. B. Guillot, who has investigated English terminology of ministry in some detail, the advantage is not merely a reduction in the tendency to think of "two separate categories in the Church, ordained and non-ordained" but that attention is shifted from ministry as "power over" or "power through" to ministry as "servanthood for" and "service to," thereby opening up common ground in ministries of different traditions and reducing the significance of differences between the historical forms in which churches have preserved their ministerial order.[6] A. T. Hanson calls this a concentration on the theology instead of on the pedigree of ministry.[7] The latter had led to a frustrating pursuit of questions put by A. M. Farrer in a now classic symposium, "Into what channels did the divinely instituted apostolic power flow? . . . Where did it go? And where is it now?"[8] Ministry, instead of power, and the verity and universality of service, instead of validation, have made that quest seem dated to many and even pointless. Both the emphasis on ministry and the resultant doctrine are modern, and, although one important factor working towards this is the traditional Protestant doctrine of the priesthood of all believers and another is a renewed appreciation of charismatic activity, probably the most influential concept has been that of ministry as "diakonia." "The Church," declares the St. Louis statement in explaining its terms, "has . . . the task of proclaiming the gospel to all, believers and unbelievers. This task or service of the whole Church is spoken of as 'ministry' *(diakonia)*."[9] The following discussion centres on that notion.

In the first decades of the Faith and Order movement, ministry meant the ordained ministry, and discussion of the subject was a set piece familiar to bishops and elders from centuries of post-Reformation debate.[10] Beyond an easy acknowledgement that ministry is necessary to the church and comprises word and sacrament, there were, as the report from the first conference at Lausanne frankly admitted, "manifold doubts, questions, and misunderstandings" in regard to "the nature of the ministry (whether consisting of one or several orders), the nature of ordination and of the grace conferred thereby, the function and authority of bishops, and the nature of apostolic succession."[11] Thus at one extreme representatives of the Orthodox Church felt obliged to enter the following grave dissenting note:

> The Orthodox Church, regarding the ministry as instituted in the Church by Christ Himself, and as the body which by a special *charisma* is the organ through which the Church spreads its means of grace such as the sacraments, and believing that the ministry in its threefold form of bishops, presbyters, and deacons can be based only on the unbroken apostolic succession, regrets that it is unable to come in regard to the ministry into some measure of agreement with many of the Churches represented at this Conference.[12]

Among the many churches referred to there were those which at the other extreme upheld the starkly simple view that "no particular form of ministry is necessary to be received as a matter of faith."[13]

Despite scholarly preparation for the next conference at Edinburgh in 1937[14] the report on ministry is a litany of the same irreducible differences, and behind the irenic but futile observation, "In every case Churches treasure the Apostolic Succession in which they believe,"[15] lay memorable and even heated exchanges.[16] Not surprisingly ministry as such was bypassed at the Third World Conference at Lund in 1952 and did not in fact form part of the Faith and Order's agenda for twenty-five years. Perhaps external events and the accidents of conference life played a part in this. Because the National Socialists of Germany proscribed the Edinburgh conference for the Lutheran Church and because schedules during the conference did not allow for debate on biblical aspects of doctrine,[17] the conference was deprived of a major sector of Lutheran opinion and in particular of discussion on the paper "The Origin of the Christian Ministry," which had been prepared by Professor Friedrich Gerke of Berlin.[18] This included one of the first formulations of a doctrine of ministry based on "diakonia" as a New Testament concept, and by also taking into account the discontinuity in the earliest Christian tradition of ministry it anticipated more recent approaches to the subject.

When the subject was resumed at Montreal in 1963 there was an immediate emphasis upon several biblical aspects, in particular on the priesthood of the whole people of God, on varieties of ministries in addition to the established ministry, and on ministry as God's service in and for the world. The central proposition put by the conference to the churches that "Christ calls the whole Church into his whole ministry"[19] was not met equably in all quarters,[20] but the position is firmly held in the report. Thus, "ministry is the responsibility of the whole body and not only of those who are ordained," and "the special responsibility" of the ordained, who are "servants of the servants of God" and "servants of the Servant of God," is "the equipment of the other members in the work of ministry"; in these ways "the whole body standing firm together is armed for its service."[21]

The doctrine of a general ministry was in no way thought to remove problems set for ecumenists by particular denominational attitudes in respect of the special or ordained ministry but it was seen as placing them in a promising new perspective and it established a pattern for all subsequent consultations.[22] The report of 1971 to the Faith and Order Commission in Louvain sought to clarify precisely this perspective,[23] and a phrase from a consultation in Marseilles in 1973 reveals how closely the doctrine of the common priesthood and the conception of ministry as service had converged to produce a picture of the ordained minister as "an associate in service in the midst of a servant people."[24] It was upon this theme that the Commission's statement from Accra opened in the following year:

> All ministry in the Church is to be understood in the light of him who came "not to be served but to serve" (Mk 10.45). It is he who said "As my Father has sent me, even so I send you" (Jn 20.21). Thus, our calling in Christ constrains us to a costly, dedicated, and humble involvement in the needs of mankind. Only so we may understand the whole ministry of the people of God, and only so the character of the special ministry of those who are called and set apart to serve and equip the Church by their stewardship of the mysteries of Christ.
>
> The ordained ministry is to be understood as part of the community. An understanding of the ministry must therefore start from the nature of the Church, the community of believers. This conviction is now shared by most of the churches.[25]

The statement then constantly resorts to the idea of service as the unifying principle amidst the diversity of ministries within and between churches. G. Gassmann has in fact noted that the draft statement was revised for the purpose of bringing this out more clearly.[26] The principle applies with particular effect to the central issue of how ministries so diverse as the episcopal and the nonepiscopal might be seen to be compatible. Service to the community is both the touchstone and the catalyst:

> The plurality of ecclesial cultures and ministerial structures does not diminish the one ministerial reality found in Christ and constituted by the Holy Spirit in the commission of the Apostles. . . .
>
> There is unity in the diversity of ministerial structures, in that the essential elements of ministry can always be identified in the very plurality and multiformity of ecclesial styles and structures. It would be difficult to imagine any structure of ministry which did not incorporate *episcope* . . . and *presbyteral* function. . . .
>
> Both the episcopal and presbyteral functions of the Church must be understood as a sharing in the *diakonia,* that is, as costly service to the community of the Church and to the world through the proclamation and actualisation of the gospel.[27]

This line of thinking came to its mature expression in the commission's next and most widely received statement on ministry, the Lima text of 1982.[28] Here ministry is presented from "the perspective of the calling of the whole people of God" (para. 6), people of "a new community" whose foundation is in "Jesus' life of service, his death and resurrection" (para. 1). On this ground a terminology is established: "The word *ministry* in its broadest sense denotes the service to which the whole people of God is called" (para. 7), with a special designation, "ordained ministry", being then necessary for that traditional role, a treatment of which forms the matter of the rest of the document. First responses of churches to this Lima text took its priority of the universal call very much for granted and, interestingly, saw a new kind of problem emerging, which is how to find room within existing church structures and practices for the ample resources of ministry now recognised. Some responses register a disquiet that the extended treatment of ordained ministry has not been matched by a similar treatment of the newer phenomenon of universal ministry. In the words of the Church of Scotland (Reformed):

> The stated importance of the ministry of the whole church (the whole people, all the people, the least people) and of the integration of ordained ministry with the ministry of the *laos* is not followed through to its implications for structured "lay" participation in ministry and government, i.e., in pastoring, liturgy, decision-making and "spiritual" functions generally.[29]

Paralleling and complementing reflection within the Faith and Order Commission are the many consultations on ministry between various confessional bodies. These have been in progress since about 1965, are usually on a bilateral basis, and are often of a national rather than an international character. Of special interest are those where pre-Reformation and Reformation theologies are represented because here the divergencies have been greatest; yet again, as F. Herzog forecast might be the case, the cutting edge has been "diakonia."[30] Thus, reporting on stages of the discussion between Lutherans and Roman Catholics in the United States, M. C. Duchaine duly noted the importance of the proposition that "Ministry *(diakonia)* is something committed to the entire church,"[31] and, in a survey of a score of various consultations, N. Ehrenstrom and G. Gassmann confirm the dominance of this principle. Their statement sums up the trend of the theology of ministry since the Montreal conference of 1963:

> In tune with general thinking on the subject today, the bilateral conversations usually interpret the ordained ministry within a multiple sequential context: the ministry of Jesus Christ, the ministry of the apostles, the Church as a ministering community with its multiplicity of services and functions, participating in and continuing the ministry of Christ to the world and being equipped for this service *(diakonia)* through a great variety of gifts of the Holy Spirit *(charismata)*, the special or ordained ministry. Rethought in this framework, several of the issues which have long been points of dissension among the churches have lost their divisiveness.[32]

The doctrine here spoken of as a breakthrough has been characterised by Claude Bridel as "the diaconal chain,"[33] and what the chain links and in what order bears emphasising. The progression is from Christ to apostles to *ministering community,* and only thence to ordained ministers, and is novel because it replaces the old "pipeline" from Christ to apostles to ordained ministers, the latter being then responsible for nurturing believers. As programmes for ministry these two are entirely distinct, and the recent one has been adopted by virtue of where "diakonia" is seen to reside, namely, in the "ministering community."

The same shift is evident in several modern translations of the passage in the New Testament to which Ehrenstrom and Gassmann allude, a passage moreover that, if there are such things as normative scriptural texts, is one such for the ordering of the church. In Eph. 4:11–12 we read that Christ on high gave teachers to the church "to equip God's people for work in his service" *(NEB)*, "so that the saints together make a unity in the work of service" *(JB)*,[34] "to prepare all God's people for the work of service" *(GN)*. This is the understanding reflected in Ehrenstrom and Gassmann's statement, and the passage was used to similar effect in the Second Vatican Council.[35] The traditional English translation, by contrast, keeps closer to the structure of the Greek: "for the perfecting of the saints, for the work of the ministry, for the edifying of the body of Christ" *(AV;* similarly *RV,* and first edition of *RSV)*; by its punctuation this translation is indicating that "ministry" pertains to the teachers. The second edition of the Revised Standard Version (1971), in keeping with the current trend, omitted the Authorised Version's comma after "saints" and thus encouraged the abandonment of the Authorised Version's meaning, which one commentator deprecates for being "aristocratic-clerical."[36] In a matter of such moment for our understanding of how the church works, however, one can only wonder how the author of the passage managed to be so unclear as to give leading divines of the sixteenth and seventeenth centuries one idea of his meaning and skilled translators four centuries later another and opposite idea. Certainly the earlier views had been that ministry here can only be the official ministry of teaching—a view supported in one comprehensive modern study[37]—and this was the mind too of the sixteenth-century reformers. Luther, despite introducing some subtleties, does not mistake in the mention of "ministry" a reference to "office,"[38] and Calvin is emphatic that the author of Ephesians is writing of the authoritative preaching ministry:

> Christ "ascended up far above all heavens, that he might fill all things" (Eph. 4.10). The mode of filling is this: By the ministers to whom he has committed this office, and given grace to discharge it, he dispenses and distributes his gifts to the Church. . . . In this way, the renewal of the saints is accomplished, and the body of Christ is edified . . . ; in this way we are all brought into the unity of Christ, provided prophecy flourishes among us, provided we receive his apostles, and despise not the doctrine which is administered to us.[39]

The common modern position forms a striking contrast with this classical position and well illustrates on what the doctrine of ministry now turns. This is, in T. F. O'Meara's phrase, on "a ministerial pleroma,"[40] which can be embodied in practice, to take a telling example from the Interim Constitution of the Uniting Church in Australia, in the principle that "ministry is a function of the whole Church to which all baptised persons are called."[41] The more vigorously such doctrine is pressed, however, the more taxing becomes the question and the more competitive becomes the field of who does what in the church. Thus David M. Gill, secretary of the Assembly of this Uniting Church, commented somewhat ruefully just eight years on from the formulation of its Interim Constitution: "At this moment we do not enjoy an excess of clarity in our understanding of . . . the calling of those who are ordained to minister to the ministering people of God."[42] The quandary is understandable. When the purpose of the God-given ministry is to bring believers up to ministry, where is ministry to be differentiated? "In a church which is completely ministerial," asked the French bishops, "on what conditions will the ministry of the priest, far from being lost in a blur of ministries, appear in its authenticity?"[43] The logic of such churchwide ministry probably issues in the "diaconal Christianity of the future" which Gyula Nagy pointed to: "In a 'servant Church' the hierarchical distinctions between members will increasingly disappear and all that will remain will be the functional distinctions."[44] Just such a prospect had led Yves Congar to repudiate the functional interpretation of ministry which he had explored in his widely read "Problems Affecting Ministry."[45]

Early in this trend of presenting the church as replete with ministry, the Faith and Order Commission itself pointed out that "there is no universally agreed language by which to describe the special ministry in distinction from the ministry of the Church as a whole,"[46] and one is led to wonder why at the threshold of a new theology of ministry traditional language should have failed theology so comprehensively. When that embittered character of fiction, the mother in D. H. Lawrence's *Sons and Lovers,* looked back to the lover she could have had, she recalled an exchange between them, one September Sunday afternoon as they walked home from their nonconformist chapel, about the path John Field would follow in life.

> "But you say you don't like business," she pursued.
> "I don't. I hate it!" he cried hotly.
> "And you would like to go into the ministry," she half implored.
> "I should. I should love it, if I thought I could make a first-rate preacher."
> "Then why don't you—why *don't* you?" Her voice rang with defiance. "If *I* were a man, nothing would stop me."[47]

In contrast with the clarity of this universally recognised usage of 1913 is the advice in 1982 of the Committee on Doctrine of the United States' Catholic bishops: "while consistency in . . . use of 'ministry' has value, it is premature to attempt a precise definition of the word. A rush to judgement would only disrupt discussion of the serious theological problems which definition raises."[48]

William Bausch, who drew attention to this odd modern perplexity, himself recommended that "if anything, ministry should be seen as a code word for the mixture of the ordained and non-ordained,"[49] but the serious problems envisaged by the bishops do not thereby go away. These are of the order alluded to by the Faith and Order Commission in collating responses to its Accra statement on ministry: "A certain distinction between general and special (ordained) ministry is recognised and accepted in practically all replies. There are, however, many diverging views on the question as to whether it is a distinction in function, degree or kind."[50] Disagreement at this level is major, and at the extremes of opinion—"function" versus "kind"—must reflect different perceptions about the nature of ministry, opening up the question of how deep accord really goes in ecumenically agreed statements that have no agreed position at this point. It is one thing to concur with the commission's Lima statement that ordained ministry is "constitutive for the life and witness of the Church" (para. 8) and another to suspect, as the Baptist Union of Great Britain and Ireland hoped it could, that the proposition is countermanded elsewhere in the document by the insistence on ordained ministry having "no existence apart from the community" (para. 12).[51]

Further to this difficulty of conceptualising ministry in a churchwide ministry are difficulties arising when the broader view of ministry is built around an older priestly or sacerdotal theology. Here there is a need to take account not only of a general and of a special ministry but also of a common and of a restricted priesthood, and indeed of a sacrament of order that has always been a dividing line between these ministries and priesthoods. There would have been no ground for surprise, accordingly, at the "discomfort" confessed to by members of the North American Academy of Liturgy when they addressed this question at an annual meeting. The members represented Roman Catholic, Episcopal, United Presbyterian, and United Methodist traditions, and their discomfort was in respect of "hierarchical and sacramental definitions of ordination" insofar as they imply power and authority for ministry and elevation of the minister to a clerical class within the community. Giving rise to their concern was the perception all members apparently brought to their discussion—namely, that "this elevation and authority tend to vitiate the function of ministry as service to the servant community."[52] Confirming what our own discussion has been illustrating, the writer of the report observes: "To make the ministry of presbyters—or of bishops, for that matter—the point of departure for understanding the ministries of the Church will yield conclusions very different from those produced when the community as a whole becomes the starting point."[53]

Such a typically modern preoccupation does not seem to have overborne the writers of the Canterbury statement on ministry, which is representative of Anglican and Roman Catholic theologies, although in presentation at least their considerations proceeded along the same line from ministry broadly conceived to the ministry of the ordained. Of the ordained they wrote, "their ministry is not an extension of the common Christian priesthood but belongs to another realm of the gifts of the Spirit."[54] To express themselves thus was to come down on the side of difference between the ministries; with the formulation proving equivocal to some, however, the commission provided an "elucidation" in 1979 which spoke of the two priesthoods as "two distinct realities", with the word "priesthood" applying to both "by way of analogy."[55] By virtue of this explanation the term "ministry" itself becomes—by an unusual stretch of language—an umbrella word for two priesthoods only analogously related, and in the light of this the commission's insistence in the introduction to its statement that the ordained ministry "can only be rightly understood" within the context of various ministries loses much of its point, for two such distinct priesthoods can hardly be considered expressions of a univocal ministry. In its response to the commission's statement, nonetheless, the Bishops' Conference of England and Wales welcomed "these very clear statements"[56]—not so the French Episcopal Conference, which would have wished for more precision at this point[57]—but went on to affirm in the document preparatory to the Extraordinary Synod in Rome in 1985, called to review the Second Vatican Council, that its thinking on ordained ministry remained grounded in the commonness of ministry: "The council . . . presented the ordained ministry in the role of *diakonia* or service, designed to enable all ministry in the Church to flourish in the unity of the one mission of Christ."[58]

With greater apparent precision than the Canterbury statement, the Second Vatican Council spoke of the difference between the common priesthood and the ordained priesthood as a difference "in essence and not only in degree."[59] A statement of the difference in such terms could once have been explained by way of the late scholastic theory of sacramental "character," but the theory—if not some of the language associated with it which survives to obscure lines of thinking—has been widely abandoned.[60] If the council had in view here a "common" priesthood and a "ministerial or hierarchical" priesthood, the difference between the two should not be sought in the notion of priesthood itself ("sacerdotium" in the council's Latin) because priesthood is what is shared, or so the statement seems to say;[61] at all events any attempt to reerect a precisely sacerdotal theory of the ordained ministry is out of the question because, whatever may have been the exact nature of official ministry in the early Christian mind, the idea of a sacrificing priesthood within Christian communities was alien to their thinking.[62] Rather, the difference should logically be sought in the qualification introduced by the council, namely, "ministerial or hierarchical." Of these two terms, "ministerial" is the novel one, reflecting a major theme of the council as we have seen in an earlier section and, because of the word "or," is presented as an alternative to "hierarchical"; it should thus of itself be an adequate designation of what characterises one who has received the sacrament of order. Interestingly, in a letter to priests in 1979, Pope John Paul II steps back from such usage in adopting "ministerial" as a characteristic of what is in his view an essentially "hierarchical" priesthood; such at any rate would seem to be the implication of his changing the phrase "ministerial or hierarchical" to " 'hierarchical' and at the same time 'ministerial.' "[63]

These Roman Catholic and Anglican statements about the difference between ordained and nonordained ministries are attempts to fit a new evaluation of ministry into an older theology of priesthood, and not surprisingly have attracted the kind of criticism that Christian Duquoc directed at *Lumen gentium* of Vatican II, namely of "the lack of articulation between the image of the ministry as service (ch. 3, para. 18) and that of the priesthood as 'sacred power' (ch. 2, para. 10; ch. 3, para. 18)."[64] Apart from a few writers who seem able to view this conflict with apparent equanimity ("Christians are hopeful beings"),[65] most assume that a new task for theology has to be done on the basis of what Anton Houtepen called "the essential rooting of the ministry in the laity,"[66] and in such writers as Duquoc, Moignt, Burrows, Schillebeeckx, Boff, Tavard, and O'Meara we meet an admixture of perplexity, hope, and inventiveness in a search for a way to balance out roles within "every-member ministry" without thereby necessarily making of Christianity "a one-caste religion."[67]

For a long historical period preceding the modern fixation on a churchwide ministry even those churches that eschewed the so-called metaphysical or ontological approach of catholic theologies in characterising ordained members had recognised that ordination—whatever its nature—is a constant of Christian tradition and that it is more plainly an induction to ministry than are rites of Christian initiation. They accordingly gave ordained members first call on the term "ministry." Some had it and others did not. This was more than a pragmatic approach to a theological conundrum. In *The Assembly of the Lord,* for example, Robert S.

Paul illustrated the resistance of the Puritan and Scottish divines, even under the extreme circumstances of the English civil war, to Erastian influences on the Westminster Assembly's "Grand Debate" on church order; these men, like their predecessors of the sixteenth century, were reformers of theological conviction. The conviction about ministry was admirably put in the Second Helvetic Confession of 1566:

> To be sure, Christ's apostles call all who believe in Christ "priests," but not on account of an office, but because, all the faithful having been made kings and priests, we are able to offer up spiritual sacrifices to God through Christ (Ex. 19.6; 1 Peter 2.9; Rev. 1.6). Therefore, the priesthood and the ministry are very different from one another. For the priesthood, as we have just said, is common to all Christians; not so is the ministry.[68]

The same conviction, redolent of "the exalted notions held as to the nature and character of the ministry,"[69] was echoed three and four centuries later in the opening statement on ministry at the conference in Edinburgh in 1937, which, ironically, was to bring ecumenical discussion of ministry to a long-standing impasse: "The ministry was instituted by Jesus Christ, the Head of the Church, 'for the perfecting of the Saints . . . the upbuilding of the Body of Christ.' "[70] Here "the ministry" means the ordained ministry and it looks to the word "ministry" in the citation from Eph. 4:12. Currently that word would be understood as "diakonia," in which form it would no longer designate the distinctive role of the ordained but the role of the community at large. Within this "diakonia" the ordained are understood to be carrying out traditional functions but are not considered to be more "diakonical" or ministerial than the nonordained.

The question is more than terminological. At root there are convictions about "diakonia" as a common and indeed a univocal Christian condition. "These people," wrote Max Thurian of the ordained in introducing some churches' responses to the Lima statement, "represent the Servant Christ in the servant church so that all the faithful may become servants of one another and servants of their sisters and brothers in the entire human family."[71] The word "servant" for these purposes in the Greek of the New Testament is διάκονος, and Thurian's view is classical "diakonia," which rests on philological judgements about the special use early Greek Christians are said to have made of terms like διακονία. The importance of those judgements in respect of the kind of theology we have been reviewing can be gauged from the precision with which they are frequently enunciated in treatments of the subject.

The Language of Office

The most influential treatment has probably been Eduard Schweizer's chapter on office in *Church Order in the New Testament*.[1] The chapter opens with the observation, "One of the most surprising consistencies of the New Testament witnesses is seen in word-statistics." The Greek language was well supplied with words for the designation of office, and in recalling some of these Schweizer points out that in the New Testament none of them applies to roles within the Christian community. ἀρχή denotes office as a position of precedence or rulership, τιμή as a position of dignity, and τέλος as a position of power. Christian writers might have been expected to adopt the term λειτουργία because it was widely current for services within the state and had both civic and religious connotations; the term had also been chosen in the Septuagint for the duties of priestly office. Schweizer proceeds:

> As a general term for what we call "office," namely the service of individuals within the Church, there is, with a few exceptions, only one word: διακονία. Thus the New Testament throughout and uniformly chooses a word that is entirely unbiblical and non-religious and never includes association with a particular dignity or position.

Having noted then some particularities of usage in Hellenistic Judaism, he provides the following "basic meaning":

> In the development of Greek the basic meaning, "to serve at table," was extended to include the more comprehensive idea of "serving." It nearly always denotes something of inferior value.

Among exceptions he notes occasional uses relating to cult and other service of God in the papyri, in Hellenistic sages, and in Josephus, the latter in particular twice designating himself διάκονος to mean "the mediator of divine prophecy."[2] In the light of general usage, however, such instances are minor and "the evidence of the choice of words" in the New Testament "is unmistakable":

> The very choice of the word, which still clearly involves the idea of humble activity, proves that the Church wishes to denote the attitude of one who is at the service of God and his fellow-men, not a position carrying with it rights and powers.

Underlying this choice is the principle that links ecclesiology with christology. The passage continues:

> This new understanding is the continuing testimony to God's action in Jesus of Nazareth. The fact that it was in lowliness that God revealed himself as God implies for the Church that through being itself prepared to be lowly it must become separated from the world, to which indeed all kinds of ceremonial associations with imposing dignitaries belong. "He who is greatest among you shall be your servant." This sentence, in its six variants, is based in Luke 22.27, in a very primitive formulation, on Jesus' conduct: "I am among you as one who serves." Special ministry takes place in the Church only in special subordination. Acts, as well as Paul, likes to describe all special activities, particularly the preaching of an apostle or some other church member, as that kind of "ministry" or "service," and the person who performs it as a "servant" or "slave," with God, Christ, or men appearing as those to whom the service is rendered.

In urging the special character of Christian usage and its particular significance for an understanding of church office, Schweizer makes numerous references to Jewish sources in both Greek and Hebrew but none directly to Greek usage as such, although several indirect references are provided by way of lexicons and

Beyer's article in Kittel. As a point of presentation this should be noted because the argument is ultimately linguistic in nature. Christian writers undoubtedly do apply the term διακονία and its cognates across a wide field of their practices, and if other writers normally apply the words to services of a lowly and nonreligious kind Schweizer could well be right to see in the pervasive presence of these words in the New Testament evidence of a Christian desire to conform all aspects of life in the community to the pattern of Jesus who had been among them as the serving one. Such at any rate has subsequently been agreed among many of those who use linguistic data in determining the general question of church office.

While for some the stress fell on the ethical value of "diakonia," reminding ministers that they are but lowly servants of God,[3] many saw "diakonia" as constitutive of institutionalised ministry or church office. When André Lemaire reviewed a significant amount of the literature to 1973, he concluded that "diakonia" had set the pattern for the modern discussion and that the few Roman Catholic writers who continued to characterise official ministry by recourse to ideas of priesthood were neglecting a basic fact of early Christian language.[4] Not all of these writers, however, took the argument so far as Schweizer to find the notion of office inappropriate for the modern church. For some it was sufficient that "diakonia" required church office to be seen and exercised as a ministry for the benefit of the community, and they would concur with Rudolf Pesch: "The ministerial structures of the New Testament communities can only be of help in that they provide a model for the organisation of ecclesiastical services in the present day; the one thing necessary is the basic structure which Jesus established for the ministry: diakonia."[5] For others the relevance lay in the seemingly nonreligious character of the Greek term so that a case could be put for the desacralisation of church office.[6] Or the idea might be used mainly to enliven the church's awareness of its ministerial capability by broadening the pool of ministers and ministries available to it.[7]

The thinking here, as Schweizer had shown, tied in with " 'ministry' as synonymous with 'gift of grace.' "[8] If ministry is a charisma, and every charisma a ministry, and if again there is no Christian who does not have his or her own charisma, then the church is thoroughly replete with ministerial powers, and the need for office becomes questionable. Smooth runnings of the church, said Schweizer, will require that a ministry of a public nature be regulated: "If we like, we can call such ministry an office; but we must be clear that this is simply a matter of order, and that an 'office' is not on principle separated from a "ministry." "[9] By force of its meaning, accordingly, "diakonia" ushers in the charismatic church; "diakonia" is a kind of ministry that is laid upon all—"everything 'that edifies' is ministry"[10]—and that engenders "a fundamental equality."[11] On such a basis the community's structure is "one of free fellowship, developing through the living interplay of spiritual gifts and ministries, without the benefit of official authority."[12] If the individual's gift is the ministry of the word, this too fits within the charismatic pattern. "There is not even a prerogative of official proclamation," according to Käsemann, so that the apostle—whose work is a "diakonia" (Rom. 11:13)—is "only one charismatic among many."[13]

On this ground writers have staked women's claim to a share in official ministry. Rosemary Ruether wrote that "the Spirit is no discriminator among persons on the basis of gender but can empower whomever it will. Ministry is proven by its gifts, not by its credentials."[14] Preeminent among women's gifts is said to be the "human dimension" they offer ministry: "Through the ages in the pain of childbirth, through dealing with personal problems which arise at every stage in the lives of loved ones, and finally through the care of the sick and dying, women have had an opportunity to centre in on all that is individual and deeply human."[15] The link between such experience and what "diakonia" says ministry is has often been made, and it was reestablished in the following reflection from the Klingenthal consultation on women and ordination, which the World Council of Churches sponsored in 1979: "Those who have long been assigned the status of servant in many cultures and societies have a unique opportunity to renew the ministry of the Church through re-examining the assumptions held about service, and about the separation of service *(diakonia)* from clerical roles."[16]

In other words, we are invited to consider ministry to be a compassionate service of others, to concede that women are better at this kind of thing than men, and then to recognise that from its earliest stages the male ministry was unfaithful to the gospel in excluding from ministry those who were more naturally fitted to meet the full range of demands in the ministerial calling. "Not the Twelve but the women followers prove to be the true disciples of Jesus," argued Schüssler Fiorenza:

> The women not only accompany Jesus on his way to suffering and death but they also *do* what he had come to do, namely, to serve *(diakonein,* cf. [Mark] 10:42–45 and 15:41). . . . In Mark's theological perspective women are the functional successors of Jesus and they represent the true intention of Jesus and his mission within the messianic people of God.[17]

Accordingly, as women gain access to ministry, they will, in Anne Carr's words, transform it:

> The ordination of women . . . would further the transformation of the priesthood: by admission of those who have traditionally only served, the sign will be clear. It will help to transform the ministry from a predominantly cultic role to a ministerial one, from a symbol of prestige to a symbol of service.[18]

Even more pointedly, Rosemary Ruether notes:

> If we seek to put some women in that historical form of the ministry which has evolved in the Roman Catholic Church, it is not because we seek something we do not have, but because we wish to give to this clergy something that they do not have. . . . We seek to help them rediscover the very nature of ministry, not as an alienated power and domination, but as the service of Christians to each other in the building up of a community of brothers and sisters.[19]

Possibly the most sensitive essay on what it might mean for women to minister was that of M. Timothy Prokes—her small book was closer to the genre of Simone de Beauvoir than to the classical theology of Schüssler Fiorenza—but she kept the reader clearly in mind, nonetheless, as to where such estimations of ministry took their rise. In the section "Ministry and the Flesh" she drew directly on Beyer of the *Theological Dictionary* to present *"diakonia* as the discharge of *any* service in genuine love which contributed to the upbuilding of the whole community."[20] Exactly here, too, Elizabeth Tetlow established the norm of authentic ministry, and by departures from the norm measured fidelity to the primordial intention about ministry:

> The ministry of servant was new and unique and had not been connected with any formal religious office in the history of Judaism. There was a greater freedom for the inclusion of all human persons in the new ministerial model of servanthood. As long as servant remained the primary model for Christian ministry women were able to minister on the same basis as men. When, at the close of the New Testament period the Christian model of servant was replaced by Jewish models of presbyter and bishop, and in the second century the Old Testament model of levitical priesthood was applied to ecclesiastical office, women came to be excluded from the official ministry of the Church.[21]

This "diakonic" mode of office has, of course, been widely advocated also among men: "Office *is* service," Walter Kasper wrote, "and that means being-for-others." "The validity of ministry," according to Kilian McDonnell, "is tied not to a given hierarchical structure, but rather to the verity of *diakonia.*" "Service," wrote Gerald Moede, "is quite definitely the criterion of office."[22] This view immediately challenges the validity of a notion of sacred authority in the forms, as we have known them, of hierarchy; thus, among "the chorus of voices insisting on the 'service' character of church government"[23] have been those urging that "hierarchy" should make way for "hierodiakonia" or "syndiakonia."[24] On this plane authority ceases to bind and loose and is replaced by "mutual subordination" and by "a spontaneity as great in serving as in obeying,"[25] which is virtually the picture of it emerging from the Lima statement on ministry:

> Because Jesus came as one who serves (Mark 10:45; Luke 22:27), to be set apart means to be consecrated to service. . . . Therefore, ordained ministers must not be autocrats or impersonal functionaries. Although called to exercise wise and loving leadership on the basis of the Word of God, they are bound to the faithful in interdependence and reciprocity. Only when they seek the response and acknowledgement of the community can their authority be protected from the distortions of isolation and domination.[26]

The views of these and other writers, among them some of the most widely read, on the nature of office in the early church rest largely on judgments about the meaning of a small set of Greek words such as the judgements we have seen here and in earlier sections of Schweizer, Beyer, and Brandt. Although the actual pattern of early church life remains obscure to this day, and unstable and variegated even where something of it can be discerned, the recurrent theme of "diakonia" in early writings has seemed to these many modern theologians to provide at least one constant; apostle, presbyter, and believer, we are to take it, found common cause with him who came to serve. They had thereby been placed in such a novel religious condition that neither the linguistic storehouse of the Jewish Septuagint nor the legal, administrative or religious language of the contemporary pagan Greek world, creative and even fanciful as that was in the religious sector, could furnish them with a basic terminology—prepacked, ready for use. Instead they had to process their own. For a starting point they are said to have turned to what Audet called "the most 'uninstitutionalised' language of the day,"[27] and they came up with a word that Hans Küng has repeatedly described in terms like the following: "a very ordinary and not a religious term, with a vague connotation of lowliness, and thus no possible evocation of any association with any public power, administration, sovereignty, high dignity or function of lords and masters."[28] The term was, as he goes on to write, " *'diakonia'*, or service (in its fundamental acceptation, *service at table*)."

Küng concluded his influential book *The Church* with a section entitled "Ecclesiastical Office as Ministry" (and the translation "ministry" for the German "Dienst" was hardly apposite, as the citations which now follow suggest). Here in a few pages he drew together some of the implications he saw in the early Christian predilection for this "very ordinary" word. Among other things we read:

> The very concrete secular sense of the word reveals what a tremendous impact it must have had not only on the Greeks but on all normal thinking people, when Jesus proclaimed: "Let the greatest among you become as the youngest, and the leader as one who serves. . . ."
>
> But it is clear that Jesus is not merely concerned about service at table. . . . His fundamental concern is with living for others (cf. Mk. 9.35; 10.43–45; Mt. 20.26–28); and the origins of the word diakonia, in contrast to other similar verbs, indicate that a completely personal service is implied. This is an essential element in being a disciple: a man is a disciple of Jesus through service of his fellow men. Jesus chose and emphasised this new conception of service. . . . This is a point where something distinctively Christian can be discerned, as the choice of a completely new word shows. The consequences are enormous. Is it possible for there to be among the followers of Jesus any kind of office which is based on *law* and *power* and which corresponds to the office of secular potentates? . . .
>
> Or can there be among the followers of Jesus any kind of office which is based in *knowledge* and *dignity*, and corresponds to the office of the scribes? . . .
>
> It is not law or power, knowledge or dignity but *service* which is the basis of discipleship.[29]

These lines are not cited for their originality—Küng did not pretend to that here and plainly acknowledged his debt to Beyer and Schelkle[30]—but because they are succint, representative of so much modern thinking in the area of church office, and because from the point of view of the Roman Catholic Church, for which they were penned, they were revolutionary. Although the consequences for official policy and thinking in regard to ministry have hardly been "enormous," as Küng had hoped, the impact of these pages on the middle and younger ranges of Roman Catholic clergy was very great indeed, just as these pages were among those most warmly received by non-Catholic writers.[31]

The purpose of reviewing what Küng and Schweizer in particular have had to say about ministry as "diakonia" has been to draw attention to the nature of the argument. This is narrowly linguistic. That is to say, the argument is strong to the extent that dictionary meanings provided for the Greek are reliable. Part of the argument's attraction is undoubtedly the plausible human story accompanying it, namely, that Jesus' radical demands on his followers could not be expressed in conventional Greek and that accordingly early Greek-speaking followers reached ingeniously into their everyday language to coin the necessary idiom. And much of the argument's force has become the fact that it is now part of the stock in trade of those dealing with church office.[32]

Deacons

The ministry totally open to a notion like "diakonia" is that of deacons. The reason for this of course is the direct link in language between diaconate and διακονία, between the deacon and his title in Greek διάκονος. This title came into English from the first Christian contact with the roots of our language in the British Isles; cognate words of the Greek term, by contrast, did not become part of English by this kind of transliteration. Thus the cognate verb διακονεῖν did not become "to deacon" but was translated. We see the difference at 1 Tim. 3:13 where we read of those "who minister as deacons": here both "minister" and "deacons" are διακον- words in Greek. This tells us that, from the first, newly converted nations—for this process repeated itself in other languages of western Christendom—were given to understand that they needed a special title for the deacon lest his diaconate be confused with other types of ministry or service designated in the Christian scriptures as διακονία. Given this intimate historical and linguistic link, a redefinition of the Greek words is inevitably going to work towards a reassessment of the deacon and his office. In an earlier section we have seen that in the nineteenth century the Lutheran deaconesses and deacons were totally oriented towards works of service, and that in this century Wilhelm Brandt prepared his linguistic study of service for the purpose of fixing this orientation. As a consequence the general assumption is that diaconate is, or should be, the institutionalised form of the notion we have been calling "diakonia."

The most far-reaching attempt to renew the diaconate along these lines has been within the Roman Catholic church. In 1964 the Second Vatican Council decreed that a permanent diaconate—that is, a diaconate in which the ordained deacon is not merely "a probationer under training for office" of presbyter (to use Thomas Smyth's derogatory phrase about "this useless order" from last century)[1]—could be restored "as a proper and permanent rank of the hierarchy."[2] By ordination these deacons would be dedicated to the People of God "in the service of the liturgy, of the Gospel and of works of charity." The council's Latin word for "service" is, significantly, "diaconia" (with immediate reference in this instance to the order of "diaconate" itself but with overtones of the then fashionable notion of "diakonia" which we have been examining). The Apostolic Letter of Paul VI *Ad pascendum* of 1972, which contained norms for the order of the diaconate, depicted its role more fully in the following statement:

> The permanent diaconate should be restored as an intermediate order between the higher ranks of the Church's hierarchy and the rest of the people of God, as an

expression of the needs and desires of the Christian communities, as a driving force for the Church's service or *diaconia* towards the local Christian communities, and as a sign or sacrament of the Lord Christ himself, who "came not to be served but to serve."[3]

Here the Latin term "diaconia" takes on the restricted meaning with which we are familiar and is coupled with the "diakonia" of the Son of man in the gospels (Mark 10:45).

That the new deacon should be a sign of the Lord's "diakonia" was thus inculcated by the Vatican's norms, was enshrined in diocesan and national regulations,[4] and has been the ideal which deacons themselves found attractive and sought to embody.[5] The Latin American bishops put the ideal comprehensively at Puebla in 1979: "The charism of the diaconate, a sacramental sign of 'Christ the servant,' is very effective in bringing about a poor, servant Church that exercises its missionary function for the integral liberation of the human being."[6]

Although this Roman Catholic emphasis upon a diaconate of service owed much to the long-standing Protestant predilection for "diakonia," the first stirrings of interest were free of that influence. Josef Hornef, a layman who wrote prolifically on the subject in Germany in the decades following World War II, was far more concerned to see the establishment of a ministry that could make up for what he accurately foresaw would be a serious shortage of priests. In this he was also influenced by Otto Pies, a Jesuit priest whose experiences in a concentration camp had fostered a highly spiritualised view of priesthood and had engendered the notion that an auxiliary ministry might safeguard the priest's commitment to prayer and reflection.[7] The early book of Wilhelm Schamoni, *Married Men as Ordained Deacons*, worked with a similar set of ideas.[8] When in 1957 the Dominican theologian M. D. Epagneul publicised this line of thinking in a leading journal, the article was referred to the Vatican's Holy Office and received a rejoinder that same year from Pius XII himself in an address to a major Congress of Lay Apostolate;[9] in the pope's view a diaconate that was not a temporary clerical state preparatory to priesthood would obscure the demarcation between laity and clergy and could undermine the received conception of priesthood. Accordingly when Winninger and Hornef returned to the subject of "an independent diaconate, that is, a permanent diaconate detached from the priestly state," they were careful not to impugn the uniqueness of priesthood ("sacrificial and priestly in the real sense") at the same time as they advocated a "levitical" ministry, "diaconal in the real sense," which would assist the priesthood at the altar, in catechesis, and in works of charity.[10]

Such early initiatives were thus pragmatic in character, and advocates of an active diaconate had to pick their way carefully between a recently developed but strongly endorsed theology of lay apostolate and a sacrosanct view of priesthood. During this period an initiative towards a more clearly defined objective for the diaconate was being undertaken in southern Germany by a young forestry worker turned student of welfare, Hannes Kramer. He envisaged a diaconate for professional married men that would represent the church's involvement in society and manifest the charity of the gospel. In Freiburg in 1951 he established what he called the first "Diakonatskreis" or diaconate group, whose aim was "to serve the needy as married permanent deacons."[11] With the support of some theologians and bishops and a deepening association with Caritas, the welfare agency of the German bishops whose work was directed from Freiburg, a movement got under way to work for recognition of such a diaconate by church authorities. Meanwhile other diaconate groups were formed in Munich, Cologne, and Trier and then in France at Lyons. A dominant idea among these men was that the church's ministry as then structured was ineffective beyond parish confines and that its scope could best be enlarged by the creation of a special ministry capable of working on the margins of church and society. Here matters of liturgy and catechesis were not of direct relevance so that emphasis fell upon works of charity.[12]

This thinking comes round to that of the founders of the German Protestant diaconate in the previous century but, interestingly, Kramer did not at first come under that influence. One reason for this was that he envisaged a diaconate by ordination, and thus as a part of hierarchy, whereas deaconesses and deacons of the Evangelical foundations were not ordained into official ministry and led an institutional life that had the appearance of being a Protestant variation of Catholic congregations of nuns and brothers. By the time Kramer invoked "diakonia" as the loving service of the needy and as the aim of diaconate it was less an exclusively Protestant conceptualisation than an idea pervading much thinking on ministry in all churches.

In the interests of gaining recognition of a permanent diaconate the two lines of thinking about what it might be came together; the one line was characterised by a traditional view of pastoral activity whereas the other looked to a specialised ministry embodying "diakonia." With the announcement in 1959 by John XXIII of the convening of the Second Vatican Council the opportunity was brilliantly seized not only to bring the notion of permanent diaconate to the church at large but to press for its implementation. The world's bishops were lobbied by means of a dossier on the issue,[13] and 1962 saw the publication before the opening of the council of the compendious volume *Diaconia in Christo* edited by Karl Rahner and Herbert Vorgrimler. This book brought together scores of historical and theological studies as well as reflections on the pastoral situation throughout the world which revealed the potential field of ministry for deacons in those lands especially where ministry was being impeded by a shortage of personnel. In the council's third session the fathers duly voted for the restoration of a ministry called diaconate. Its inauguration in individual nations was left to the decision of national conferences of bishops acting under approval of the Vatican. Action was in fact slow and sporadic, due in part to the Vatican's tardiness in issuing guidelines, and in part to the reluctance of some bishops to incorporate married clergy within the hierarchy, but increasingly to theological uncertainty about the need for and the nature of an ordained ministry whose functions are virtually all within the competence of the nonordained.[14] Thus, although by 1988 there had been almost 14,000 permanent deacons ordained since the first in 1968, 8,500 of these were in the advanced and well-serviced society of the United States, with another 1,200 in the similarly situated church of Western Germany, but a mere 500 and 200 had been made available to the vast churches respectively of Brazil and Africa.[15] And significantly, in reviewing the experience of diaconate on the occasion of the twenty-first anniversary of its renewal, writers from half a dozen countries echoed one another in recording doubts about the identity of the deacon and in reporting tensions between deacons and priests or deacons and nonordained pastoral assistants.[16] The basic problem was in the underlying notion of "diakonia," because when "diakonia" is service of others and is a mandate from which no Christian can be excluded one will have difficulty finding a place in the church for specialised servants like deacons of the "diakonia." Within another closely allied tradition this was precisely the ground upon which a working party once recommended that the diaconate—and they were writing of the traditional diaconate—should be allowed to lapse.[17]

Strangely, during the period when the renewal of the Roman Catholic diaconate was being implemented, any new linguistic work relating to deacons and their origins did nothing to support the notion of "diakonia" by virtue of which the renewal was proceeding. Jean Colson, who wrote much on ministry in early sources, produced a small book on "the diaconal function," which was the first of the period to look in any way closely at the evidence of the New Testament.[18] Here, acknowledging what he called "an original 'mystique' " engendered by the consistent play on "diakonia,"[19] he nonetheless declined to accept that as the mainspring of the hierarchical diaconate. He saw the title "deacon" originating rather in a Greek translation of a title current in the conservative Jewish settlement at Qumran and carried over into Christian circles. Debatable as this proposition is,[20] it allowed him to place the diaconate within a harmoniously structured and clearly defined order of early Christian ministry which was ministerial strictly insofar as it dispensed religious salvation.[21] This was to place the diaconate, alongside the episcopate and the presbyterate, in the realm of the sacral. Colson repeated his views in his contribution to the volume *Diaconia in Christo,* which was a main instrument in bringing the diaconate to the attention of the Second Vatican Council, but they are views that do nothing to illumine the rationale of "diakonia." Perhaps for this reason his essay was replaced by one that aligned itself with "diakonia" when a selection of the studies in *Diaconia in Christo* was published with acts of a congress on the diaconate held in Rome in 1965.[22]

About the same time Manuel Guerra published a comprehensive study called "Greek and Biblical Deacons," which similarly called into question a basic tenet of the modern position. The main feature of the title "deacon" in non-Christian Greek, as depicted by Guerra, is not what it says about service of others but what it says about service of the gods and of the city-state. This is a major shift of meaning. Although Guerra's presentation is marred by undue religious connotations that he attributes to talk about service of the state, his extensive collection of sources does point to the highly literary character of the usage, and in the area of service of the gods, in which he is able to adduce references to cultic "deacons," the usage indisputably exhibits a religious character. On turning to the New Testament Guerra fails to see what this could imply, perhaps because in seeking to concentrate on the deacon of church order he pays only cursory attention to Pauline usage, which is the bulk of the matter and the most telling from a linguistic point of view. He concludes that the word "deacon" (Phil. 1:1 and 1

Tim. 3:8–13) occurs in a nontechnical sense of assistant to community leaders in the service of the faithful, a use that established itself by virtue of the assistant's role at the eucharistic table.[23] If this was a disappointingly minor conclusion from so major a study, Guerra appreciably broadened the field in which it is necessary to consider the original diaconate and reduced the likelihood that the narrow base of "diakonia" is its correct setting. Unaccountably Guerra's work seems to have been almost entirely neglected.

The same neglect on the part of students of church order has been noted earlier in respect of the work of Dieter Georgi.[24] Georgi openly contested the evidence for the existence of a Christian notion called "diakonia." "The NT term," Georgi maintained, "almost never involves an act of charity."[25] For him, in line with his interpretation of Pauline usage, the deacon is a preacher. For André Lemaire, who unlike Georgi found no helpful precedents for Christian usage in general Greek usage and wrote on evidence in the New Testament and early church fathers, the original deacon was an itinerant officer for liaison between churches.[26]

These few opinions about the role of the early deacon and the language by which he was designated are born of the little linguistic work on the matter that has been done independently of Brandt and Beyer, and it must be seen as working against the tendency, set in motion by them in the 1930s, towards a diaconate of service. Even in their day the judgement of Hans Lietzmann was available—and it was reflected in Moulton and Milligan's standard work of reference—that the Greek term $\delta\iota\acute{\alpha}\kappa o\nu o\varsigma$ commonly occurred in "more elevated language."[27] If this opinion is reliable, it suggests that early Christians may have had more in mind when they adopted the title "deacon" than a fellow Christian engaged in the kinds of service to which they were all in fact obliged. Theology of the diaconate, it would seem, was carried forward in recent decades with almost total reliance upon a small school of opinion which may not be compatible with other specialist opinion about the meaning of the word "deacon." One of the most eloquent writers on diaconate, Claude Bridel, also an ecumenical colleague of those working in the International Centre for the Diaconate in Freiburg, expressed the following grave disillusionment about what we have been calling "diakonia":

> In the first place . . . we have the inflation of the term and its erection into a veritable myth. To such an extent does every one speak of serving—baptising his administrative, parish-pump or philanthropic activity with a word that has become banal—that Christian declarations in this style appear merely to be following in the wake of the spirit of the times without any expression being given to just where the service of the church is to be distinguished from various humanitarian projects unless this is by way of a vocabulary that is obscurely technical (ministry, diakonia) and of a pious phraseology which attempts to give substance to it.[28]

2

The Servant Son of Man

How confident can we be that the modern notion called "diakonia" corresponds to a notion entertained by early Christians? The best way to approach this question is to go to the only point in the early tradition where they have written expressly of Jesus as "serving." At this point expert opinion should crystallise, and we can hope to arrive at an authentically Christian view of the service of Jesus. The following are the two familiar passages concerned, and the relevant Greek words about serving are in parenthesis. The translation is from the Revised Standard Version (RSV).

> You know that those who are supposed to rule over the Gentiles lord it over them, and their great men exercise authority over them. But it shall not be so among you; but whoever would be great among you must be your servant [διάκονος], and whoever would be first among you must be slave of all. For the Son of man also came not to be served [διακονηθῆναι] but to serve [διακονῆσαι], and to give his life as a ransom for many. (Mark 10:42–45)

> The kings of the Gentiles exercise lordship over them; and those in authority over them are called benefactors. But not so with you; rather let the greatest among you become as the youngest, and the leader as one who serves [ὁ διακονῶν]. For which is the greater, one who sits at table, or one who serves [ὁ διακονῶν]? Is it not the one who sits at table? But I am among you as one who serves [ὁ διακονῶν]. (Luke 22:25–27)

In a strikingly similar way both of these passages bring ethical teaching about disciples comporting themselves like servants around to the attitude and behaviour of Jesus himself. Equally striking, however, is the difference in context of the two passages. The second passage is from Luke's account of the last supper, and "one who serves" (expressed in Greek as a present participle, "the serving one") is appropriately a designation for the table waiter. In the first passage, by contrast, reference to dining is not explicit, and the words "servant," "slave," and "to be served" refer in a more general way to minor functionaries and lowly activities in a royal or lordly household. Because the similarity of ethical teaching and language suggests that one passage may have owed something in its formulation to the other, commentators have often looked for signs that might clarify the relationship. Among these, two are commonly said to be telltale signs. First, Mark uses διακονῆσαι, "to serve," a word said to refer in the first instance to serving at table. Second, in Mark the Son of man moves on from the idea of serving to the idea of saving, that is, from ethics to theology or soteriology, and this is a variation which makes of Mark 10:45 one of the most discussed statements in the gospels. Nowhere else, apart from the parallel passage in Matthew, does Jesus speak of himself so expressly as a saviour. Because Luke makes no explicit mention of this aspect of doctrine, and because he uses the word "serving" in a context that is understood to be natural to it—namely, with reference to serving at table—is it possible that Mark has changed the context so that he could turn an original small body of ethical teaching into a proposition about Jesus as saviour and use it at the climactic point of his gospel where Jesus is about to go up to Jerusalem?

In 1913, in his influential *Kyrios Christos,* Wilhelm Bousset judged that this was the case. He wrote:

> One of the most important observations of synoptic criticism . . . may be made in a comparison of Mark 10.41–45 with Luke 22.24–27. For on the basis of this comparison it is established that the logion of Mark 10.45, which is so heavily freighted with dogmatic import . . . , has its simple original form in the saying, "I am among you as one who serves" (Lk. 22.27).[1]

In an explanatory footnote Bousset then outlines the stages by which the simple original form developed into the form with dogmatic import. First, the title "Son of man" was inserted in place of an original "I"; second, the formulaic word "came" was introduced; third, the contrasting clause "not to be served but"; fourth, "to serve" was glossed by the addition of a phrase ("and to give his life as a ransom for many") referring to sacrificial death.

In the more recent writing cited in the following pages we shall see that some such history of the saying is often taken for granted. Some commentators will say that the glossed "ransom" phrase originates in reflection among Greek Christians on the mission of Jesus, others that it originates in Palestinian or Semitic circles. On either view Mark 10:45 splits at the middle, so that it is customary to speak of verse 45a and mean "the Son of man came not to be served but to serve," and of verse 45b and mean "and give his life as a ransom for many." This division is usually adopted also by those who are less interested to discuss the provenance and formation of the saying than to explain the meaning of the saying as a whole. The division also, of course, emphasises the role of the notion of service in the verse and leads us to ask what Mark, or the tradition behind Mark, could have intended by that notion if in this new context and expanded saying it could no longer have been intended to convey the notion of service at table, as in Luke 22:27.

The purpose in reviewing an amount of modern commentary, then, is to assess the level of agreement about the kind of service attributed to Jesus. In particular we will want to know whether scholarly opinion supports the view that this service is the "diakonia" so widely discussed in other areas of theology. In other words, by speaking of his own service, is the Son of man giving divine sanction, as one writer puts it, to "the new economy of service and sacrifice for others in action,"[2] or, as another puts it, is he teaching that "greatness in the Kingdom is according to the amount of service rendered"?[3] Or is the service something else again?

Service as a Saving Action

Wilhelm Brandt was the first in more recent times to draw special attention to the presence in these sayings from Mark and Luke of the verb διακονεῖν rather than of some other Greek word for serving.[1] Because his book is not easy to come by and because of his influence on Beyer, whose treatment will be reported next, his approach is outlined rather fully. We note that he leaves aside the question of how the sayings in Mark and Luke are related to one another and prefers to treat them as witnessing independently to what διακονεῖν means in statements about the mission of Jesus.[2] In approaching this question he considers the meaning of the word elsewhere in the gospels and finds that it designates household activity, especially that associated with the table. In this connection Brandt passes a judgement that determines the outcome of his enquiry: although the word denotes service at table, it does not point so much to the activity of serving as to the help that is provided in the act of serving.[3] He instances one passage where the latter aspect of the verb's meaning crowds out in effect any idea of serving at table so that it is to be taken as designating any helping act.[4] This is at Matt. 25:44 in the parable of judgement where the wicked answer the king, "Lord, when did we see thee hungry or thirsty or a stranger or naked or sick or in prison, and did not minister (διακονήσαμεν) to thee?" Brandt proposes that in this passage we are able to discern the precise aspect of service that early Christian writers intended to convey by this verb, and he writes about it in the following way:

> Διακονεῖν designates an act of assistance pure and simple which, plainly, loses nothing by being directed at an external need. In all the passages mentioned[5] the word designates a service rendered to one's fellow-man. It does not express the service in terms of a relationship to the master whom one might be serving. An exception here would be John 12.26 ["If any one serves me, the Father will honour him."] but even so the idea in this instance is that in rendering service to a master one is advancing his interests; it is a διάκονος of this kind that the Father will honour. At all events we can say in general: the verb διακονεῖν does not denote service under the aspect of obedience but as a relationship to one's fellow-man. Διακονεῖν is one of those expressions which suppose a "Thou," not however a "Thou" to whom I am free to relate myself in any way I choose but a "Thou" to whom I subordinate myself as a διακονῶν.[6]

In this attempt to get to the meaning of the word in the gospels one cannot mistake Brandt's intention to elevate the notion of goodwill and personal commitment above any connotation of servility that a word meaning "to serve" might be expected to carry in the Greek language of the time. And it is indeed a cardinal point of this section of his book that an orientation to the other person, to the "Thou," is the element in the word's meaning that Christian writers were seeking to elicit in order to reflect faithfully the teaching of Jesus. That the conception of service delineated in the preceding citation corresponds to and in fact originates in the teaching and attitudes of Jesus Brandt shows by way of Luke 22:27 and Mark 10:45.

An indication that the saying in Luke, "But I am among you as one who serves," expresses a refined sense of selfless dedication to the care of others is the emphatic use by Jesus of the expression ἐγὼ δέ, "But I." Because he is speaking here in his capacity as Messiah, the expression "as one who serves" represents, beyond any particular act of Jesus, "the sum total of his actions in regard to the disciples."[7] Brandt writes:

> διακονεῖν is used as a designation of Jesus' life-work. That a word from the language of banqueting should take on this meaning was not thought to be out of place. Indeed the passage in Luke reflects an awareness of the exceptional character of the development. By the question with which it begins ["For which is the greater, one who sits at table, or one who serves?"] the passage establishes that the act of διακονεῖν was normally held in low esteem. With the ἐγὼ δέ, however, we come to realise that Jesus was fully conscious of this attitude but nonetheless uses διακονεῖν to express the meaning of his life.[8]

The other thing worth emphasising in Brandt's understanding of the phrase "as one who serves" is that the service is directed exclusively towards people. Luke 22:27 "speaks of Jesus' relationship to the disciples." Brandt observes: "In that relationship we have the meaning of the saying but also its delimitation. Parallel to that side of Jesus' life which concerns people is the side by which he is related to God. Of the latter relationship, however, the saying in Luke has nothing to tell us."[9]

Against this background of usage in the gospels Brandt turns to consider what the verb means in Mark 10:45, "For the Son of man also came . . . to serve, and to give his life as a ransom for many." He finds that the sense the word carries in Luke 22:27 is also immediately recognisable here but that by force of the title "Son of man" and of the pregnant expression "came" the word "to serve" states more clearly than in Luke the meaning of Jesus' life.[10] To this a profound perspective is added by the phrase about giving his life as a ransom, because the phrase determines what Jesus means by serving: "Mark 10.45 gives a quite specific meaning to the διακονεῖν of Jesus: his service is authenticated in the laying down of his life; even such an action is service."[11]

In comparing the two sayings, accordingly, Brandt is able to point to the following difference in the kind of service comprehended by the Greek word: "In one saying the meaning of the word extends to every helping act performed in Jesus' life; in the other before anything else it is the laying down of life for others."[12] On either presentation, however, service so characterises what Jesus does that it reveals what his messiahship consists in: "In statements about his messiahship service is never renounced; rather service is the expression of messiahship: *the Christ* serves."[13] By a unique combination of this service to men and women and of obedience to the will of the Father—an obedience which on Brandt's understanding is notionally distinct from the διακονία of Jesus—Jesus achieves his saving work: "It is for the sake of his mission that he is the servant."[14]

Brandt's approach through this kind of service to an interpretation of the saving mission of Jesus is already familiar from the review in the preceding chapter of modern attitudes towards various aspects of the mission of the church. The central role played by what we called there "diakonia" in the formation of those modern attitudes is however to be attributed less to the immediate influence of Brandt's exposition than to the treatment of the διακον- words by H. W. Beyer, who drew substantially on Brandt's ideas.[15] Even though Beyer interprets Luke's phrase "as one who serves" of a particular action by Jesus at the supper rather than of his general dealings with the disciples, he concludes with Brandt that Jesus is here "instituting in fact a new pattern of human relationships." "He makes this no less clear," Beyer continues, "in terms of the specific process of waiting at table than by His own action in washing the feet of His disciples."[16] Beyer's judgement at this point deserves further comment.

Because he sees a parallel of sorts between John 13:1–11 and Luke 22:26–27, Beyer is drawn to attribute to the Jesus of Luke's supper the moral attitude he discerns in Jesus at the washing of feet, which is in effect to endow the notion of waiting at table with an overlay of concern or care for others. As will appear from the following pages, other commentators also make use of the passage in John to clarify the passage in Luke, but in Beyer's case the outcome is to reinforce ideas he already holds about the meaning of the Greek verb in Christian sources. Brandt himself, it may be noticed, did not use the comparison for such narrowly lexical purposes but mainly saw in the action of washing feet the kind of lowliness that should characterise mutual service among Christians.[17] At all events Beyer accepts Brandt's view of a gradual enrichment of the meaning of διακονεῖν within Christian circles from "to wait at table" to something like "to care for others." We see Beyer calling Martha's activity, for example, "solicitous" (fürsorgende)[18] when a more straightforward reference to her role at the meal (Luke 10:40) would depict her as "waiting on" her guest. In the statement about serving the king in the parable of judgement (Matt. 25:44) the development of the word's meaning within the field of Christian ethics has reached its peak. Here, as Beyer puts it, the word "comes to have the full sense of active Christian love for the neighbour and as such it is a mark of true discipleship of Jesus."[19]

This developed sense is basic also to Mark 10:45, but because it applies there to Jesus himself we recognise the unique kind of care that transformed Jesus' relationship with people into a saving work:

> διακονεῖν is now much more than a comprehensive term for any loving assistance rendered to the neighbour. It is understood as full and perfect sacrifice, as the offering of life which is the very essence of service, of being for others, whether in life or in death. Thus the concept of διακονεῖν achieves its final theological depth.[20]

The phrase "being for others" is noteworthy. It is the succinct expression of Beyer's and Brandt's perception that "to serve" at Mark 10:45 establishes the serving character of the whole of Jesus' ministry. Indeed it is by virtue of this moral stance, in life as in death, that Jesus saves. Beyer states that the value of Jesus' suffering is in the service that is rendered there; the service is a condition of the suffering being acknowledged as sacrificial.[21]

Such an elevated assessment of the place of service in the theology of Jesus' mission is not often reflected in subsequent discussion of Mark 10:45. Even those who interpret verse 45a as a saying about service to others and who then examine the "ransom for many" of verse 45b in the light of this interpretation refrain from giving "to serve" an inherent soteriological value. Their views will be recorded principally in the subsequent section on "The New Ethic" where it will frequently appear that these authors find Beyer's strikingly simple conception of a ransoming service incompatible with rather complex views about the formation of verse 45a–b of the kind referred to under the name of Bousset earlier in this chapter. Perhaps the writer who has maintained Beyer's approach most closely at the same time as he has taken into account opinions about verse 45a–b that run counter to that approach is Karl Kertelge. He writes that "it is only in the light of the range of Jesus' deeds *for* people, seen as a laying down of his life in service [dienende Lebenshingabe], that we come to understand the saving power of his death."[22]

Quite different is the use which P. H. Boulton made of Beyer.[23] Unconvinced by the attempts of those whom he calls "the 'community sayings' connoisseurs" to distinguish a saying of Hellenistic origin about a "ransom for many" from a saying of simple ethical character about service of others, he finds sufficient material in Beyer for "a radical revision of the all too common practice of ignoring the place held by διακονέω and its cognates in the theology of the Gospels—and indeed in the mind of our Lord himself."[24] A consideration of this material leads him to propose that by virtue of the association of service at table with the theme of the messianic banquet (Luke 12:37; Matt. 22:13) an eschatological connotation attached itself in the gospel tradition to a "primary Greek meaning of 'to wait at table.' "[25] The theme is represented in Luke's reference to Jesus "as one who serves," and in Mark's use of "to serve" it as adapted to a theological statement "expressing the complete fulfilment of a complete sacrifice": "the Son of Man Himself has come, not to be treated as one sitting at meat, but to be identified with the classificaion διάκονος in relation to those He must redeem."[26] The theme of the messianic banquet is actually lacking in Beyer, just as Beyer's characteristic motif of care for others is not taken up by Boulton. Boulton feels nonetheless that he is restoring to the mention of service in Mark 10:45 a soteriological dimension attributed to it by Beyer but lost subsequently to view through a haze of dubious critical opinions.

Many besides Boulton accept the same primary Greek meaning—as against a primary meaning in the gospels—of "to wait at table," W. Grundmann for the apparently sound reason that if a sense of service other than service at table had been intended the author would have chosen some other word than διακονεῖν from the ample supply of Greek words meaning "to serve" in one way or another.²⁷ Grundmann, however, is singular in then taking Mark's "to serve" as one of the "picture words" that Jesus uses to describe the banquet in his Father's house. This metaphor embraces the activity of his life and extends also to include his death, so that the verb assumes in this instance, as with Brandt, Boulton, and Beyer, an inherent soteriological value. In saying that he has come to serve, the Son of man is saying that he has come to save.

Jesus' Service at the Supper

J. Wellhausen understood "to serve' at Mark 10:45 as an explicit and direct reference to "the diakonia of the last supper," that is, to the distribution by Jesus of bread and wine.¹ The reference is obscured of course by the present context of the verse but is sure because, as he says, $\delta\iota\alpha\kappa o\nu\epsilon\hat{\iota}\nu$ means "to attend upon, to wait at table." Further, the reference is supported by Luke's use of the same word in the context of the supper and by the incident of the footwashing in John 13. Then, just as the two latter passages say nothing of the saving work of Jesus, so the service Jesus speaks of in verse 45a does not lead into the doctrine about a ransom in verse 45b. In fact Wellhausen calls this transition a $\mu\epsilon\tau\acute{\alpha}\beta\alpha\sigma\iota\varsigma\ \epsilon\grave{\iota}\varsigma$ $\check{\alpha}\lambda\lambda o\ \gamma\acute{\epsilon}\nu o\varsigma$ or jump from one category to another, but he finds a justification for it in the fact that at the supper "Jesus used bread and wine in giving out his body and blood."

G. Dalman took a more restricted view of what "to serve" referred to. This was to the footwashing itself,² an action that was symbolic of "the renunciation of one's rights for the sake of others." In Jesus' case such renunciation was "absolute" and "the essential thing in His life work," but again it did not necessarily imply the offering of himself in death. This thought is expressed by Mark in the separate statement of verse 45b.

H. Schürmann's approach to "to serve" is determined by the fact that he sees in the verse a motif of the servant figure from the Book of Isaiah. Other writers, like those in a subsequent section on the Isaian servant, see the motif in "to serve" itself, but Schürmann has difficulty conceding such a direct dependence because Mark and the Septuagint use different words for "serve" ($\delta\iota\alpha\kappa o\nu\epsilon\hat{\iota}\nu$/ $\delta o\upsilon\lambda\epsilon\acute{\upsilon}\epsilon\iota\nu$).³ Mark's words in fact, especially the Greek of "to be served," are reminiscent of service at table, and are judged by Schürmann to be relics of an earlier saying that would have shown a more explicit reference to waiting at table, much as Luke 22:27 still does. Although he then points out that service at table could well bear a soteriological connotation (cf. Luke 12:27; John 13) and need not be the vehicle for merely ethical teaching,⁴ he argues that the soteriology of Mark 10:45 does not derive from reference to the table and that reference to the table in verse 45a has in fact been blunted by the attachment to it of verse 45b, where the soteriology of the verse becomes explicit through the different and unrelated idiom of Isa. 53.⁵

The greater originality of Luke 22:27 argued in these studies by Schürmann is decisive for E. Schweizer, who then suggests that Luke's "as one who serves" could well reflect "a reference by Jesus to his serving" at the last supper.⁶ Mark

10:45, by contrast, was probably produced in the Greek-speaking Jewish church on the basis of the Isaian doctrine of the servant. Such doctrine, however, is confined to verse 45b,[7] and its formulation was occasioned by the tradition about Jesus serving (verse 45a) at the supper.[8]

J. Roloff has combined a use of Schürmann's analysis of these sayings with some assumptions that lie behind the studies of Brandt and Beyer. His purpose is not merely to find a connection between "as one who serves"/"to serve" and the supper but to arrive at the soteriological value of these expressions in the light of what the supper itself meant.[9] His paper is a very close examination of the material. He agrees with Brandt and Beyer that "$\Delta\iota\alpha\kappa o\nu\epsilon\hat{\iota}\nu$ is one of those terms which in the course of being taken over by early Christians developed a new specific meaning."[10] Where those writers however turned in the first place to the gospels, discerning in "care for others" an adequate expression of the saving work of Jesus, Roloff turns first to the writings of Paul and the Acts. He is in fact critical of Beyer for forming judgements on the basis of usage in the gospels without giving due weight to the tradition that formed the usage.[11] Paul and Acts, on the other hand, in passages like 2 Cor. 3:7–9 and Acts 1:17, provide instances of a settled usage, "a specific use" for "functions within the community and their authentication."[12] This usage is so foreign in his judgement to non-Christian Greek, where "the basic meaning of the verb $\delta\iota\alpha\kappa o\nu\epsilon\hat{\iota}\nu$ is of course waiting at table,"[13] that it becomes a base from which to trace back the development through Christian sources.

From the material available in the synoptic gospels, accordingly, Roloff seeks to determine at what point the original profane meaning begins to change by virtue of the word's currency in Christian communities. Traces of the change are discerned in Mark 10:43–45, but meanings there only approximate to the later specific meaning.[14] Turning to Luke 22:26–27, he takes from Schürmann that verses 24–27 are not a redaction of Mark but of an independent source in which these verses were attached immediately to verse 20; also that the redaction was carried out in the interests of producing a code for leaders of eucharistic assemblies. He concludes that in Luke 22:26 "the transition . . . to the specific ecclesiastical meaning is complete."[15]

Further to this, however, Roloff envisages the possibility that Luke's source originally lacked its present ethical orientation. The consequence would be that $\delta\iota\alpha\kappa o\nu\epsilon\hat{\iota}\nu$ carried a meaning more closely in tune with the context of the supper, that is, it would refer more directly to an activity around the table. Roloff pursues the thought by way of verse 27. He does not accept that the verse is merely a Hellenistic formulation. Nor is the verse pointedly ethical in intent, as a comparison with verses 24–26 and especially with Mark 10:45a shows. Its terms and antitheses are rather to be seen as realistic traits of an actual meal, just as the emphatic use of the Greek "I" and of the phrase "(I) am among you" suggest the presence of a real person. The verse points in fact to the supper itself and to Jesus there as "serving."[16] In the circumstances, however, serving cannot be any such action as the breaking of bread because Jesus was doing that by force of custom as head of the household, nor can it be the washing of feet because that incident is unknown to the synoptic tradition. "Serving" is rather an interpretation of Jesus' attitude at the eucharistic meal: "Jesus' giving of himself and the shedding of his blood ὑπὲρ ὑμῶν [for you]—for the community that is taking shape in the persons of the disciples—are presented here as a beneficent service to those gathered about the table."[17] By contrast with Luke, Mark 10:45a does not suggest the actuality of the original eucharistic meal, and the verse's explicit theology ("came"/"ransom") could even be a compensation for this. Nevertheless, "to serve" there does stem from the tradition in which Luke's "serving" expressed in reference to Jesus "the central motivation of that role of his which was being actualised at the meal."[18] As it stands in Mark's saying, however, "to serve" is midway in the process of passing from that tradition where it received its original Christian determination into the church's vocabulary as a term for specifically community functions.

> In Mk. 10.45 we are at the precise point of tradition at which recoining began of the general and non-specific root διακ- in the form of specific ecclesiastical terminology. . . . Thus, through being adopted as the indication of Jesus' attitude on the occasion of the last supper, the general term for waiting at table—διακονεῖν—would have become the key term for designating any service performed in the community after the manner and under the mandate of Jesus.[19]

The Community's Eucharist

J. M. Robinson sees no reason to look beyond the tradition itself for the explanation of the origin and meaning of "to serve."[1] The word is simply one of the "eucharistic motifs" that possibly indicate the influence on formation of Mark 10:43–45 of difficulties experienced by communities in relation to the Eucharist (cf. 2 Cor. 11:27–34).

The tradition is credited with a more positive role by H. E. Tödt,[2] who is followed rather closely by H. Kessler.[3] According to Tödt the service of Jesus at the supper (John 13; Luke 22:27 "as one who serves") took on a new significance when the supper was interpreted by the Palestinian tradition in the light of Isa. 53 (cf. Mark 14:24). When the quite different question was raised of the authority of the Son of man and of favours disciples might expect through his exercise of it (Mark 10:35–45), a word was required that would unmistakably define the mission of the Son of man as "to become lowly."[4] "To serve" was chosen because its link with the eucharist (Luke 22:27) provides the opportunity to amplify the statement about mission ("to become lowly") with the eucharistic theme of "ransom."

For W. Popkes[5] not only the term for service but the notion itself derives from tradition about the supper, as the Lukan context and catchword ("serving"; cf. John 13) indicate. The Hellenistic community combined the saying about service with the saying about ransom, which had its own earlier history within the tradition of the supper.

Mention might best be made here of Bo Reicke's investigation into the connection between liturgy and care for the poor.[6] On Reicke's reading, the primary reference of διακον- in the New Testament is not to a service of care but to liturgical service. Only in a subsidiary study, however, does he say what this might mean for an understanding of Mark 10:45.[7] The saying exposes the motive for the Christian practice of "diakonia" as "service to one's neighbour" and, in the Greek term used for "to serve," exposes also the liturgical connection which is more clearly in evidence in Luke 22:27 and John 13.

The Isaian Servant

Despite the frequency of the judgement that Isa. 53 lies behind Mark 10:45, there has not always been a readiness to equate the activity of the servant figure there with the service of the Son of man. Thus in the second section in this chapter we saw Schürmann could not accommodate the Son of man's service within an admitted Isaian motif because the words for "to serve" are different, διακονεῖν in the one case and δουλεύειν in the other. The same objection is voiced by writers who do not admit an Isaian influence at all; Hooker, from the next section, and Barrett, from later in this chapter, might be mentioned as being especially widely read in the English-speaking world. Hooker makes the further point that the service of the Isaian figure is directed to God, not as with the Son of man to people. C. E. B. Cranfield has responded to the latter point by insisting that "the actual content of the [Isaian] Servant's service of God is a service of man,"[1] and to the former by observing that since the kind of service designated by δουλεύειν lacks a "manward aspect" διακονεῖν fills a gap in so far as "it recalls, and sums up in a word, the *whole picture* of the service to men which the Servant of Yahweh is to render." P. Benoit treats the matter more lightly, attributing the emphasis upon different kinds of service to differences of context.[2] A. J. B. Higgins concedes that the difference in terms could pose a problem but argues that διακονεῖν is required to balance the term διάκονος in verse 43.[3] He adds that δουλεύειν would not have been convenient in any case because, apart from having no passive by which Mark's "to be served" might be expressed, that word has an "over-servile" meaning.[4] In the same way F. H. Borsch explains Mark's term as "an attempt to soften the harshness" of the word in the Septuagint.[5] S. Lyonnet, on the other hand, practically eliminates the problem by taking διακονεῖν to mean "service to the community," which allows him to appeal directly to the Septuagint where "the Servant of Yahweh is said to exercise his office 'by serving well the community.' "[6]

Higgins is not so precise as to who it is that receives service; it is "congruity of thought" with the passage in Isaiah that matters.[7] This suffices also for R. H. Fuller, for whom the three phrases of Mark 10:45 are "generalised summaries of the fate of the Suffering Servant," with "to serve" being "equivalent to 'fulfilling the mission of the Servant.' "[8] Hardly more precise are L. L. Carpenter,[9] T. W. Manson,[10] J. Schniewind,[11] P. Carrington,[12] S. E. Johnson,[13] A. Jones,[14] and A. Lemaire.[15]

While C. H. Dodd is similarly satisfied to see the verse as "no bad summary" of the Isaian servant's career, he realises that the recipient of the service in Isa. 53 is God.[16] F. H. Borsch is one of the few other proponents of an Isaian motif in Mark's "to serve" who acknowledges that fact; but although he then discerns a shift of emphasis in Mark from the idea of service to God to the idea of being "a servant of the people," he recognises "the same background of beliefs" in both passages.[17]

In contrast to such broad views of the service of Jesus in his role as the Iasian servant, other writers see it in precise terms as referring to "one definitive act of self-surrender" (W. Manson),[18] or as the "service . . . of vicarious and representative suffering" (V. Taylor).[19] By drawing on Boulton's study mentioned in the first section in this chapter, L. Sabourin also considers that "to serve" expresses "the idea of a sacrifice pressed to its furthest limit."[20]

The most persistent advocate of an Isaian servant Christology in Mark 10:45 has been J. Jeremias, but it is of interest that in earlier studies he had been reticent in respect of Mark's "to serve" itself.[21] Later however he brought the word more closely into line with the verbal parallel he had long proposed between Mark 10:45 and Isa. 53:10–11 by relating "to serve" to the verb $‛bd$ attested in the versions (as against the noun of the Massoretic text: $‛μbdy$). Further, the word "and" after "to serve" is epexegetical, so that the verse is to be understood as saying that Jesus "serves by surrendering his life."[22] In regard to Luke 22:27 he credits "serving," in the phrase "as one who serves," with a certain parabolic character, which appears exclusively in ethical teaching but which arises in this instance nonetheless from specific activity of Jesus at the supper.[23] While Mark's reference to service, therefore, is due to a biblical interpretation of Jesus' mission, Luke's is due to an equally old tradition about what Jesus did at the supper.[24] The parabolic quality in Luke's saying, however, gives it a meaning in regard to Jesus' mission as a whole:

> Common to both strands of tradition is the fact that each presents Jesus as a pattern of serving. . . . The difference between them is that the way in which Jesus serves is illustrated differently; in Luke by his waiting at table (cf. John 13), in Mark by the surrender of his life (cf. Isa. 54).[25]

What remains unsaid by Jeremias is whether the use of διακον- in Mark (instead of δουλ-)—and the former is "very closely connected with the eucharistic words"[26]—is due to a connection between Mark's saying and tradition concerning the eucharist. The stress on the distinct traditions of Mark 10:45 and Luke 22:27[27] and on the Hellenistic character of Luke's ὁ διακονῶν ("the serving one")[28] would seem to indicate that in Jeremias's view no connection exists.

The New Ethic

E. Lohse relies largely on Jeremias's investigations into the Semitic background of verse 45b in presenting this as a doctrinal formulation from the Palestinian tradition cast in terms of the Isaian servant.[1] He also finds with Jeremias that Mark is free of the Hellenising tendencies discernible in Luke. Because Mark's theme of service (verses 41–45a) is however an ethical instruction about how disciples are to live, he finds there is a conflict between that and the theme of eschatological salvation (verses 35–40, 45b) into which it has, understandably enough, been introduced.

Earlier A. Schweitzer had taken advantage of the eschatological context to give a profounder meaning to the ethics.[2] For Mark, in contrast to Luke, service is not simply an "inversion of the ideas of ruling and serving" but the law of *"interim-ethics . . .* in *expectation of the Kingdom of God."* [3] The disciples will reign only if by abasement in service they provide a counterpart to the suffering by which Jesus acquires and confirms "the messianic authority to which he is designated."[4] In J. Schreiber's view,[5] as important as the eschatology and doctrine of salvation is the way Mark's passage reflects the community's sacramental activity, an aspect of the passage to which G. W. H. Lampe has also drawn attention.[6] For Schreiber, service to one's neighbour, even to the extent of sacrificing one's life for him, is the expression of the faith in Jesus' mission that is celebrated in baptism and eucharist.

In verse 45b E. Best is as ready to see the theme of Maccabean martyrdom as the Isaian motif; in contrast to the less absolute standard laid down in verse 45a for disciples, death "particularises" the kind of service Jesus is called to.[7] Like Tödt, Best is impressed by the contrast here between the idea of Jesus as the "humble servant" and the otherwise usual Markan emphasis on "a strong Son of God who has authority."[8] H. Anderson, who also tends to favour the influence of Maccabean idealism over a possibly discredited opinion that Jesus is presented as the Isaian servant, sees verse 45a as "the divine sanction given in the Son of man's own service" for the teaching in verses 40–44 about "service and sacrifice for others."[9] The doctrinal element in verse 45 is, as comparison with Luke 22:27 suggests, "a later theologising expansion,"[10] showing the influence perhaps of Pauline thinking. B. H. Branscomb would understand a similar development from a saying about service to doctrine about a ransom after the manner of Paul.[11] G. Bornkamm,[12] F. Hahn,[13] and N. Perrin[14] also see the origins of the passage in ethical instruction but attribute the gloss (verse 45b) to reflection on the Isaian servant; for Perrin this was "very probably in a Eucharistic setting."[15]

The ethical context is stressed to such an extent by M. D. Hooker that it is considered to inform verse 45b with the soteriological thought evident there and to eliminate the necessity of invoking the figure of the Isaian servant for its inspiration.[16] The theme of service is ultimately to be traced to tradition about the Son of man.[17] C. Colpe speaks no less forcefully of the ethical value but makes explicit allowance for Markan redaction, which through the addition of the saying about a ransom "brings out the depth of Jesus' serving."[18] A more detailed analysis of tradition behind the saying leads H. T. Wrege to conclude that the final statements in each of the independently transmitted groups of sayings—Mark 10:42b–45a, and Luke 22:24–27—reveal the beginning of a process in Christian ethics whereby commands to disciples are made obligatory by exemplifying them in the person of Jesus.[19]

A few writers bypass questions that might be raised about the tradition behind the verse to concentrate on the ethics. Approached in this way, according to M. J. Lagrange, the verse offers no difficulty; it presents "the idea of serving others even unto death."[20] L. J. Cameli on the other hand has considered the verse in the full light of subsequent criticism and fails to find the disjunction between its parts *a* and *b* which is so commonly accepted as the key to its interpretation.[21] Analysing the structure of the section Mark 10:35–45, he depicts "a 'mounting' technique" with a climax in verse 45 as a whole.[22] Far from being an untoward addition, verse 45b is so naturally "the specification of Jesus' *diakonia*"[23] that the whole statement must have originated in its present form, whether in the community or with Jesus himself.[24] Though not a direct citation of Isa. 53, the verse is "a living assimilation" of the passage.[25] At the same time the kind of service referred to, though not precisely designated by the verb διακονεῖν, belongs to the field of "humble and costly service of man,"[26] and it demands that "those who would share in His death must join Jesus and live His death in lives of service."[27] Commenting then on the parallel statement in Matt. 20:28 and in the light of Matt. 25:44, Cameli writes, "Anyone who would contribute towards shaping and building the community to greater conformity with its ideal in Jesus must do so through service."[28]

Rule in the Community

Cameli's investigation into this section of the gospel was undertaken for the purpose of determining what it might contribute to an understanding of the role of the ordained minister in the church of today. He finds its ethical teaching of the closest relevance to the exercise of ministry. By contrast R. Bultmann has found that the ethical content of the saying as it must originally have been has been dissipated by its application in its present form to the actualities of life in an early Christian community. For Bultmann Mark 10:45a is a secondary form of the saying in Luke 22:27,[1] which with its "I" form is related to the inner life of the early church.[2] Even the independent saying about "greatness of service" (verses 43b–44) has been edited to a second-person form in the interests of community regulation.[3] The effect of this attitude upon an assessment of the ethics of the section has been bluntly stated by E. Klostermann when he writes that "the matter relates to actual rank in the community itself and not to deeds of service in the hope of future glory,"[4] And E. Haenchen gives the following description of the kind of tensions that produced the saying: "Here we have a view of the problems affecting the community after the resurrection. The community had grown large and was already engaged in struggles about influence and authority."[5]

Functions in the Community

H. W. Kuhn describes the same transformation under the impact of community problems of a rule of humility (Mark 9:35) into a rule for the governance of the community (Mark 10:43–45).[1] The Son of man's "to serve" is not only exemplary but, as the paradoxically humble activity of that heavenly figure, is the act enabling community leaders to hold office as servants. In drawing attention to the paradox Kuhn acknowledges Tödt, but where Tödt attributes the use of διακονεῖν to a eucharistic milieu Kuhn finds that the choice of terms has been determined by the presence of the words "servant" and "slave" in the preceding verses. The term belongs to an idiom, represented especially in the writings of Paul, that purposely expressed functions within the community by words denoting lowly activity. K. G. Reploh had earlier written in a similar vein,[2] and part of the thinking surfaces in P. S. Minnear's brief comment, "Only the slave of all is qualified to govern all. And slavery to others is embodied in the act of dying for them."[3]

Ecclesiastical Office

M. Guerra y Gomez also recognises the language of community function in Mark 10:45 but attributes it to ecclesiastical usage.[1] Although of itself the verse commits Jesus to a quite general service of all people, the use of διακονεῖν in this connection associates him indirectly but of set purpose with the hierarchical function in the church. This association, noted also by W. D. Davies,[2] is maintained elsewhere in the New Testament when the terms "bishop," "pastor," "teacher," and "deacon" are applied to Christ.

Attendance on the Rabbi

Commenting on the women's attendance on Jesus (Mark 15:41), W. D. Davies writes that "ἀκολουθεῖν [to follow] . . . is interpreted as διακονεῖν," and he regards this as an "admirable translation of the rabbinic šmš . . . used of students being in attendance upon a scholar as a disciple."[1] He sees the sense also at Mark 1:31 and raises the possibility ("unlikely") that it might affect the interpretation of Mark 10:45. "Jesus," he writes, "seems to be repudiating the reverential treatment meted out to the great: his terms may suggest rabbis."[2] H. J. B. Combrink has also been attracted by the same possibility.[3] He notes that the Aramaic šmš corresponds to šrt as used in the Hebrew of the Old Testament for service of a personal or intimate kind and could thus underlie the use of διακονεῖν in the New Testament as a term for serving.

In Fealty to God

Where most writers have seen in Mark's "to serve" what Cranfield called the "manward" aspect of Jesus' service, an aspect that this Greek verb is commonly understood to convey more clearly than other terms for serving, H. G. Swete saw in the "new conditions of social life" that Jesus inaugurated by precept (Mark 9:35) and example (Mark 10:45) a statement about δουλεία (slavery) or "absolute submission to God." Mark's "to serve" manifests that "form of a slave" (Phil. 2.7) under which Jesus conducted his whole life.[1] H. Gross has been as little impressed by semantic differences between δουλ- and διακον-. For him Jesus' notion of service is inherited directly from the Old Testament command to love one's neighbour and is a Christian "slavery."[2] In a study of the writings of Josephus, A. Schlatter had proposed that words of the διακον- group had already made the transition there from being terms for service at table to denoting "every activity that takes place according to the will of God."[3] Some such meaning underlies C. K. Barrett's understanding of Mark's "to serve" as referring to the general ministry of Jesus seen in the light of the great servants of God from the Old Testament.[4] Since Barrett speaks of the ransom undertaken by Jesus as "service to the mass of his people,"[5] and is inclined to relate John 13 to Mark 10:45a,[6] Mark's term would include service to others under the idea of service to God. B. Gerhardsson also rejects the narrow conception of Jesus as the Isaian servant in favour of the broader tradition of Jesus as one among the servants of God,[7] but pays more direct attention to what this implies for "to serve." Noting that it is by the word λατρεύειν (to serve) by which Jesus is related to God (Matt. 4:10) and διακονεῖν by which he is related to people, he points out that Mark's use of the latter word expresses nonetheless the essence of Jesus' obedience to God:

> The moral standard upheld here by Jesus is really the fundamental law of the "rule of heaven": *not to seek one's own but the interests of the other person.* . . . Jesus thus undertakes to be "the servant" par excellence; he "humbles himself" and carries out the will of the heavenly Father for the sake of those to whom he is sent.[8]

A plain statement along these lines is that by W. L. Lane; he is ready to see in verse 45a an allusion to the Isaian servant but also allows for the influence of the Maccabean ideal of martyrdom and writes of this verse as a whole that "the death of Jesus is presented as his service to God."[9]

In Summary

From the summaries presented under the preceding ten headings we can see that the various approaches to the interpretation of Mark 10:45 have not arisen in the main from disagreement about the lexical meaning of the word used for "to serve." The majority of writers work on the assumption that Mark's verb means in a quite general way "to serve" and that this service is directed either to one's fellow (the first, fifth, sixth, and seventh sections) or to God (the tenth section). Such general senses are taken also by those who see a reference to the Isaian servant (the fourth section). A few writers suggest specific areas of service relating to ecclesiastical (eighth) or rabbinical custom (ninth). Many others, under the influence particularly of the meaning in Luke's phrase "as one who serves," see behind Mark's use of "to serve" a reference to serving at table (second and third sections). It should be clear also that writers more often adopt a particular meaning within this narrow range under the influence of attitudes they have adopted to the formation of the saying in Mark than for strictly lexicographical reasons. Both the general meaning of the saying and the context in which it occurs have long suggested that its closer interpretation will depend on the correctness of the attitudes taken. Thus, in regard to the weight of the saying by virtue of context, N. Perrin has written that "whether Mark is composing, redacting or simply using the saying, its position as the climax of the section 8:27–10:52 alone guarantees its immense importance for Markan theology."[1]

In regard to the general meaning of the saying the picture is less simple, and the interpreter may be swayed now by one factor and then by another. The two most influential factors are the presence of the phrase about a ransom for many in verse 45b—Wellhausen's jump from one category of thinking to another—and the context in which Luke has placed his version of the saying, where it lacks an explicit soteriology and appears to be an ethical statement cast in terms of service at table. For most writers these factors imply that separate elements may have been joined to form the present verse with the result for interpretation that verse 45a is considered in isolation and is variously related to general Christian ethics or to the specific environment of community life or church order. As against this, numbers of those who support an interpretation on the lines of the Isaian servant maintain that the unity of the saying is original, although they may sometimes feel obliged to account for διακονεῖν instead of the Septuagint's δουλεύειν by reference to a eucharistic tradition. The point here is that investigations into and surmises about the saying in which διακονεῖν occurs in an ambiguous sense seem to take precedence over enquiries about what διακονεῖν might be expected to mean in a saying of this importance.

Of those who have pursued the meaning of this word and its cognates in Christian and other sources, Schweizer has considered Mark 10:45 only incidentally, being mainly concerned to discuss the significance of the word group within the terminology of church order. Aspects of church order have also been the main interest of Lemaire and of Guerra y Gomez. Brandt, it is true, engages in a close examination of Mark 10:45, but his purpose was less lexical definition of διακονεῖν than the delineation of contrasting estimations of service—expressed by whatever words—in Christian and non-Christian circles. Nonetheless his contention that from being a word meaning "to wait at table" διακονεῖν came to be a vehicle for expressing the worthiest Christian concern for others was taken up by Beyer, and its influence outside of commentaries on the text of Mark has been immense. As expounded in the preceding chapter, Brandt's and Beyer's conviction that we are dealing, in Mark 10:45 in particular, with thinking at the apex of Christian existence—with a notion comprising loving assistance to people in full and perfect sacrifice—has provided churchmen and theologians with an instrument for the refashioning of much thinking in ecclesiology. Contrasting with the success of their ideas in this area, however, is the negligible effect they have had among commentators on Mark, an area where one may have expected the impact of a specialised word study to have been greatest. With some justification Roloff has attributed this neglect to Beyer's not having taken tradition into account, but in attending to that very aspect of the matter he has himself not gone behind the proposition with which Beyer began that διακονεῖν means "to wait at table."

As well then as having no agreed opinion among commentators as to what the service of the Son of man consists in, we have reason enough to doubt that his service is the kind comprehended under the modern notion of "diakonia." But if we are not confident that the notion is formulated at Mark 10:45, are we perhaps to find it more explicitly in evidence in other areas of early tradition and church life? That is the question of the next chapter.

3

The Early Servant Church

Mark's statement that the Son of man came "to serve" might be the most noticed instance of διακονεῖν in early Christian literature but it is only one of some 150 instances of this word and its cognates in the New Testament, the Apostolic Fathers, and the Apologists. If the instance in Mark bears but dubious testimony to the existence among early Christians of the modern notion of "diakonia"—and leaves us with an intriguing problem of interpretation—to what extent do the others instances echo, to use Lemaire's phrase, the loving attitude of the master?[1] As we look at their number we realise that by comparison with Jewish Greek literature occurrences in early Christian literature are strikingly more frequent. Thus the Greek Old Testament provides no instance of the verb διακονεῖν, only 7 instances of the common noun διάκονος, mainly in the book of Esther, and three of the abstract noun διακονία,[2] whereas these words occur in the New Testament, a much smaller body of text, in the order of thirty-seven, thirty, and thirty-four.[3] As against this, however, the spread of the words in the New Testament is uneven. 2 Corinthians for example provides twenty instances, Galatians one, while none occurs in 2 Thessalonians, James, Jude, 2 Peter, or 1–3 John. There is also the fact, probably requiring explanation, that the common noun διάκονος does not occur in Luke or Acts. Of instances in the Apostolic Fathers a third occurs in Ignatius in the sense "deacon"; the verb occurs only in Hermas, while no instances occur in Barnabas, 2 Clement, or in Romans of Ignatius. Does an examination of these instances show that the words were among those that early Christians, as Nigel Turner has put it in a slightly different context, "filled with new meaning beyond recognition"?[4]

Works of Mercy

A third of the early instances relate to the preaching of the word of God. Thus, when Paul, for example, calls himself διάκονος, the designation has customarily been taken as the expression of the humility of the apostle. In this connection J. F. Collange wrote of "the strictly Christian colouring of the term where the teacher thinks of himself as a servant."[1] Collange was in fact countering the contention of Dieter Georgi that in such a case the designation is neither an expression of humility nor one peculiar to Christians. Rather it would be a borrowing from the idiom of religious propaganda current especially among the wandering Cynic preachers of the Hellenistic world. For them a διάκονος θεοῦ, a formulation used by Paul at 2 Cor. 6:4 ("servants of God"), is not merely God's servant but his plenipotentiary envoy.[2] This view has proved of interest to a number of scholars and will be examined in the later chapter on the language of diplomacy. Its interest here is that it marks a clear departure from conventional interpretation; moreover it would illustrate a Christian dependence on non-Christian usage at the very point where so many like to see a uniquely Christian character in the usage. The view also leads to Georgi's further contention that "The NT term almost never involves an act of charity."[3] This judgement is particularly relevant to the interpretation of the several passages in Rom. 15 and 2 Cor. 8-9 where Paul writes of the collection for the poor in Jerusalem. Where others interpret instances of the words in this context on the lines of "aid" for the poor (thus *RSV* at Rom. 15:25) and have difficulty in reading any other meaning into the context because, as one has put it, "service means service,"[4] Georgi would see meanings—although less clearly in some of these instances—related to that of his "envoy."[5] If the case could be sustained in relation to these passages that the notion of "mission" (see *RSV* at Acts 12.25) should replace the notion of "assistance" (see *RSV* Acts 11.29, "relief"), few places would remain in the New Testament where the words might unequivocably express the idea of service of the needy.

C. E. B. Cranfield sought to isolate such passages.[6] He writes of "a specialised technical use . . . to denote the practical service of those who are specially needy 'in body, or estate,' "[7] and adduces ten instances from statements about the collection just referred to, five others in regard to which he expresses reservations,[8] five about deacons, and then Acts 6:1-2 and the locus classicus Matt. 25:44. These are not many, and it is fair to think that, in the light of Georgi's work and of other work on deacons and the passage in Acts, Cranfield's reservations could extend to more of them. Lemaire has since argued, for example, that the deacon was not instituted in the first place for works of charity but to be an officer liaising between communities.[9] So much for being named a "servant" of the needy. In effect, then, we could be left only with the usage in the parable of the last judgement with which to illustrate a unique connection between these words and works of mercy. And even here an interpretation will now be ventured that has no bearing on any such connection.

Towards the end of the parable the wicked ask the king, "Lord, when did we . . . not minister to thee?" (Matt. 25:44) Throughout this parable those words have been emphasised that designate in a matter-of-fact way a series of good works: "you gave me food, gave me drink, welcomed me, clothed me, visited me, came to me." In all, the list is repeated in full three times. The words do not, however, occur on the lips of the wicked. On the other hand neither the righteous nor the king use the word "minister." When the king uses verbal shorthand for the list of good works he says simply "you did it to me" or "you did it not to me." In avoiding the list and introducing the word "minister" the wicked might also seem to be falling back on shorthand, but there could be more to it than that. Before translating the word as "help" *(GN)* or in the phrase "when was it that we . . . did nothing for you" *(NEB),* might we not take a clue from the awesome solemnity of an occasion when a devastating judgement has been handed down by a king and anticipate some deviousness, born of despair, on the part of those against whom the judgement has gone? In other words we should perhaps view the wicked at this moment as fawning upon the king and in their appeal using an appropriate term from the language of the court, "minister." After all this is one of the words attributed to Jesus speaking elsewhere of the courts of rulers (Mark 10:42–45). To the wicked it is a declaration of their abasement before the king and of their readiness to carry out his wishes. To the king, however, it is further damning evidence against them. They appeal in abject terms of the kingdoms of this world, having failed to comprehend despite the lengthy judicial procedure that the kingdom exists in others. Such obtuseness of itself would justify the king's rejection of their appeal. Far then from inculcating a range and style of good works under the banner of "diakonia," the word "minister" here might be a minor but effective literary device finally to discredit the wicked.

To move from the New Testament to any sort of detailed consideration of the language and attitudes of early church fathers in respect of service of the needy is beyond us. Harnack did in fact open his account of early Christian philanthropic activity with the citation of Mark 10:45 as the new language of "ministering love,"[10] but he found it necessary immediately to observe that while the literature of the period frequently summons the Christian to unconditional almsgiving the summons is not often put in terms of love. Still less was it put in terms of ministering. And those, as J. V. Taylor has noted, who "hope to build a service-theology . . . on the actual use of words like *diakonos"* will be disappointed.[11] The statement in Mark and Matthew about the Son of man's service is virtually ignored.[12] That early Christian attitudes in these matters were focused elsewhere is quickly apparent. At root almsgiving is just another commandment of the gospel, as the Shepherd of Hermas expounds it, and is to be observed by Christians in the interests of salvation; to harbour possessions is to forfeit one's citizenship of the distant city: "invest not in land but in people in distress" (Sim. 1.1–11). In the middle of the third century the same arguments are advanced in the widely influential treatise of Cyprian of Carthage concerning *Works and Almsgiving.* When Cyprian concludes on the theme of divine imitation, it is not with reference to Jesus but, as in the earlier *Epistle to Diognetus,* to the bounteous God. In early literature, as A. D. Nock has put it, Jesus is "a saviour rather than a pattern."[13]

We cannot be surprised accordingly that in the course of surveying this kind of literature G. W. H. Lampe furnishes no instance of these cognate words in the sense of serving the needy.[14] Even the passage from Dionysius of Alexandria (Eusebius, *Ecclesiastical History* 7.22.6–10), which is recounting assiduous service to the plague-ridden by deacons and lay people, uses words of service from a different background. In Clement of Alexandria's tract *Who is the rich man that shall be saved?* we meet our cognate words only in reference to the service of God (29; 30) or in allusion to the gospel (10); in fact Matthew's parable of the judgement is cited with his word for "minister" omitted and a paraphrase substituted (30). In the first volume of his sources for the history of "diakonia" Herbert Krimm provided in translation sixty passages from literature of the first three centuries in illustration of loving Christian service,[15] but what the theologian requires is evidence that such service was conceptualised and expressed as precisely "diakonia." When the Greek words supposedly underlying this concept are traced through Christian works of the time, they introduce us not to works of service but to worlds of angels, revelation, prophecy, hierarchy, and to some of the stranger corners of cosmology.[16] Of course a presentation of such material cannot be undertaken here when not even Paul's usage has been reviewed, but the question can be brought down to something manageable, more or less in the manner of a test case. This will concern the origin of the designation "diaconiae" for institutions in Rome that are known to have been centres for works of charity.

"Diaconiae" at Rome

Since the twelfth century the cardinal deacons of Rome have taken their titles from ancient churches that have the word "diaconia" in their name, as in Diaconia San Teodoro. Much earlier, according to the *Liber Pontificalis,* Pope Fabian (236–250) had divided the city into seven administrative regions under seven deacons who were responsible for temporal administration and for the implementation of the "frumentatio" or relief of the poor. Because the words "deacon" and "diaconia" are cognate, the opinion was, at least from the time of Baronius and as late as Hatch, that the "diaconiae" were the centres from which the deacons had distributed this relief.[1] The connection would be favoured of course by those who see in "deacon" and "diaconia" designations for persons and activities related to the care of the needy. At the end of the last century, however, L. Duchesne showed that a connection is not sustainable.[2] He pointed out that with the establishment of the Lateran at the peace of Constantine the deacons operated from those quarters and continued to do so even after the fifth century. It was not in fact until the seventh century that the word "diaconia" first appeared in connection with Rome. This is in the phrase "monasteria diaconiae" from the life of Benedict II (684–685),[3] and it points to a connection between Roman "diaconiae" and monastic establishments. The word next occurs under Hadrian (772–795), who raised the number of "diaconiae" from sixteen to eighteen. To each of these a chapel was attached, but none was associated with the deacons, who had remained seven in number and had long been known by the titles "deacon of the first region, . . . of the second region" and so on ("diaconus regionis primae").

Two facts in particular indicate the late foundation of the "diaconiae." No less than eight of them were concentrated in the area of Rome's forum; since only one of the twenty-five "tituli" of the presbyters, all dating from the fourth and fifth centuries, is located here, the inference is that the forum was not at that time the centre of a population large enough to attract the foundation of so many "diaconiae." Second, twelve of the "diaconiae" were adaptations of public buildings, which none of the "tituli" or of the seven basilicas had been and which would not have been a building method permitted to Christians prior to the Constantinian peace. H. I. Marrou has presented the further evidence to show that even so the "diaconiae" were not a native development but a Byzantine importation, which would again place them late.[4]

The word "diaconia" is itself an obvious borrowing, being known previously in Latin only from Cassian (ca. 360–435) and Gregory the Great (590–604). Significantly Cassian is referring in his *Conferences* to monastic institutions in the Greek east, whereas references in Gregory's letters are to Pesaro, Ravenna and Naples, all ports in Italy with strong Byzantine connections.[5] Six at least of the "diaconiae" in Rome bear the names of Greek saints, which would point to their foundation some time after Justinian's conquest (535–555). Richard Krautheimer has in fact dated S. Maria in Cosmedin, which seems to be the earliest of the type with at first a hall and later a church, around 550.[6] He describes Greek influences in its construction, possibly also in its design; he notes its location in what was probably the Greek quarter, and its Greek title. Because S. Maria is first recorded a "diaconia" only in an eighth-century inscription, and because the date 550 would anticipate Benedict's "monasteria diaconiae" by over a century, Krautheimer calls the establishment "a building that was in fact a *diaconia* even if the name had not been invented."[7] We can however come closer to the name than that.

The "diaconiae" are known to have been for a period centres for the distribution of alms and even for the reception of the poor, being endowed by popes and lay people for this purpose, and they seem to have been basically monastic in character, which is consistent with the fact that they are mentioned in official documents almost only on the occasion of papal benefactions. Marrou has attempted to trace their origin to monastic foundations of fourth-century Egypt, and takes the occasion to append a history ("strange and touching") of the word itself.[8] Like most writers of his period he finds that in early Christian literature the meaning of the original Greek word extended into a new field covering the loving care of others, and to this usage he attributes its adoption as a designation for monastic centres where charity was disbursed. Because we have already pointed out that Christian literature of the period leading up to the establishment of monasticism in Egypt in the fourth century does not provide uses of the kind that would support the provenance attributed to "diaconia" by Marrou, it may be valuable to look briefly at some available records of the Egyptian monasteries themselves.

Most of the records are Greek papyri of the fifth and sixth centuries, and they do instance διακονία as a designation for the monastery or, as Jean Maspero interestingly put it in dealing with one of the papyri, for the whole landed estate.[9] None of the documents, however, gives the impression that these monasteries were distinguished for their involvement in works of charity. Rather the records are of dues levied by the bishop, of land disputes, of building rights on the estate, of rents paid into the monastery, and of purchases from it. The Greek designation occasionally employed there carried over, like many another Greek term, into Coptic records of monastic life. The following seventh-century petition illustrates this as well as opening a little further the question of what the word means in a monastic context.

> The unworthy, humblest one, the monk Pshêre the son of Daniel, writeth, entreating and praying and casting himself before God, thereafter casting himself before your Lord brotherhood which is in Christ, (namely) the whole διακονία, from small to great; in order, that, if the thing please your revered lordships, ye would do charity with my humility and be charitable unto me, as if ye would redeem a captive and even as ye do lend your aid to many a good deed for the Lord's sake. For God knoweth, I am a wretched poor man.[10]

In this petition, the exact purpose of which is unknown, we notice that the word would seem to indicate the assembled brethren (rather than, as Crum suggested, "the body of officials, the staff of the monastery, here in their charitable capacities").[11] In fact, taking a lead from the Septuagint where in a variant to the phrase "the king's attendants" the word means "body of royal attendants" (Esther 6:3, 5), and in line with the expression "brotherhood," we could think of "servanthood," that is, all those pursuing the commands of the Lord in monastic life. The passage illustrates in any event that the word had more than one application in the area of religious establishments and institutions. The same would be indicated by the brief dedicatory inscriptions in Greek on two religious artefacts of the fifth–sixth centuries.[12] The interpretation of one of these considerably exercised Maspero's ingenuity. He found that "from the treasure of" made the best of ἐκ διακονίας, as if the little perfume or incense box were part of the church plate under the charge of the presbyter Praepositus,[13] whereas the phrase, which occurs in a similar construction on the other item also, fits more easily as "[dedicated/made] at the behest of" the presbyter.

Our purposes will be met without closer exegesis of such inscriptions and papyri. It is sufficient to draw attention to the varied uses of the word and to propose that they belong not within the area of charitable activity but within that of mission and mandate. Some of our English uses of the world "mission" are helpful here. In theological writing we might apply this word to Jesus and mean his divinely appointed task, but in a wider and related Christian usage we might also speak of "the Methodist mission" and mean a building down the road, or we might say "The mission has bought a plane" and mean that a purchase was made by the authorities of an evangelising organisation, or say "The mission extends beyond the plantation to the coast" and be referring to the extent of an estate. Accordingly when the word διακονία occurs in the mosaic floor of an annexe to a sixth-century church building in Gerasa,[14] we might be mistaken to infer that this part of the building was the special preserve of deacons or that it was an office from which distribution was made to the poor or that the building and its functions are linked historically with institutions in Egypt. The person who commissioned the Gerasa mosaic may have been recording something simpler like "This sacred building was established by the grace of God on such and such a date." If the meaning of the inscription is thus doubtful, and if Gerasa is the only known instance of an ecclesiastical establishment so designated outside Egypt and Italy,[15] it would be straining the evidence to suggest that a chain of charitable institutions linked Egypt, where some establishments were called διακονίαι, and Italy, where some were called "diaconiae." What we are dealing with, rather, is the fact of language that Greeks found διακονία a suitable designation for monastic foundations in Egypt and that other Greeks found it a suitable designation also for monastic foundations in Italy which may or may not have had the same role as the Egyptian monasteries.

Returning then to the "diaconiae" of Rome, we recall that for historical reasons they could not have been so named by reason of their association with the city's deacons. We may now suggest that neither were they so named because they duplicated a type of Egyptian institution, and add that in either case there is virtually nothing to suggest that the title designated these places as centres for works of charity. What does this imply for the attitude of the church of Rome towards the terms διακονία and "diaconia" as the expression of the Christian ideal of charity? For centuries the Roman church had operated works of public assistance through the agency of its deacons ("diaconi"). It had seen no reason, however, to refer to this work or its institutions by a term so ready to hand as the Greek διακονία. When this did appear as a loan word in Latin, it denoted an institution that was foreign and distinct from its own. Thus at a point where the Greek word group is intimately associated with the practice of service to the community, Roman churchmen—and indeed Greek monks—did not attach to it connotations of a Christian ideal of service.

Deacons of Old

If we were to look in detail at the kind of language churchmen used in connection with the charitable activities of the deacons themselves, we would be led to the same conclusion, namely, that διακονία and its related words were not designations for this part of the deacon's function. We will not take up such detail here, however, because the matter butts upon questions arising from the very first sources of the diaconate in the New Testament and in the Apostolic Fathers, and these are handled later in chapter 13. A few points may be taken nonetheless from the *Apostolic Constitutions,* a church order of the late fourth century which has the advantage for us over Hippolytus' *Apostolic Tradition* of the early third century that the Greek text is extant. The charitable role of deacons (and of deaconesses) is touched upon at various points but a chapter of particular interest from the linguistic point of view is 3.19. This chapter aims largely to protect the church's works of charity against the avarice and sloth to which the diaconate was much exposed. It does so by reminding the deacons where their primary responsibility lay, and this was to the bishop, whose "mind and soul" they were. The instruction refers frequently also to their role in charity, and the various words denoting service are common (δουλ-, θεραπ-, υπηρετ-, εργατ-, επισκεπτ-). The first phrase to speak expressly of "serving the needy" uses ὑπηρετεῖν (3.19.3), not διακονεῖν. The latter word does then immediately occur by way of a citation of Mark 10:45, where the Lord himself establishes the standard of the deacon's ministry. It then recurs in a phrase similar to the preceding and meaning "to minister to the needy" (διακονῆσαι τοῖς δεονένοις, 3.19.4), but that this occurrence is an allusion to the gospel and not a customary ecclesiastical expression for helping those in need is clear enough from the citation of the gospel, from the form of the verb, which is identical to that in the gospel, from the uses of this verb elsewhere in the Constitutions[1]—as from the use of other verbs when the idea of serving others is intended—and from the tenor of this chapter itself. Thus the passage proceeds to urge deacons to become imitators of him who endured slavery and the cross; in other words it is taking from the διακονῆσαι of Jesus as stated in Mark not the idea of helping others but an absolute standard and exemplification of subservience. In Jesus' case this is to the will of the Father, as references to the faithful Isaian servant make plain (3.19.1, 5); in the case of deacons it is before anything else to the demands of their office, which include "being enslaved to the brethren" (δουλεύειν, 3.19.4). In his death Jesus showed the extent to which fidelity in one's ministry must be carried.

The thought that deacons must carry out their ministry within the terms of their mandate is an important one for the Constitutions and has carried over from the earlier *Apostolic Tradition* of Hippolytus. The reasons for the emphasis are well known. Deacons were not always exemplary men and could take advantage of or neglect their office. More important, an influence deriving from their association with the bishop incited rivalry and even hostility on the part of the presbyters. Gregory Dix has gone so far as to describe the ninth chapter of the *Apostolic Tradition* as "an attack on deacons under cover of an outline of their duties."[2] One phrase here is as interesting as it is famous. It states that the deacon "is not ordained to the priesthood but to the bishop's service, to do what the latter tells him" (non in sacerdotio ordinatur, sed in ministerio episcopi, 9.2). This makes it perfectly plain where Hippolytus was placing the ministry of deacons, even if at the same time he was making room for presbyters. But we have only to omit the word "bishop," as the version of Hippolytus known as the Egyptian Church Order did, and his thought is obscured. Bernard Botte considers in fact that the phrase then becomes practically unintelligible.[3] We notice that it is this shortened phrase that is cited in the Second Vatican Council's Constitution on the Church (art. 29), and this encourages some to interpret the phrase's "ministry" in the modern sense of "diakonia" or service to the people.[4] According to Botte's reconstituted text of Hippolytus, however, in which he incorporated remnants of the original Greek, the word for "ministry" was not διακονία at all but ὑπηρεσία, a word with a clear bureaucratic ring and reinforcing the orientation of the ministry intended by Hippolytus.[5] This orientation is even more firmly marked in the *Apostolic Constitutions*. There, as Christ carries out only the will of the Father, so the deacon is "to do nothing of himself" but is "to serve the bishop" (λειτουργεῖν, 2.26.5). It is in this context that we are to understand what διακονεῖν means within the ministry of deacons according to a document that represents a long-standing tradition. At one stage it even proffers the following terse definition, "to carry out the orders of bishop and presbyters, this is διακονεῖν, not to be busy with all the rest" (3.20.2).

Conclusion to Part I

The preceding two chapters have sought to discover what kind of support could be adduced from the New Testament and other early Christian sources for the notion of "diakonia" described in chapter 1. The central statement relating to the notion in the New Testament would be Mark 10:45, but a survey of opinion in chapter 2 revealed widespread uncertainty about the character of service here attributed to the Son of man. A more cursory look in the present chapter at other material in the New Testament revealed scholarly opinions in some important sectors that do not fit with the notion called "diakonia" and raise doubts about the relevance to any such notion of the rest of the usage there. After then alluding to patristic literature, which really lies beyond the area we need to examine, we chose two areas from church history—the Roman foundations known as "diaconiae" and the diaconate itself—which are areas most likely to evidence an ancient conceptualisation similar to the modern "diakonia," and argued that no sign of it was to be discerned.

These negative conclusions, however, raise questions about the Greek term διακονία to a more significant level. Words of the group like "deacon" have gone with the gospel into many languages through many lands; statements using other words of the group, like the words in Mark 10:45 and Eph. 4:12, contain doctrinal elements basic to an appreciation of both Jesus and church. At such a level inaccuracy or uncertainty is not good enough. Accordingly the chapters that follow will try to disclose the linguistic background of the group as a whole. They will do this not by way of Christian sources—because the signs are that of themselves these have not been adequate—but by way of non-Christian sources. This approach could be all the more necessary in that a principle underlying the modern lexicography of διακονία asserts that the meaning of the word underwent a "sea change" in the new stream of Christian thought. Without a careful look at what other Greeks were saying by this and cognate words we cannot really be sure that Christian Greeks were saying something different. Admittedly the field has been examined before, but most closely by those who advocate or support the notion of "diakonia." They would concur with the view expressed by a classical scholar that "though the word διακονεῖν is Greek, there is little help to be gained from a study of its usage and that of its cognates in the classical period when our task is to elucidate its meaning in the Christian Church."[1] Against this, however, we already have the elucidation that Georgi has tried to bring to bear on the title διάκονος in the very first Christian churches from some such background. And perhaps in dealing with this small set of words we have to bear in mind the conclusion drawn by Gerhard Friedrich, the second editor of Kittel's *Theological Dictionary,* that "seen overall Christianity did not have that capacity for developing language which has been attributed to it. . . . With every single word we have always to be asking what it means and where it came from."[2] Conducted on such lines, a successful enquiry may produce insights into early Christian thinking which are of interest from a linguistic and historical point of view but which may also be of value to the processes of Christian life today.

II

NON-CHRISTIAN SOURCES

Herodotus, who died about 425 BCE, is the first writer of prose known to have used words of the διακον- group (in his Ionic dialect the words appear as διηκον-). His *Histories,* the first book of its kind—and telling of a century of struggle between Greece and Asia—is a lengthy work built up from sources but stacked also with anecdotes and wondrous reports from far places. In it he uses the verb from our set of words once only and the common noun three times, two of these in the same context. Such a spare use of the words is typical of other writers also, the seventy instances in the voluminous works of Josephus five centuries later being only an apparent exception. Let us look at what Herodotus intended by the verb. This exercise will introduce us to the kind of question we will be asking throughout part II and to the approach we will be taking to resolving it.

Etearchus, a ruler in Crete, entertains the merchant Themison as a guest and "made him swear that he would διηκονήσειν for Etearchus whatever he required" (οἱ διηκονήσειν ὅ τι ἂν δεηθῇ, 4.154.3).[1] Translations of this statement might say that Themison was "to do a service for" his host. He was in fact to dispose of a girl by throwing her into the sea; Themison declined. But details of language and context suggest that "service for" might not be to the point. First, the verb has a direct object as well as a dative of person. Second, Themison is a person of substance and a guest, and is taken into the confidence of Etearchus. Third, the request is put under oath. That Themison is not bound in servitude to Etearchus is clear; thus no menial connotation attaches to the verb. But neither does the verb express a relationship arising from Themison's kindness towards Etearchus or concern for his interests; otherwise the oath would hardly have been necessary. The sentence suggests rather that Themison is to "undertake" or "effect" something for Etearchus; the dative "for him," as is normal with Greek expressions of that kind, need say no more than that the intention to act originates with Etearchus and that the action will be morally his. The real question raised by the sentence is why Herodotus does not use the verb elsewhere. The best clues here are that the story is from Crete and that the sentence might reproduce the formula of a religious oath.

Two of the nouns occur in reports of customs in Scythia beyond the Caspian Sea. When Scythians bury their king, Herodotus writes, they "strangle and bury with him one of the concubines, the cupbearer, the cook, the groom, the διήκονον, and the courier" (4.71.4). Again, on the anniversary of the king's death "they take fifty of the most reliable servants [θεραπόντων] left. . . . they take these fifty διηκόνους and the fifty finest horses and strangle them" (4.72.2). In the first instance the person referred to is one of those in closest contact with the king; since he is not listed with the cook and the cupbearer, he is not likely to be the butler but is rather the valet. In the second instance the term is used more widely as an alternative designation of servants in the royal household. The third instance occurs in the account of an incident closer to home. Pausanias, regent of Sparta and commander of its forces at the battle of Plataea, entered the abandoned tent of the Persian commander and was amazed at the luxury of its appointments and of the food; "for a joke he ordered his own διηκόνους to prepare a Spartan meal" (9.82.2). Here attendants on a high personage are designated; it is unlikely, given Pausanias's lighthearted and Spartan disregard for the luxuries of the table, that we are to understand he was accompanied by waiters.

Common to these uses of the noun is that it designates those attending on a royal person or in a royal household. Neither they nor the verb necessarily connote service at table. The nouns suppose subordination to a person of higher standing and are part of menial terminology. The same cannot be said of the verb. The correlation of such disparate uses will be the main subject of the following chapters but first some information and general observations on the sources.

As just noted, and as illustrated earlier in respect of the Septuagint Bible, the incidence of these words is not high. The following study covers only some 370 instances from about 90 authors for the period 500 BCE to roughly 300 CE; in addition some 20 and 30 other instances from inscriptions and papyri respectively, and some 45 instances of cognate words that do not occur in Christian literature of the period. Coverage of the inscriptions and papyri is to my knowledge complete. That of literary sources of course is not. Perhaps half of these instances are recorded in lexicons, and most of the rest have been taken from conventional indexes and concordances; reading and chance account for the remainder. Among these in particular are words compounded of prefixes (ὑπο-, παρα-), which have not otherwise been pursued. Naturally many more authors than 90 have been consulted but nothing near the 1,800 (covering twenty million words) under study in the "Thesaurus linguae graecae" of the University of California.[2] The object of building up a stock of instances is purely to have a large view of the usage; a variety of contextual and idiomatic uses benefits, at the same time as it complicates, the process of definition. One thing may be said of the usage as a whole because it affects the presentation that has been adopted. It is that the usage shows a remarkable consistency over nearly a thousand years, and at no stage did it seem necessary to devise a presentation that would show differences from one literary period to another. The study is thus synchronistic and not diachronistic.

An opinion that kept recurring in modern comment on these cognate words, as seen in part I, is that διακονεῖν means "to wait at table." This was even called "the basic meaning." Others may be able to say whether words have such things. Certainly a word may be used to designate a particular action, object, idea, or feeling more frequently than it is used to designate anything else, and in the case of this word group it is true that a reference to service at table is the most common single reference. The reference occurs in fact in about a quarter of all instances. Of itself, however, this need not make of the reference the most significant factor from a semantic point of view, for it is not telling us what understanding of service at table the Greek user of one of these words had in mind—Is the διάκονος, for example, really a "waiter"?—let alone is it telling us why a word used for service at table is also used by Herodotus as in the passage we have just glanced at (4.154.3). Further, if we think of waiting at table as a menial occupation, and add to occurrences of this sense those referring to menial attendance on a person or around a household, we arrive at something like half of all instances referring to menial occupations. Even such frequencies, however, are not the criterion by which to judge the inherent usage of a word because they are easily affected by incidental factors. Thus of 32 uses of the verb in the *Antiquities* of Josephus, 20 occur in books 17–19; 9 of these occur in the idiom διακονεῖν κελεύματι, rare elsewhere, suggesting perhaps an author's passing idiosyncrasy, but since 5 phrases of the latter type fall in a running controversy between Petronius and the Jews we are also reminded that an expression tends to recur if its original context is extended. Similarly the high frequency of references to attendance at table could be affected by the fact that the ancient banquest was the occasion of much humour, crime, discourse, religion, diplomacy, and sheer extravagance, and thus merited the attention of writers in all branches of literature.

The source in fact of half of the uses referring to attendance at table is Athenaeus's *Deipnosophistae*, a hotchpotch of a book reporting the supposed conversations of cultured men at a number of dinner parties in Rome. In the course of it, and in addition to much interesting and sometimes diverting information about life in the ancient world, Athenaeus anthologises literary and current opinion on Greek and barbarian eating customs, religious festivals, the origins, functions and orders of slavery, the wiles of parasites or professional diners-out, and the artifices of cooks.[3] Such topics invite the use of any word that bears a reference to service at table. What is noticeable in this work, however, is that διακονεῖν and its cognates occur less frequently than other terms. Thus in books 8 and 9 (sections 331–410) verb forms of τιθ- and φερ- abound in citation and in general discussion,[4] while διακον- occurs only four times, always in a citation from comedy's stock character, the cook, and three of the instances from the same speech.[5] When Athenaeus's own usage is considered, it suggests that our cognate words might be restricted to dining situations of a special character. In the interlude between discourses, when Athenaeus sustains the fiction of a banquet by introducing slaves with refreshments for those participating in the discussion, no word of the group occurs; the phrasing is invariably of the kind, παρεφάνη πλῆθος οἰκέτων τὰ πρὸς τὴν ἐδωδὴν εἰσκομίζοντες (262b, "a group of slaves appeared bringing in something to eat"). διακονεῖν and διάκονος, by contrast, occur only in his idealised description of a banquet (192b, f; cf. 420e), in reference to the religious festivals of Delos and of Aphrodite Pandemus (173a, b; 659d), and to a formal custom of the Spartans (139c). Something similar is observable in the usage of authors whom he cites, passages that will be examined in chapter 7, "House and Table," and it suggests that the incidence of the word group in passages about dining might be determined by the public, official, or religious nature of the occasion. If true, that would be a useful piece of information; it would certainly run counter to the common view that the words of this group were taken up by Christians because they were part of ordinary everyday language. While that possibly tells us something of the colour of these words, it does not bring us much closer to what they were saying to the Greeks. To arrive there more help might be gained from the use of the words in passages that are plainer. In the first chapter of this part, then, we will begin with a familiar passage from Plato's *Republic*.

4

The Go-Between

In the second book of the *Republic* Plato writes a few paragraphs on the place of trade and merchandising in society (R. 370–71). His use of our cognate words there will be examined in itself and then in association with usage in two other passages from the *Politicus* (*Plt.* 290) and *Gorgias* (*Grg.* 517–18). Although the three passages discuss social functions, they use our words in quite different ways. Thus in the first the word group provides a colourless term in the definition of a nonpolitical function, in the second it denotes a function that is highly compatible with political activity, and in the third a function that exemplifies political ineptitude. If despite the differences of meaning that these uses imply, an aspect of meaning is nonetheless observable that is common to all three, a valuable insight will have been gained.

The passages are also likely to be instructive in that in each one a clear emphasis falls on the words, to the extent at times that the argument turns on their meaning, and because they are sometimes set in contrast to words like $ὑπηρετεῖν$ and $θεραπεύειν$ which are often seen as synonymous with $διακονεῖν$ and as meaning, like it, "to serve." The treatment that follows is detailed because such equivalence of meaning, which is represented in the traditional commentaries, is not considered to come close enough to what Plato intended by the respective usages. An understanding of $διακον$- in terms broadly of "go-between," as indicated in the heading of this chapter, is argued to make for a fuller appreciation of Plato's argument and to open the way for the interpretation of other uses in literature. Of incidental interest is the fact that theologians occasionally refer to one or other of these passages to illustrate some aspect of service, usually for the purpose of drawing an unflattering contrast between non-Christian and Christian ideals. The chapter will conclude on a more technical note in gauging the extent to which insights from etymological studies coincide with interpretations offered.

A Self-Sufficient Community

The first passage, *R*. 370–71, comes early in Plato's investigation into the nature of justice. Plato suggests that the nature of justice is more likely to be discovered in a study of the relations between citizens than in the examination of a lone individual (369a). Accordingly he undertakes to construct an ideal community where the essential natural relations between individuals may be viewed in isolation. The basic principle of the construction is that no individual is self-sufficient but relies on cooperation with others to meet his basic requirements in food, shelter, and clothing. All inhabitants of the community are thus "helpers" ($βοηθοί$, 369c) of one another and put their work at the disposal of the community (369e). This leads at once (370a) to the desirability of a high degree of specialisation within the community, for it is better for the farmer to keep to the plough that he knows than to attempt also to build his dwelling and weave his cloth. It issues eventually (433a, b) in the realisation that justice is "to do one's own business."

On these principles it is highly unlikely that what is described initially as an environment of mutual help should become at any stage an environment in which an ethic of mutual service might flourish. In fact the possibility is excluded by the overriding necessity for the individual always to pursue his own needs. "Everyone," Plato had concluded at 369c, "who gives or takes in exchange, whatever it might be, does so in the belief that it is to his own advantage." Men cooperate for their economic needs but do so "entirely selfishly," as Cross and Woozley put it. "The first city is no high-minded community, fired by ideals of brotherly love."[1]

Trade

Along these lines a small community of farmers, builders, weavers, and some others[1] develops to a stage where it experiences needs it does not find convenient or even possible to meet from its own resources. Barter between one community and the next must begin. "It is impossible, I said, to locate the city where there will be no need of imports" (370e). Because to allocate food producers and artisans to the conduct of this side of the community's affairs would be to compromise the principle of sepcialisation, the community must be expanded to incorporate a special class "who will bring what is needed from another city" (370e). These will in fact be called the "merchants" ($ἔμποροι$), but at the abstract level on which the enquiry is being carried forward generic descriptions like the preceding are used until the exact nature of the function under consideration has been clearly stated and its necessity established. Only then is the specific or technical designation introduced. To state the trading function Plato needs to add the idea of exporting to that of importing. In doing this he introduces a generic term as follows (370e): "If the $διάκονος$ goes empty-handed, taking nothing which those people want from whom are to be brought the things which the community has a need of, he will certainly come back empty-handed."[2]

This sentence, where phrase runs into phrase as trade is described in nontechnical terms, achieves its coherence largely by the use of the word διάκονος, which replaces what would otherwise have been yet another phrase. This indicates at once that the meaning of the word is to be found in what that phrase is likely to have said, namely, something about a person who takes things from one place to another. An English substitute would accordingly be in the line of "courier" or "go-between." The word "servant," in its ordinary sense, would be wide of the mark because at this stage in the community's development there are no masters. Nor may one suppose that the importer and exporter is a servant in a more general sense of the rest of the community because such a notion would cut across the notion of justice, which is the right and the obligation of the individual, including the διάκονος, to do in his own interests what he is best fitted to do in a communal situation. As διάκονος, the exporter-importer performs a task which is complementary to the tasks of others in the community, and he is no more their servant than they are his. Equally he is not a servant of the state because this state is subservient to the needs of each individual within it. Nor is he being called διάκονος of the state, but is an individual residing within it who is designated by that term because it is a general nontechnical term suitably indicating what his avocation involves.

The specific technical term "merchant" is not introduced until the final factor affecting the growth of trade has been taken into account. This is that once intercity trade has begun goods will have to be produced for the foreign as well as for the home market, which will mean an increase in the numbers of food producers and artisans, also therefore of couriers. "And so we will need yet more διακόνων to bring in and take out the goods. These, of course," concludes Plato, "are our merchants" (371a).

A Colourless Term

Once introduced to the discussion in this way to designate the trader, διάκονος or a cognate word appears wherever a general term is required for commercial functions. There are three further such occasions involving retailers (371c, d) and hired labourers (371e). In these cases one element of the notion of trading so far under consideration is lacking, namely the movement of the merchant from one place to another. Retailers remain in the marketplace to do their business, which is also, let us say, where manual labourers will gather to await the nod. But movement is only one aspect of the merchant's function. He is also the bearer of goods and the middleman in transactions between city and city. All three aspects are expressed, and in a particularly clear way, by the word διάκονος at 370e. The essential element of commerce lies however for Plato in the idea of exchange. This is what engages the retailers sitting in the market place "exchanging goods for money with those who want to buy and money for goods with those who want to sell" (371c). They thus form with others the whole commercial class described in the *Politicus* as "those who transfer the products of husbandry and other arts to other hands and bring them to an equality, whether in the market-place or by land or sea traffic between one state and another, whether by exchange of coin for commodities or of coin for coin, —the class we know as money-changers, merchants, ship-captains, small tradesmen" (*Plt.* 289e).[1] Because transference or conveyance, expressed here by διακομίζοντες, is central, Plato retains for the retailing function the kind of word used at 370e in defining the merchant's function as conveying (cf. κομίζονται) goods, namely διακονία. This word designates retailing as an operation carried on between two parties by a third party. Producers, he says, will neglect production if they sit around the market with their goods when no one is there to exchange with, "but there are those who will see the situation and involve themselves in the procedure [ἑαυτοὺς ἐπὶ τὴν διακονίαν τάττουσιν ταύτην]" (371c). Thus, he goes on, the city gathers to itself the retailers—now for the first time named specifically κάπηλοι—in addition to its merchants. Where the latter wander from city to city, these "sit in the market operating as middlemen in buying and selling [τοὺς πρὸς ὠνήν τε καὶ πρᾶσιν διακονοῦντας]" (371d).

The final component in the population of any natural community will be the unskilled labourers. Plato does not say they are needed; they are simply there, people of no worth but for their physical strength, and this of course, like the attainments or qualities of every one else, is to be put to use. They are fitted into the scheme not by virtue precisely of their dull avocation but by virtue of the way they enter upon it. This is by "trading the use of their strength." From this point of view they belong to the commercial sector of the populace, which is why Plato takes up the topic of hired labour with the statement, "Yet there will be still other διάκονοι" (371e).[2] As in the preceding cases a general term is thus used to designate those who may not be producers but who hold their place in society on some middle ground by trading in a commodity. For διάκονος as applied to the labourer, a suitable English word is hard to find; "agent" is likely to be best because it has a use in commerce and we can think of the labourer as the agent in the sale of his own strength. His function once satisfactorily presented as a commercial function, Plato comes back to the specific term by which labourers are normally designated, "wage earners, because they call the price they receive a wage."

Within this section of the *Republic* (370–71), then, our cognate words have occurred only in connection with commercial activities. Within this context the idea of service—either as a beneficent or as a menial activity—to a person, community, or institution does not enter. What we might call services to the state is hardly a consideration anywhere for Plato. In *Laws* he writes of artisans and the military as "serving the country and the people" (*Lg.* 920e) but his word is θεραπεύοντες and such service is seen as part of reverence to the gods.[3] More significant in the light of modern attitudes to the word group is the lack of a menial connotation. This merits a separate comment.

At the stage of its development that Plato is depicting, the community is still what he calls "healthy" (372e), that is, no one belongs to it who is not doing a necessary task. Human nature being what it is and inclined to indulge in luxury, he will later reluctantly have to cater for such tastes by the inclusion of individuals employed in superfluities (373). For the moment however he has a natural community where equity reigns, where individual skills are all part of community life (cf. 369e), and where those who engage in a thing called the διακονία of trade are beholden to no one but share in the idyllic content which a life according to nature brings (cf. 372b). This holds true even for the labourers; despite the fact that their intellectual attainments hardly make them worth associating with (cf. 371e), they are free men, and the "wage" from which their normal Greek designation derives is not to be confused with the fee charged by Greeks in hiring out slaves. The absence of any derogatorry connotation in the word group is the more noteworthy in the light of the disfavour in which commerce was held by Plato and the philosophers generally. Plato's attitude is reflected in the way he bypasses traders in the *Politicus* in his search for an avocation that might exhibit the mark of statecraft (*Plt.* 289e), and is plainly exhibited in the constitution drawn up for the republic in *Laws*. Here, despite provisions made to rid trade of the dishonour accruing to it by reason of widespread malpractices, the first trade law makes involvement in commerce an indictable offence for landowners; this is complemented by the second, which stipulates that commerce is open only to resident aliens and visiting foreigners (*Lg.* 918–20).[4] Even so Plato allows that commerce is not a naturally discreditable occupation. He asserts in fact that the trader would be a benefactor of society if only he could be relied upon to bring about that equitable distribution of goods which a truly natural community supposes (*Lg.* 918b); this has inspired the legislation, which aims precisely to introduce as much equity into the republic as the rapacity of the distributors might allow. The *Republic* supposes honourable commerce and finds in the term διάκονος and its cognates suitable designations for its operatives and operations. They are this because they say something about the middleman's function in the process of exchange that trade is: the merchant travels, carries, exchanges; the retailer and the labourer merely exchange. A loose analogy is the English word "trade" itself. Meanings like "merchandising" or "skilled handicraft" do not reveal the word's connection with "tread" or suggest that an obsolete meaning was "path." The analogy is loose because the English word has lost its earlier connotations to have now only specific applications. The Greek words by contrast are not technical commercial terms, and retain various connotations. Being generic, they are applicable to trade as to other functions, as the following examination of parts of the *Politicus* will illustrate.

Functions of the Subordinate

In the *Politicus* instances of the word group are at 290a,c,d and 299d. The first three will be examined as falling within the section 289c–290e, the last as within a different but related section. The Passage *Plt.* 289c–290e makes a useful study for three reasons. The words again convey the notion of go-between; the functions designated are civic, not commercial; and the words occur in parallel with those of the ὑπηρετεῖν group, which on this occasion apply to commercial activities among other things. The difference in the usage between the *Politicus* and the *Republic* is to be explained by the different purposes of Plato's two reviews of society. In the *Republic* he was reviewing social functions in the first place to determine which ones are necessary for the material well-being of individuals. In the *Politicus* he begins to review them at 287 to discover which one exhibits the quality of statecraft (τέχνη/ἐπιστήμη/πρᾶξις βασιλικὴ καὶ πολιτική). What the latter approach implies for an understanding of the word group is the present topic.

The first social group under review is one whose role is to provide society with instruments and means; its role is clearly subsidiary and thus cannot be political. Next to be examined is "the class of slaves [δούλων] and any other functionaries [ὑπηρέτων]" (*Plt.* 289c). Now if we are to understand what words like διακονεῖν contribute to the ensuing discussion it is important to know how far discussion extends under the heading just cited. Its limits are marked both by the structure of the paragraphs and by the vocabulary and ideas that characterise them. First, the structure.

Plato takes up his consideration of the new class with the words, "Now I forecast that precisely among these we shall see the ones who claim to rival the king in his intricate task" (289c). Because the group includes slaves this must seem an unlikely prediction, but as the review proceeds through several kinds of functionary it appears with the mention of civil servants that these were the ones Plato had in mind. "So I was not imagining things," he writes after indicating their role, "when I said that somewhere here we would come across a particular group claiming to play a political role" (290b). Briefly discounting their claim, however, he says next, "Now let us come to closer grips with those who have not yet been investigated" (290c). The two last statements are signposts. The latter is introducing a new line of enquiry, the former is the definitive conclusion of investigation into slaves and other functionaries.

If the structure thus makes of 289c–290b a plainly marked unit, what of the idea that informs it? As one would expect, the idea is introduced in the treatment of the first component, the slaves. The statement is that slaves cannot be exercising statecraft because they are "functionaries to the highest degree" (μεγίστους ὑπηρέτας) and are thus the exact "contrary" of what is to be looked for in a statesman (289d). In other words slaves are so completely "functionary" as to have lost freedom of action, and are thus situated at the opposite pole from the one at which the statesman operates. The implication is clear: to the degree that functions of other social groups exhibit "the condition and conduct" of a functionary they cannot be said to be political. Because the question is one of degree,[1] however, there may well be activities where the mark of the functionary (τὸ ὑπηρετικόν) needs to be pointed out before it is discerned. That this should be the case in regard to some who are free men is to be anticipated, so that Plato proceeds to scrutinise these.

In regard to two groups at least the issue is easily decided. These are the commercial class and the hired labourers. Of the former it is interesting to observe that while their function is described in terms closely resembling those used in the analysis of the same function in the *Republic* the term ὑπηρετ- is used instead of διακον-. It would be misleading nonetheless to infer that the former means here what the latters means in the *Republic*. The context of the *Politicus* establishes on the contrary that the word groups are serving different purposes. Instead of defining the commercial function in itself, ὑπηρετ- indicates the relationship of those engaged in commerce to other persons in the state, and determines the point at which commerce falls short of being a political activity. This is that as distributors of produce those engaged in commerce put themselves at the disposal of primary producers and manufacturers (εἰς ὑπηρετικὴν ἑκόντες αὑτοὺς τάττουσι, 289e). In so doing they surrender something of their quality as free men and, consequently, of the qualification for statecraft. The labourers do this more blatantly, being ready to hire themselves out to anyone who will pay the fee (290a).

The third group is what is called the herald class (290b) but comprises much of what we would understand as the civil service. Of them it is said that they are "functionaries [ὑπηρέτας] who do not really exercise authority in the state." We should not accordingly look for statecraft among them, and yet it was with this class in mind that Plato had forecast at 289c, "we shall see the ones who claim to rival the king." The claim, Plato makes clear, is made in all seriousness and is not merely to be related to the ambitious manoeuvrings of petty officialdom. He abstracts from that, pointing out that these men are gifted, diligent, and close to the source of power. The power however is not their own but is exercised in the name of higher authority. From the point of view of statesmanship, therefore, they belong essentially like the slaves "in the category of functionaries" (ἐν ὑπηρετικῇ μοίρᾳ τινί, 290c).

This phrase marks the last use of ὑπηρετ- in this part of Plato's exposition.[2] It closes his observations on the herald class and indeed on all groups with a functionary role in society. τὸ ὑπηρετικόν or the idea of functioning as a subordinate has thus been the controlling idea in the consideration of slaves, traders, labourers, and civil servants, and makes of 289c–290b a unit of the exposition. (The unit in fact overlaps slightly with 290c.) This demarcation was also indicated by the structure of the paragraphs. As will shortly be pointed out, other commentary on this part of the *Politicus* has tended to overlook the factors that are considered here to establish its strict unity. This tendency, allied with that of regarding διακονεῖν as synonymous with ὑπηρετεῖν, has led to the misinterpretation of the sole occurrence of διακονεῖν (290a). Before suggesting what the latter word should be taken to mean in the passage, however, a further observation on ὑπηρετ- will be helpful.

A word of this group first appears in the heading of the unit that I have translated "the class of slaves and other 'functionaries' " (289c). Instead of "functionaries" other translations might have "personal servants"[3] but that would be to set up a category to which traders, labourers, and civil servants could not belong. Certainly the Greek word commonly has this meaning, but the usage is far broader. As Skemp notes, the word had a technical use in government circles,[4] and we can often translate it as "officer." It was widely current also to designate officials in military, legal, religious, and club circles.[5] The instance at *Plt.* 291a in reference to priests' attendants exemplifies one application. In the passage just considered we notice that the noun itself is applied only to slaves in a customary use and to civil servants in a technical use; it is not however applied either to traders or to labourers because they are not ὑπηρέται in any normal sense of the noun. For these Plato uses an adjective so that he can say that their relationship to other members of society is of the kind to be seen in a ὑπηρέτης. What he is meaning of all social types covered by any of these words is that they fill their role by acting at the behest of another. This characteristic does not of itself carry a derogatory connotation. Plato is merely being analytical. He aims to give a due account of functions within society so that those that are not carried out autonomously may be discovered and discounted as irrelevant to statecraft.

Functions of the Go-Between

The instance of διακονεῖν in the unit 289–290b may now be examined. After any claim the labourers might have to political potential has been met, the Stranger, who is one of the characters in the dialogue, directs the attention of Socrates to a new group, subsequently to be identified as the herald class, by means of a question that for the moment we can translate only in part: "What would you say of those who are forever διακονοῦντας for us certain matters?" (290a) To which Socrates at once replies, "What matters and who are these people?" From this we see that Socrates needs more information before coming to a judgment about the political or nonpolitical nature of the function designated διακονοῦντας; in other words, for the purpose of the enquiry, the term does not place these people within the category of functionary. As διακονοῦντες they are not ὑπηρετοῦντες; the words are not synonymous. Further, the dative "for us" or "to us" points to a relationship with the general populace that is likely to be different from that between functionaries of the kind Plato has been referring to and parties who authorise their actions; the public as such does not make functionarties of anyone. This is not to deny that the διακονοῦντες might not also be acting in someone's name but to assert that attention is not being drawn to that aspect. The Stranger is rather using the word to designate a certain social group's function that Socrates would like to look at more closely before pronouncing on it one way or the other.

The answer to Socrates' query "Which matters and what people?" must be supposed to develop naturally from the meaning of "διακονοῦντας certain matters for us." The answer provided by the Stranger identifies only the people, namely the herald class, comprising those dealing in official documents and "certain others most accomplished in performing many other processes connected with government" (290b).[1] Of the matters referred to he has nothing explicit to say and we infer that they are ordinances of government. The Stranger is thus speaking of "those who pass commands on," as Skemp refers to them,[2] or to "administrative middlemen," in P. Friedländer's phrase.[3] While neither of these commentators is translating the word διακονοῦντας—Skemp translates that by "those who render other kinds of service"—the conception they have of the function of the herald class must in fact be what the Stranger had in mind when he first used the word. Accordingly we translate his question, "What would you say of those who are forever communicating certain matters to us?" Once these people are identified as people in government we appreciate the Stranger's comment, "I am convinced it was no mere dream which prompted me to say that somewhere hereabouts would appear the men who above others put in an opposing claim to the political art" (290b).[4] The claim is discountenanced, however, not because they are διακονοῦντας but because the power exercised in that function is not their own.

For two other instances of the word group we move on to a separate unit of the discourse, one where functions are no longer subordinate or ὑπηρετικός but are indeed διάκονος itself. Of these Plato will say in his search for the authentic sign of statecraft, "at last we seem to be on the trail, so to speak, of our quarry" (290d). The persons whose roles he examines are diviners and priests. He begins by saying that divining or soothsaying is "part of a 'diaconic' skill" (τινὸς ἐπιστήμης διακόνου μόριον, 290c). And we repeat that if this word were synonymous in this context with ὑπηρετικός the claims of diviners to statecraft would be summarily dismissed. The word is used here as an adjective specifying the kind of skill or science possessed by diviners. The skill is called διάκονος "because we know diviners are held to be interpreters for the gods to men." The role of the priests is outlined with greater precision: "They know how to give to the gods in a manner pleasing to them gifts from us by means of sacrifices, and to win for us from them by means of petition the bestowal of good things" (290c–d). Both aspects of this dual priestly function are said to be part of a "diaconic" skill (290d). What the skills of diviners and priests have in common, of course, is that they are exercised in the middle ground between gods and men, so that the word is designating functions of diviners and priests as those of a go-between.

For our purposes it is not necessary to go beyond this to know why Plato is now on the track of his quarry; it is sufficient to know that insofar as they are "diaconic" these functions are exercised independently of other men's control. The aspect of meaning the two instances reflect, reflected also in the preceding instance at 290a, is the important thing. The same aspect appeared in instances in the *Republic*. One further instance from the *Politicus* will conclude the matter.

Later in the dialogue, with a view to the autonomy that should characterise the skill of the true statesman, Plato is considering the deleterious effect the making of laws has on other skills. To take up the threads of earlier discussion he presents a list of fourteen such skills or sciences, which he pairs off. The last pair itemised is "the science of soothsaving or whatever the 'diaconic' [διακονική] science includes" (299d). Because of the connection between soothsaying and "diaconic" science at 290c the same ideas must be in the air. Confirmation of this is to hand in the relationship between sciences in each other pair of the list. With each pair a specific science is mentioned first and then a general related area of science. As Campbell observed many years ago, "The Stranger takes every opportunity to exercise his pupil in recognising the 'kindred' between divers arts."[5] The list works perfectly well in respect of the first six pairs,[6] and the relationship within the last pair will be consistent with this. Plato is thus placing the function of soothsaying within the larger context of mediation, perhaps better, of the function of communicating between heaven and earth.

The Functions as Menial

Now that we have examined a number of uses in both the *Republic* and the *Politicus,* the virtue of an interpretation centering on the notion of go-between can perhaps be better appreciated if we notice a few inconsistencies arising from interpretations based on a different understanding of διακον-. In the case for example of the "diaconic"' science at *Plt.* 299d we meet renderings like "the art of serving"[1] or "branch of personal menial service"[2] which distort the balance of Plato's classification at this point to no perceivable purpose. If the same idea of menial service is introduced to the discussion about priests, diviners, and heralds, his thinking is difficult to follow at all.[3] The resulting incongruity has not gone unobserved. Thus Campbell, after concluding that the phrase "servants to the highest degree" implies contempt for the function of διακονία, necessarily finds it strange that rivals to the king should appear within this category.[4] Others attempt to explain how this might be so. Skemp sees behind an idea of "personal 'services' " Plato's own "aristocratic sense of *nobless oblige"* which led him to conceive of the Statesman as "the Prime Minister."[5] On this reading the motif of personal service makes an extended section of 289–290e, which Skemp somewhat arbitrarily divides into slaves, labourers, merchants, and then clerks, soothsayers, priests, thus overlooking Plato's intention to include the clerks (or herald class) with the slaves. But a larger consequence follows because by the connotation of meniality attributed to διάκονος the priests also are cut off from the developing consideration of kingship, and this represents a serious dislocation of the argument. By contrast Taylor applies a lighter touch. Aware that the whole section (beginning 287) is manifestly constructed to draw attention to the importance of 290, he explains a pervasive connotation of meniality as "an ironic pleasantry" on Plato's part. "It tickles his humour to think that the function of the 'king' in a society is more like that of a menial or 'gentleman's gentleman' than like any other more honoured occupation."[6]

Such attempts to make meniality compatible with statecraft read the more strangely in the light of the accurate perception commentators have of the true role of civil servants, diviners, and priests. Thus Skemp refers to "the service of the priest as the official mediator,"[7] and Friedländer depicts the direction of the argument from the point where civil servants are mentioned as follows: "From these administrative middlemen in society, we move upward to the intermediaries between men and gods, soothsayers and priests."[8] Some of these ideas—"mediator," "middlemen in society," "intermediaries between men and gods"—are, according to the exposition given above, precisely what Plato is expressing by means of διακον-.

In the *Republic,* on the other hand, διάκονος has on occasion been treated more circumspectly. As applied to the merchant, the word has been translated "trader,"[9] "agent,"[10] "servitor,"[11] or in the French "le commissionnaire."[12] For the verb at *R.* 371d, translated previously "operating as middlemen" in buying and selling, Chambry has given "as intermediaries."[13] Given the affinity between what Plato is saying of intermediaries both in commercial and in civic and religious life, it is curious that the special meaning occasionally attributed to the word group in the former instance has not affected interpretation of the latter.

Functions of the Attendant

The third and last passage to be considered is from Plato's *Gorgias,* a dialogue that analyses the art of rhetoric and compares it unfavourably with the art of the philosophic statesman. Rhetorical skills may please the public but only the true statesman makes citizens better and wiser. The word group occurs at *Grg.* 517b, d; 518a, c; and 521a. It is applied to the Four, some of the most eminent men in the political and military life of fifth-century Athens (Pericles, Miltiades, Themistocles, Cimon), and again in a way to show the aspect of the group's meaning, which we are calling "go-between." In this connection, however, a derogatory connotation arises from the figurative use of the meaning "domestic attendant." The question is why the words are pejorative as applied in this way to politicians. The answer is stated more easily by making use of the contrast in the passage between διακονία and a kind of service called θεραπεία.

At *Grg.* 464 the term θεραπεία is adopted to designate the highest form of care of the body (therapy), and embraces the arts of gymnastics and medicine. Each of these arts, however, is liable to come under the influence of a harmful habit, in one case mere adornment of the body and in the other the pampering of the body with food. The habits are called distortions or parodies ("flattery": κολακεία, a very strong word) of therapy. This doctrine is put to use at 517–18 in the assessment of Pericles and others of the Four. In general esteem these politicians are held to have contributed much to the well-being of Athens, but in Plato's judgment their contribution consisted of "such nonsense" as harbours, docks, and walls, in the provision of which, he maintains, they have neglected the therapy of the body politic. To call them therapists is like calling the pastry maker, the gourmet cook, and the liquor merchant "good trainers of the body in gymnastics" (518b). What they are in fact is διάκονοι and "caterers to men's appetites, stuffing and fattening their bodies" (518c). To expect the ideal statesman to pursue a similar policy of public works is to ask him to become διακονήσαντα and the city's flatterer (κολακούσαντα, 521a).

The phrase "servants of the state" (διακόνους εἶναι πόλεως, 517b) as applied to politicians has in consequence a meaning that is narrowly determined by the rhetoric of the passage. It draws on the image of the household attendant who fetches food and the articles used in daily life. Essential to the image is an appreciation of the way a domestic goes about these tasks. He is at the master's beck and call, and attends any whim without regard for the master's true welfare. In a similar way the Four are presented as having pandered to the whims of the Athenians. The phrase is thus wholly prejudicial to the estimation of what the Four accomplished, and those who would detect a grudging concession on Plato's part that these men were in any laudatory sense "servants of the state" overlook the connotation imposed on the phrase by the irony of the context (517a–b):

CALLICLES: Surely, Socrates, none of the present generation has ever done anything like such deeds as one of these others, any one of them you please.

SOCRATES: My dear sir, neither do I find any fault with them as precisely attendants [διακόνους] on the city. On the contrary I think they have shown themselves more attentive [διακονικώτεροι] than the men of our time and ampler providers of what the city wanted.

Socrates here seizes on the issue raised by Callicles because it provides the occasion to make a statement on what is service to the state and what is not. To be a διάκονος of the state is the negation of therapy; it is to be the city's lackey.[1]

This seems to be clear enough. It is necessary to point out further, however, especially in the light of the attempt made previously to eliminate a derogatory connotation from the usage in the *Republic* and the *Politicus,* that Plato's invidious purpose is achieved not so naively as by attaching despicable tags to famous men—calling them names—but by illustrating that their activities are the wrong kind of activity for politicians. Having given themselves to providing things, they can be said to have performed an activity that is διακονική (517d). The usage suggests that of *R.* 370e where the merchant—with whom these politicians are also compared—as courier brings in articles from other places. In the *Gorgias* however the words are not used generically but with reference to the household domestic, and the mechanical aspect of supply is emphasised to allow a contrast with therapy, where the dominant aspect is care or concern. When the Four therefore are called διακονικώτεροι (517b), Plato means that they are prompter than contemporary politicians in getting to the public what the public wants. And while this is undoubtedly taken to mean that they fawned on the public like menials on their master, they are ridiculed in terms of the word group not specifically as servile but as mindless purveyors of civic baubles. Thus we would be mistaken to take a servile connotation from the word merely from the fact that it occurs with others meaning ''slavish'' and ''unfree.'' When the arts of cooking and so on are so designated (δουλοπρεπεῖς τε καὶ διακονικὰς καὶ ἀνελευθέρους, 518a), the first adjective refers to status of the activity; the second to its style, which is characterised by the fetching of things commonly performed by slaves but not proper to them and not in itself servile; and the third to the moral quality of the action. Because the action is mechanical Socrates declines Callicles' invitation to become διακονήσαντα in the politics of Athens. To be ''at the beck and call'' of the city in this way would mean ''saying only what pleases'' and giving the city a ''flattering'' opinion of itself; he would rather ''struggle to make the Athenians as good as possible'' (521a).

Plato's Usage

These passages from the *Republic,* the *Politicus,* and *Gorgias* have features that make them advantageous starting points for the study of usage elsewhere in literature. Of special value is the analytical character of the discourse in the *Republic,* where familiar social functions are closely inspected but defined in general terms only before the specific terms of everyday life are invoked. This has meant that the role of a broadly commercial class has been presented in a way to show the notion of exchange lying at the root of the various functions. That notion has been expressed by διακον- because words of its kind speak of go-between. The words have not elicited any connotation of meniality or of beneficent service. Similar generic uses have been observed in the *Politicus* in the dipiction of civil and religious functions. Of interest here was the contrast between the notion of go-between and that of action in subordination, expressed by ὑπηρετ-. In regard to the politicians of *Gorgias* connotations of meniality do arise by way of reference to the activity of domestic attendants, but the activity itself is seen to be that of one who provides or brings things on call; it is this aspect of the servant's role, his fetching, his being a go-between, that is elicited, and not merely his servile or menial status. The contrast with θεραπ- indicated that the servant who is διάκονος is understood to go about his chores in an uncritical, unthinking, and mechanical manner. Varied as are the uses of words like "service" in English, "service" may often be a misleading word to express the Greek. Insofar as it commonly implies a relationship to a master or to a beneficiary, and is used to the latter effect especially by those who speak of "diakonia" in modern theology, it may overlook the underlying notion of go-between and misrepresent the nature of a function in question.

In the following chapters what has been taken from Plato's usage—and is now perhaps discernible also in that of Herodotus as presented in the introduction to part II above—will be related to uses in other written sources. The present chapter will be concluded by relating it to possible etymologies suggested for the root and to more general observations by philologists and lexicographers.

Speed

The *Etymologicum Magnum*, a late Greek compilation of philological opinion and edited by Thomas Gaisford in 1848, offers more than one explanation of διάκονος. It gives the meaning as ὁ διαφέρων τὰ κελευόμενα ("the one who carries out orders"), and relates the word to ἐνέκω, τὸ φέρω καὶ προσενέγκω (it is pointless to translate this and a few other comments here), but records also ἐγκονεῖν, τὸ ἐπείγεσθαι (268.25–31; cf. 92.18). Of these ideas that of carrying out orders is perhaps recognisable in Herodotus 4.154.3, that of getting something from one place to another in Plato, *R*. 370e, and that of promptness or speed in Plato, *Grg*. 517b. Related ideas appear in the phrases of Hesychius:

διακονῆσαι· Κατεργάσασθαι. . . .
διακονούμενον· ὑπηρετούντων ἢ ἀποκρίσεις τινὰς ἐκπληρούντων

The importance of speed to the ancient understanding of the word group is illustrated in the opinion that -κον- in these words represents the dust (κόνις) raised by hasty movement (*Etym. Magn.* 311.11). To the modern scholar this opinion has appeared "amusingly naive."[1] The scholarly problem is that δια- ends in an unaccountably long -α- (reflected in the even longer Ionic διη- as used by Herodotus) whereas the presumed prefix διά ("through") is short. The opinion has the virtue nonetheless of indicating that the word group was understood to say something about the style of an activity rather than about a relationship to the person on whose behalf the activity was carried out. The most recent opinion cannot account for -κον- but retains a connection between διακονεῖν and ἐγκονεῖν ("to hurry"), Frisk assuming a development from the idea of "hurry" to that of "serve."[2]

In the earlier modern opinion of the German philologist Phillip Carl Buttmann, the development was rather after the following fashion: διάκονος - διάκων - διάκω/διήκω - διώκω, the last word meaning "to run," and the second word, which is in fact the usual form in the papyri and is reflected in the Latin dative "diaconibus" of Phil. 1:1 (cf. "diaconos" as an accusative from "diaconus" at 1 Tim. 3:8),[3] would be a present participle of the verb meaning "to run". Buttmann writes: "Διάκονος, therefore, derived from this διώκειν, to run, . . . properly means *the runner;* whence *a messenger, a servant.*"[4] In addition he makes the firm conjecture that ζάκορος, meaning "temple attendant," is another form of the same word, and more interestingly asserts that the German words "jagen" ("hunt", "drive," "chase") and "Diener" ("servant") come from the same root as the Greek. He continues that the word derived from διώκειν is "always retaining the free and honourable idea implied in the original word." We see this in the English derivative "thane," which can mean "royal servant" but can also designate a member of the nobility, as in the twelfth-century Peterborough Chronicle where thanes are listed at court after archbishops, bishops, abbots, and earls.[5] There is an irony in this because today's writer of theology in the German language—those writing in other languages do something similar—prefers to use the word "Diener" rather than more traditional words suggestive of office and dignity in designating ordained ministers on the ground that "Diener" comes closer to what is held to be the ordinary everyday Greek word διάκονος, the lowly servant or waiter at table. Of course, the modern German word has lost the honourable connotation it once had, but Buttmann was saying that the connotation was always retained by διάκονος. His observation is echoed in one or two works widely consulted in the study of Christian origins[6] but cannot be said to have had any influence in the modern thinking on the nature of early Christian ministry. In regard to etymology, on the other hand, other philologists of the nineteenth century concurred with him in seeing "to hurry" as the root idea, notably A. Goebel,[7] J. H. H. Schmidt, and R. C. Trench. The opinions of the last two will be noticed shortly. Meanwhile, a further contention of Buttmann's, supporting ideas he has already advanced, will be followed up.

Hermes

Buttmann was commenting on the word διάκονος in the course of a study of the rarer and archaic διάκτορος, proposing that the two are etymologically the same. The latter is known as an epithet of the god Hermes, hence Buttmann's further observation that "the free and honourable idea" retained in διάκονος "became still more honourable in the other antiquated form διάκτορος, and so was an epithet well suited to the messenger or herald of the gods."[1] The question of the etymological relationship need not concern us, but well worth illustrating is the close association of these words over a long period in reference especially to Hermes. The association says much for the estimation of διάκονος.

The older word[2] was already an epithet of the god in Homer but its meaning in this connection has been variously explained by scholars ancient and modern: Hermes conveyor of the dead, of funeral offerings, guide, god of eloquence, messenger of the gods.[3] As designating a messenger role, it is applied by later poets to other gods also.[4] In this application it seems to be interchanged with διάκονος at Aeschylus, *Pr.* 942 in reference to Hermes on an errand from Zeus.[5] For Epictetus it means that Hermes is "most excellent messenger" (κάλλιστος ἄγγελος, 3.1.37).[6] A subtler indication is to be taken from the altercation between Cario and Hermes in Aristophanes, *Pl.* 1146–70. Here Cario is reviewing the various divine titles by virtue of which Hermes is hoping to have a share in the sacrificial offerings. Cario finds none of the titles—hinge, commerce, guile, guide—an adequate qualification for the god until Hermes suggests "god of games." Cario accepts this last, comments on Hermes' good fortune in having so many titles, and lets him in. As he does so, however, he indulges in a final quip, telling Hermes to lug some guts to the well for washing "so that you may at once be seen to be διακονικός" (1170). One meaning of the instruction is clear: Hermes is to work for his supper, filling the role of the διάκονος who attends on the cook at religious festivals, but Cario is also punning on the most famous title to which Hermes has not appealed: διάκτορος/διάκονος, whether in the sense of messenger or of conveyor.[7] That the latter served as a title there is no doubt. An Athenian curse inscription of the third century BCE, which will be mentioned in the following chapter, consigns thieves to the gods of the underworld and to "Hermes διάκονος."[8]

In the second century CE, Lucian also makes something of this connection. *Cont.* 1 depicts Charon, ferryman of the dead, asking Hermes to act as his guide on a tour of the world, but Hermes replies that he is too busily engaged (διακονησόμενος) for Zeus. Charon presses his request on the grounds that Hermes is "friend, shipmate and συνδιάκτορος," his expressions indicating that he recognizes Hermes as an associate, in particular a "co-conductor" of the dead. It would seem that in the word διακονησόμενος Lucian has deliberately anticipated this theme and prepared his reader for a daring appeal by Charon in terms of the god's own prestigious title, cast too in the rare form of "co"-conductor.[9] This passage will be reconsidered later with other passages in Lucian that refer to the roles of Hermes. We will see that for Lucian διακονεῖσθαι is the mark of Hermes both as the gods' go-between with men (διάκτορος) and as the cupbearer of the gods. On this occasion, in proffering his excuses to Charon, Hermes does in fact refer to his cupbearing duties but does not use διακον- because Lucian has already earmarked his use of that word to allow Charon to pick up an allusion to the other role.[10]

To judge by indications in the *Etymologicum Magnum* these terms were taken as equivalent also in scholastic Greek. *Etym. Magn.* 268.20 interprets the older word as applied to Hermes by the phrase ὁ διάγων τὰς ψυχάς ("conveyor of souls"), in a note to which F. C. Sturz (1818) reports the variant reading ὁ διακονῶν, observing for his own part, "and that also could be said." Similarly 268.21 records Callimachus's application of the ancient epithet to the owl, and the *Etymologicum* interprets, Διάκονον καὶ διαγογέα τῶν ψυχῶν ("porter and conductor of souls").

The fact that διακον- has formed an intimate part of the tradition of διάκτορος indicates how well suited the word group must have been to covering the roles comprehended under the latter term, at least those to do with message and conveyance of the dead. Interplay between the terms also suggests that the commoner word group might share something of the numinous character of the archaic and more poetic term. On such a ground we are perhaps to account for Plato's use of the word group in connection with diviners and priests (*Plt.* 290c, d; 299d) and for the occurrence of the verb in the context of a religious oath in Herodotus 4.153.3. Again the notion of "go-between" underlying the instances just mentioned and those at Plato, *R.* 370–71 fits with the idea of errand, just as the notion of promptness at Plato, *Grg.* 517b, fits with the idea of speed, both being ideas central to most judgements, ancient and modern, about the etymology. These impressions of the character of the word group can be set in a broader context by a consideration of what some philologists have had to say about Greek menial vocabulary in general.

Greek Slaves

In a work then in progress, F. Gschnitzer pointed out that although the Greek language was liberally supplied with designations for those who fill menial roles in society no one term originally expressed the idea of "slave" as it has come to be thought of—namely, a notion comprising "servant who is not personally free, work animal, and a breathing asset."[1] The terms were applied in fact to many who today would not be considered slaves, and even as applied to persons who can be recognised as slaves they did not express the comprehensive condition of slavery but only one or other aspect of that condition or else one or other function performed under that condition.[2] Gschnitzer suggests that around 400 BCE the terms reflected the situation of a preclassical society where they designated various functions that were eventually to come together and be performed by a special social type designated δοῦλος, a word that in late classical and Hellenistic periods took on the more or less complete connotation of "slave."[3]

These observations can be considered in conjunction with those of an earlier philologist, J. H. H. Schmidt. From examination of a similar vocabulary Schmidt recognises in one group of words, δουλ-, ἀνδραποδ-, οἰκετ-, expressions for "slave" in the ordinary sense, although in his opinion the words are distinguishable in that they connote respectively the slave's subjection, his low social status, and his relationship to a household.[4] Other terms, which sometimes connote a slavish condition and at other times do not, Schmidt considers under two heads. Under one of these he includes ὑπηρετ- and ὑπουργ-, which express assistance or cooperation without connoting the kind of activity involved.[5] The other group, by contrast, which comprises θεραπ-, λατρ-, διακον-, does in his view have connotations. Thus the activity designated by θεραπ- is characterised by concern and trust, observable especially in medical and religious contexts. The nuance arising from the religious context is often said by others to attach also to λατρ-, whereas Schmidt maintains that in classical literature there is little to distinguish λατρ- from δουλ- apart from the connotation of hired labour given often by the former.[6] In the case finally of διακον- he observes that although the word group frequently signifies attendance on a person, especially a person at table, an earlier usage concerned message, and here διάκονος denotes a servant who is "on the move" (Laufbursche).[7] In line with this, and drawing on a few literary examples, he attempts to show that the verb basically expresses the activity of the servant as he goes about his task. Some support for this is found in an etymology proposed by Vaniček which relates διάκονος to διάκτορος through a root DJA meaning "to hurry."[8] This support is however ambiguous for Schmidt, who sees a relevance for uses in respect of message but little if any, strangely, for other uses. His observations are helpful nonetheless in the light of the possibly special character of the word group suggested by its literary association with διάκτορος. Taken with Gschnitzer's views, Schmidt's suggest further that menial service, which is so commonly accepted as being the word group's basic field of reference, may be peripheral to the notions conveyed by διακον-. The shifts in the way the meanings of the words have been presented in some of the standard dictionaries will form a final brief topic for this chapter.

A Crossroads in Lexicography

With infrequent exceptions the definitions provided in the great *Thesaurus* of Stephanus are in terms of "to wait on table, to be a slave, to perform the work of an attendant or slave."[1] In one case the exception is only tentative, the passive verb in Demosthenes 50.2 (a passage to be examined later) being rendered in the first instance by Estienne himself in terms of "to manage and conduct," a judgement to which the following nineteenth-century gloss is added: "Perhaps [Necessarily], however, διακονεῖσθαι is to be taken here also in the sense "to be attended upon, to be presented with": a metaphorical use with reference to those who minister drink or food to those needing or requesting it." Such a shift from Estienne's idea of getting something done to the idea of service—especially emphatic in the bracketed "necessarily"—and then by a further stretch to service at table is forced in a passage that speaks of the execution of responsibilities by a public person. The English lexicon by Liddell, Scott, and Jones *(LSJ)* renders here by "to be supplied," which marks almost its only departure from the ideas of "minister, serve, render a service."[2]

The way in which an understanding of the word group other than one closely tied in to the notion of menial service could emerge only to be overlooked can be illustrated from the sphere of New Testament lexicography. In the manner of Buttmann, R. C. Trench favoured the view that διάκονος is related to διώκω and "thus indeed means 'a runner' still." He observed: "διάκονος represents the servant in his activity *for the work* . . . not in his relation, either servile, as that of the δοῦλος, or more voluntary, as in the case of θεράπων, to a person."[3] Trench, who was a professor in New Testament and highly reputed as a philologist (for twenty years also archbishop of Dublin), published this view in 1854 in a work on synonyms in the New Testament that was to be frequently reprinted (ten times even in a modern reprint between 1953 and 1978). His view came to be reflected in the writings of a few scholars of the next generation,[4] and in particular in the German "biblico-theological" dictionary of Hermann Cremer, a forerunner to Kittel's "theological" dictionary which was to contain Beyer's quite different treatment of the word group. But whereas Trench emphasised the notion of work to the exclusion of servile or voluntary relationships to a person, Cremer registered a shift of emphasis from the notion of the work itself to that of the relationship arising from the work:

> While, however, in δοῦλος the relation of dependence upon a master is prominent, and a state of servitude is the main thought, in διάκονος the main reference is to the service or advantage rendered to another [serviceableness], even as ὑπηρέτης refers to *labour* done for (serving) a lord (villenage). . . .[5]

This connotation of benefit then allows Cremer to see in a phrase at Rom. 15:25—which the New English Bible translates dispassionately and functionally as "on an errand to God's people"—"a beautiful expression for compassionate love towards the poor within the Christian fellowship" (compare the Revised Standard Version, "with aid for the saints").[6] At a later stage Beyer, who drew on Cremer as well as on Brandt, went on from there to overlook entirely the aspect of the work itself in favour of connotations of benefit that the work might bring, asserting that the comparison of synonyms in non-Christian usage reveals in διακονεῖν "the special quality of indicating very personally the service rendered to another [die ganz persönlich einem anderen erwiesene Dienstleistung] . . . a stronger approximation to the concept of a service of love."[7]

As is clear from much of the discussion in the first chapter, what Beyer calls an approximation to a service of love is the notion now usually taken from those passages about διακονία in the New Testament that have a bearing on functions in the church; indeed on the modern view the mere presence of a word of this group in any statement tends to give the statement an ecclesiological significance. When this happens we have a nice exemplification of the process described in general terms by James Barr: "the value of the context comes to be seen as something contributed by the word, and then it is read into the word as its contribution where the context is in fact different."[8] To go further at this stage and assert that the modern conceptualisation of "diakonia" exemplifies also what Barr calls elsewhere "premature theological evaluations of biblical linguistic data"[9] would not be satisfactory from the point of view of method, but the few samplings of lexicographical opinion already supplied do at least indicate the point at which the theological evaluation we have been calling "diakonia" emerged. This was when a notion of benefit was admitted to the word group's field of meaning and ultimately excluded aspects of work and movement of which earlier philologists like Trench and Buttmann had been aware.[10] Had scholarly influences worked differently so that perceptions about work and movement prevailed, the idea of a lowly service of others could hardly have exerted its strong influence in modern theology. The correctness of perceptions about work and movement cannot be said to be established on the basis of the few passages examined in this chapter, but in the notion of a go-between the passages have opened a field of meaning close to what earlier scholarship had taken for granted. This finding invites a detailed examination of other non-Christian usage to see to what extent notions of work and movement might be expressed there, and indeed to state the meanings and to depict the colour of the word group over as broad a range as possible. By arriving at the Greek mind on the use of these words we will be in a better position to assess what writers of the Greek New Testament and of similar early literature intended in their use of them.

5

Word

In a familiar phrase Luke speaks of "the ministry of the word," the διακονία τοῦ λόγου of Acts 6:4. The term "word," which means so much to the Christian church, has made the phrase memorable, but was an emphasis on "word" Luke's main intention? Or, for him, could the main thing have been διακονία? In this chapter we will put ourselves in a position to assess what the term "ministry" contributed to the phrase "ministry of the word" in Luke's mind. Would the term have come naturally to a Greek writing about a preaching ministry or would a Greek have needed to be a Christian to use it? To resolve these and other queries arising from similar usage elsewhere in early Christian sources we will look at those parts of non-Christian usage that deal with the handling of messages, bearing in mind that this field covers something like a third of all instances of our cognate words. And in examining these occurrences I hope not only to determine what ancient writers meant when they wrote about the διακονία of message but to get an inkling as well as to why they occasionally wrote of message in these terms.

A Servant Girl and a Scholar

Lysias, the Athenian orator of the fifth century BCE, and Aelius Aristides, a man of letters of the second century CE who falls somewhere between the aesthete and the mystic, provide useful passages with which to begin. In the one case an instance of the word group appears at first sight to be little more than an alternative designation of a servant girl and to be of only doubtful value to the argument, whereas in the other a phrase occurs that on customary readings is simply opaque. By taking a lead from the notion of a go-between, however, we find on the one hand that Lysias's word makes a specific point in the presentation of legal evidence, and on the other that Aristides is being clear in his rather elegant way.

Both passages remind us that an author's meaning can easily elude us if we think of our cognate words in terms of service.

In Lysias the Athenian Euphiletus is before the court on a charge of having murdered Eratosthenes. The defence claims that the dead man was a known philanderer and had been involved in an affair with the defendant's wife. As evidence Lysias adduces information that an old woman had volunteered to Euphiletus:

> "Euphiletus, do not think it is from any meddlesomeness that I have approached you; for the man who is working both your and your wife's dishonour happens to be our enemy. If, therefore, you take the servant-girl [θεράπαιναν] who goes to market and waits on you [διακονοῦσαν ὑμῖν], and torture her, you will learn all. It is," she said, "Eratosthenes of Öe who is doing this; he has debauched not only your wife, but many others besides; he makes an art of it." (Lysias 1.16).

The translator[1] has designated the servant girl as the one who attends on Euphiletus at home. This is a possibility. The value of this part of the statement would then be that the girl presumably observed Eratosthenes' visits to the household. This would mean that the phrase "who waits on you" duplicates an idea already contained in "servant girl" without adding to the evidence. As it stands in the Greek, however, this phrase is more naturally alluding to an activity connected with the market. The old woman would then be indicating how she came to know of the girl, namely, on her errands for her mistress, and so identifies her as "the servant girl who goes shopping for you in the market." This in fact is what the eighteenth-century scholar Reiske made of it, and shortly afterwards his Oxford editor T. Mitchell took up the idea when he noted that the participle designates a servant who is employed *"for the purpose of sending on errands."*[2] Later we will find an unambiguous instance of such a use in Theophrastus (*Char.* 2.9), but the passage from Lysias already suggests that the ambiguity with which it presents us could be due less to Lysias than to our unfamiliarity with his idiom.

In the instance in Aristides meaning is more elusive. In the funeral oration in which he extols the superiority of his tutor and mentor Alexander over learned contemporaries, Aristides observes that scholars who are expert in their discipline often have no flair for communication and that others who lack nothing in eloquence are often suspect in their learning; he proposes that Alexander's success lay in bringing eloquence and learning together. Further to this, his unique worth is to be seen in the fact that he did not use his gift for language in writings of his own but in interpreting the writings of his precedessors. It is in the expression of this thought that Aristides' Greek is unusual. It reads: ἀλλ' ἑλέσθαι τὴν τοῖς παλαιοῖς Ἕλλησι διακονίαν (1.82). Commentators are agreed as to the thought here, that Alexander devoted himself to the interpretation of writings by ancient Greeks, but are not clear as to why διακονία should be used in the designation of such a literary pursuit. Does Aristides mean simply that Alexander used his talents in the service of classical literature, the reader being left to perceive that the service consisted in interpretation? Jebb apparently thought so.[3] But a consideration of style and rhetorical balance can help here, and Aristides was normally meticulous in both. This section of the oration shows a distinct attempt to balance words against each other within the same field of meaning, so that when a word meaning properly "writings" occurs earlier in the sentence we do not expect Aristides to finish the sentence with a word meaning properly "service." That is to say, we might expect διακονία to be a proper designation for an activity within the literary field. If we then recall that Plato conceived of the interpretation of heavenly knowledge as the exercise of a skill he called διάκονος (*Plt.* 290c), we see that Aristides also may have been intending to designate an interpretative function. Thus, as seers engage in a διακονία from heaven to earth, Alexander would be presented as passing on to men of his own time the ideas of an earlier generation, and Aristides would be expressing himself not under any figure of service to the ancients but in a phrase which might be translated, "Alexander made it his task to interpret the ancient Greeks."

Both the instances just examined are best seen as falling within the field of message, that is, "doing" a message from one place to another and conveying a message from one period of time to another. The two passages also exemplify how tenuously the meaning of the cognate words can be tied to context, contrasting in this with a passage like Plato, *R.* 370e, where context demands the meaning "courier," and suggesting that other passages also might need to be examined closely before a suitable meaning is found. In one important respect the use of different English words to get at the meaning of the Greek has its disadvantages. Because the verb in Lysias was translated satisfactorily and idiomatically as "to go shopping," and the noun in Aristides was translated more formally and with a touch of paraphrase as "the task of interpreting," the reader loses sight of the fact that we are dealing with one and the same Greek word group. The difficulty here is a familiar one to translators but is a little more awkward than usual when the aim is not merely to state the meaning of a word but to build up the range of connotations that the word gathers to itself in its varied uses. In the use of different and unrelated English words to express various aspects of a central idea underlying the Greek, the connotations are unavoidably separated out and we lose touch with the feeling of the word in the original. We should bear this in mind as we run through a series of passages surprisingly rich in their suggestion of message from heaven.

To Heaven

In the Greek *Testament of Abraham* (Recension A), a work probably contemporary with the New Testament, the patriarch approaches death with one desire yet to be realised on earth: to be granted a vision of the whole of God's creation. He is visited by Michael, God's "chief-captain," whom he asks to take the request to God.

> The chief-captain . . . went down unto Abraham; and when the just one saw him he fell upon his face to the ground as one dead; and the chief-captain told him all that he had heard from the Most High. Then did the holy and just Abraham rise up with many tears and fell at the feet of the spirit, and besought him saying: I beseech thee, chief-captain of the powers above, since thou has deigned

to come thyself altogether unto me, a sinner and thine all-unworthy servant, I implore thee even now, O chief-captain, *to be the medium of my word* yet once (again) *unto the most High,* and thou shalt say unto Him: Thus saith Abraham, thy servant, O Lord God.

The italicised phrase in this translation[1] corresponds to τοῦ διακονῆσαί μοι λόγον . . . πρὸς τὸν ὕψιστον (*Test. Ab.* 9.24). A more recent translator has similarly given "be my interpreter" but has noted that the "literal" meaning is "serve for me the word."[2] The note is typical of the kind of comment elicited by some uses of the verb but is not enlightening from a linguistic point of view because the context does not invite a distinction between literal and actual meanings. Thus Abraham's request is the culmination of the scene so that our interest centres on what it is he is asking; his phrase contains two objects and a qualifier to leave no doubt that it expresses the idea of transmitting a speech for one person to another through the offices of a third. The author would not thus be using the verb in an applied or figurative way but rather, in keeping with the elevated sentiment of the passage, would be placing on the lips of Abraham an expression appropriate to the office of the "chief-captain of the powers above" who communicated with God himself.

In Lucian's story of Menippus we have a similar incident recounted with less reverence. The grammatical structure is also the same as in *Test. Ab.* 9.24 although the object of the verb is not determinate as there ("word"), a detail that allows Lucian to introduce a touch of humour to an interchange between the mortal Menippus and the moon goddess Selene. Discontent with the conflicting theories issuing from the schools of philosophy, Menippus aspired to reach heaven where he was confident of being better advised. To this end he equipped himself with wings. He does not travel far before encountering Selene, who is herself fed up with the scientific theories by which philosophers would explain her existence away, and who avails herself of the heaven-bent voyager to lay a complaint before Zeus. She hopes that the god might "pulverise the physicists, muzzle the logicians, raze the Porch, burn the Academy" and eradicate philosophy from the minds of men. Menippus relates the encounter in the following terms: "I had only flown a couple of hundred yards, when Selene's feminine voice reached me: 'Menippus, *do me an errand to Zeus,* and I will wish you a pleasant journey.' 'You have only to name it,' I said, 'provided it is not something to carry.' 'It is a simple message of entreaty to Zeus.'"[3]

The italicised sentence is the translation of διακόνησαί μοί τι πρὸς τὸν Δία (*Icar.* 20) and catches nicely the ambiguity intended by Lucian. The object is indefinite ("deliver *something* for me"), leaving Menippus to fear that he could be laden down with a parcel on his precarious flight. Selene, however, had not intended to arouse any such anxiety. To her mind her first words were clear enough, and in the Greek her second statement is grammatically a continuation of the first, a sign that "message of entreaty" (like "word" in *Test. Ab.* 9.24) is as fitting a complement of the verb as another word that might signify an object to be carried.

If in the end Selene's meaning is clear to both Menippus and the reader, a further point about the usage may be taken from her attitude. Lucian's skill as a writer requires us to suppose that the moon goddess expected to be accurately comprehended when she first accosted Menippus; otherwise he would be using an ambiguous expression so blatantly as to lose the minor comic effect he was after. The irony that permeates the whole essay and has several truly comic peaks finds an expression in this instance only if Selene and the informed reader are momentarily caught short by the naivety of Menippus in assuming that in his and Selene's circumstances διακονεῖν could refer to anything else but the taking of a message to heaven. The words Selene uses for "message" and "entreaty" are πρεσβεία and δέησις, both fitting for an address to God. Suspended as she is betweeen earth and heaven and thus accustomed to seeing heavenly messengers pass her way, she might be expected to ask them to act in her behalf in terms just as fitting. In other words, to her the circumstances of her meeting with Menippus created a context sufficient for what she meant—namely, that Menippus was to be a medium (like Michael in the *Testament of Abraham*) between her isolated station and the realms above.

In Lucian, *Icar.* 20, and *Test. Ab.* 9.24, therefore, we have clear instances of the same use of the verb. Context indicates also that the use was customary. Indeed it was so established that in the second century Lucian played on his readers' familiarity with it. Centuries earlier, as we have seen, Aristophanes had done the same (*Pl.* 1170). We also recall once more that in discursive writing Plato was satisfied that the whole area of dealings between gods and men could be designated διάκονος (*Plt.* 290c, 299d), a linguistic fact that takes on a larger significance in the light of stories about Hermes, "messenger most excellent" as he was to the ancients, which employ this and cognate terms. Instances of such message from heaven will now be drawn together in the following pages, and they are to be complemented by usage in respect of other aspects of dealings with the gods which will be examined in succeeding chapters.

From Heaven with Hermes and Iris

Hermes' role as διάκτορος/διάκονος has already been sketched in illustration of the relationship between these two terms. Among literature cited there was Hermes' reply to Charon after being asked to act as a guide in a tour of the world (Lucian, *Cont.* 1), "I am bound on certain errands of the Upper Zeus, certain human matters."[1] The Greek is in fact much closer to the two phrases just discussed than this excellent translation by Fowler suggests (ἀπέρχομαι γάρ τι διακονησόμενος τῷ ἄνω Διὶ τῶν ἀνθρωπικῶν), and we can represent it as "I am leaving to deliver something concerning mortal affairs for the God above." Whereas, however, the two previous examples contained a phrase meaning "to God" which established the direction of the errand as from earth to heaven, this sentence has only "to deliver something for God." The dative "for God," corresponding to "for me" in Lucian, *Icar.* 20, and *Test. Ab.* 9.24, indicates the person in whose name the errand is being done, accordingly also the direction, which is now from heaven to earth. Thus the participle διακονησόμενος of itself implies movement. At the same time, as was argued earlier, Lucian's purpose in employing this word was less to designate movement than to introducce a conventional sign of Hermes' office. In declining to accompany Charon, therefore, Hermes was saying he was on earth not for one of his escapades but in an official capacity. Of other passages already cited as referring to an office, two are now recorded in detail.[2]

In Aeschylus Prometheus refers unmistakably to the office of Hermes in designating him "the messenger of the new tyrant" (this is Zeus; τὸν τοῦ τυράννου τοῦ νέου διάκονον, Pr. 942). The designation falls between phrases meaning "courier of Zeus" and "he has come to announce news," and the passage is important as providing perhaps the earliest known instance of the word group, which is thus also one enshrining the notion of messenger (and recorded as such in LSJ).

The second passage is the curse inscription from the Athens of the third century B.C.E. containing the expression Ἑρμεῖ διακόνῳ. The exact meaning of this is difficult to determine but G. W. Elderkin, who edited the text and whose translation is used here, rendered "messenger."[3] The lead tablet contains the prayer of a householder who had been robbed and who hoped to recover his property and bring divine retribution upon the thieves. We read from line 4:

> I inscribe and consign to Pluto and Fates and
> Persephone and to Furies and every evil one,
> I consign also to Hecate eater of animals,
> I consign to underworld goddesses . . . and gods
> and to Hermes messenger,
> I consign the thieves who take their name from
> the little house of a certain slum quarter.
> Bid the thief, O Hecate, restore [?] three coverlets,
> a fleecy white cloak. . . .

These are gods of the underworld. Hermes is last mentioned. He was not in fact a god who was confined to the underworld nor a naturally inimical power. The term added to his name is clearly intended to be a title. A title is also given to Hecate, probably because she is the plaintiff's chief hope. Perhaps Hermes is included because he conveys the dead to their place—he was also well known as a thief, and even the ancients may have gone in for setting a thief to catch a thief. The title itself, however, would not bear specifically on the role of conveying the dead but would be affixed by force of custom. It may also have served to distinguish him as a god of heaven from gods of the underworld whose company was alien to him.

A versatile, unpredictable, but much loved god, Hermes was the centre and even the butt of many stories. From his infancy, as Lucian recounts in *Dialogues of the Gods,* he was marked out as the future messenger of heaven by two qualities in particular, his nimbleness and his glibness of tongue. Apollo amused himself pointing out these qualities to Hephaestus, heaven's smith, who saw in the infant only "such a pretty little thing, with a smile for everybody." The child had already, however, stolen Poseidon's trident, Ares' sword, Apollo's own bow and arrows, and, before Hephaestus could credit such naughtiness, the smith's tongs. "Ah," continues Apollo, "and you don't know what a glib young chatterbox he is; and, if he has his way, he is to be our errand-boy!" The last phrase is again Fowler's translation of ὁ δὲ καὶ διακονεῖσθαι ἡμῖν ἐθέλει (*DDeor.* 7.3),[4] where the infinitive is an allusion to Hermes' classic title διάκτωρ, although Lucian is referring only to the doing of messages and not to the god's other characteristic roles which that title might suggest and with which Lucian rounds out this portrait:

> Even at night, Maia was saying, he does not stay in Heaven; he goes down poking his nose into Hades—on a thieves' errand, no doubt. Then he has a pair of wings, and he has made himself a magic wand [a gift in fact from Hephaestus], which he uses for marshalling souls—conveying the dead to their place.

That essay includes also a reference to Hermes outsmarting Eros in a heavenly tumble, a feat applauded by Zeus because the god of love was of course mischievously quick. It was to mean that Hermes would liaise for Zeus with earthly paramours. No less than three of these are mentioned in one day's round of duties in Lucian's conversation piece between the god and his mother, Maia.

> HERMES: Mother, I am the most miserable god in Heaven.
> MAIA: Don't say such things, child.
> HERMES: Am I to do all the work of Heaven with my own hands, to be hurried from one piece of drudgery to another, and never say a word? I have to get up early, sweep the dining-room . . . then I have to wait on Zeus, and take his messages, up and down, all day long: and I am no sooner back again (no time for a wash) than I have to lay the table; and there was the nectar to pour out, too, till this new cup-bearer was bought. . . . Leda's sons [Castor and Polydeuces] take turn and turn about betwixt Heaven and Hades—*I* have to be in both every day. And why should the sons [Heracles and Dionysus] of Alcmena and Semele, paltry women, why should they feast at their ease, and I—the son of Maia, the grandson of Atlas—wait upon them? And now here am I only just back from Sidon, where he sent me to see after Europa, and before I am in breath again—off I must go to Argos, in quest of Danae, "and you can take Boeotia on your way," says father, "and see Antiope." I am half dead with it all. Mortal slaves are better off than I am: they have the chance of being sold to a new master; I wish I had the same!
> MAIA: Come, come child. You must do as your father bids, like a good boy. Run along now to Argos and Boeotia; don't loiter, or you will get a whipping. Lovers are apt to be hasty. (*DDeor.* 24.1–2)[5]

The topic of love will interest us later. The main point of this passage, its facetiousness aside, is the picture of the messenger of God at work. The god of eloquence as he is, Hermes has no time to put in a word of his own to indulge himself but is forever expending his energy and talents at the behest of Zeus. The journeys are arduous, and he must be quick about them. At the same time we notice that Lucian does not here call Hermes διάκονος, and it is tempting to surmise that by this omission of the word in an irreverent account of the messenger god's hectic day Lucian intends to convey the impression that Hermes sees small virtue attaching to his role as διάκονος/διάκτορος. Taking messages is διαφέρειν τὰς ἀγγελίας. On the other hand Lucian uses διακονοῦμαι to mean "I wait upon the sons of Alcmena and Semele" (*DDeor.* 24.2). We thus have the reverse of a point of usage noted in Lucian, *Cont.* 1. Each of these satires refers to Hermes' duties of waiting at table and carrying messages, but although διακον- is a proper designation of either duty Lucian employs the verb in one sense only on each occasion: in *Cont.* 1 for the delivery of message, because the encounter with Charon requires a play on the idea of διάκτορος, and in *DDeor.* 24.2 for waiting on table, because here the other sense would be inappropriate when the hallowed role of διάκτορος is being gently ridiculed. The pattern does at least suggest a sensitivity on Lucian's part to a word that is, after all, comparatively rare.

The story of Hermes is too long to follow here, and the theology developing around his role as messenger of the gods too elaborate to trace, but we do need to appreciate that the word διάκονος lent itself naturally to both story and theology. The connection began in an etymology—true or at least popular, as proposed earlier—that related the word to διάκτορος, and then developed because, as Plato has illustrated, the word's meaning fitted well with the passing of knowledge from one realm to another. When Plato says that divination is διάκονος because its practitioners are the gods' interpreters (ἑρμηνευταί, *Plt.* 290c), we can be sure the word attached itself to the "herald of the gods and most trusted messenger"— in the phrase of Diodorus Siculus (5.75.2)—because of "his clarity in expounding [ἑρμηνεύειν] everything given into his charge." Indeed, as Diodorus goes on to say, "he perfected, to a higher degree than all others, the art of the precise and clear statement of a message,"[6] and in the opinion of the Neoplatonists, as represented by Iamblichus, was not only "the initiator of speech" (ὁ τῶν λόγων ἡγεμών) but "the guardian of true knowledge of the gods" (*Myst.* 1). Against this background of popular and critical esteem the people of Lystra saw Hermes in Paul of Tarsus, because he was, as the translations have it, "the chief speaker" (ὁ ἡγούμενος τοῦ λόγου, Acts 14:12).

This seemingly strange incident in Acts puts us on the level of the ordinary Greek mind. We have another and delightful view of it in a scene from Heliodorus's romantic novel *Theagenes and Chariclea.* A rogue merchant named Nausicles has organised a sacrifice in honour of Hermes, "the patron of merchants and traders" (5.13)—which reminds us perhaps of Plato's choice of terms at *R.* 370e— and in the course of the ceremony comes by a windfall in the shape of a precious ring. This he attributes to Hermes (we shall look at the incident again because the narrative includes at this point an interesting instance of διακονεῖν). As the banquet progressed, the men were drinking and singing to Dionysus, with the women dancing in their own quarters for Demeter, until Nausicles prevailed on the priest to resume his tales. A visitor is duly impressed, and offers the following address to Nausicles:

> "Though you have provided every kind of music for our entertainment you are willing to forgo ordinary kinds of amusement and are eager to listen to arcane matters tempered with a pleasure truly divine. In my opinion you show admirable judgement in coupling Hermes with Dionysos, thus mingling the pleasures of discourse with those of wine. I am filled with admiration at the sumptuousness of your sacrifice, and I cannot think how anyone could render Hermes more propitious than by making a contribution to the feast of the thing which is appropriate to Hermes—discourse" (5.16).[7]

Hermes and good conversation, in other words, went together—in a startlingly real way, for the one stimulated the other, and the god was manifest in the talk.

Because our cognate words are never far away in accounts of these matters, it would seem that through the term διάκονος—and despite the double standards of a Nausicles or the frivolity of a Lucian—we are in touch with one of the most sensitive areas of the ancients' experience of the other world. Further evidence of this, drawn from statements of a more philosophic kind, will be examined in the next chapter.

That the association bears emphasising is clear from the difficulties of interpretation over a reference in Aristophanes to the gods' other messenger, Iris the Swift (*Av.* 1253). Peisthetaerus, annoyed at being pestered by messengers from Zeus, threatens to rape her. Now usually translated "messenger," the instance of διάκονος here once posed commentators a problem. F. H. M. Blaydes, for example, understood it to mean "handmaid," a designation he judged to be out of keeping with the standing of Iris. Accordingly he favoured an emendation of the text that would allow him to suppose that the term referred not to Iris but to a handmaid accompanying her.[8] Similar narrow assumptions about the meaning of this and related words have occasionally influenced the choice of reading elsewhere in non-Christian literature and epigraphy, and, more interestingly from a theological point of view, on at least one occasion in the New Testament (1 Thess. 3:2).

Messengers Less Than Gods

If Iris and Hermes are themselves gods and preeminent among the messengers of God, the διακονία of message from or to heaven is not a function reserved to divine beings. We have already seen the mortal Menippus engaging himself in a flight to heaven (Lucian, *Icar.* 20), and Michael, a leading figure in the divine entourage but not himself divine, being prayed by Abraham to mediate a request (*Test. Ab.* 9.24). Writing of Abraham's encounter with angels (Gen. 18), Philo calls them "the servitors and lieutenants [ὑποδιάκονοι καὶ ὕπαρχοι] of the primal God whom He employs as ambassadors [πρεσβευτῶν] to announce the predictions which He wills for our race" (*de Ab.* 115).[1] Here the connection with message is not as direct as the passage at first sight suggests because ὑπο/διάκονος is one of Philo's terms for the agent or executive of a superior power (see such passages as *de spec. leg.* 1.66, *de decalogo* 178—both to do with angels and God—and *de gigant.* 12). At the same time, as *de spec. leg.* 1.66 itself will show, as also *de spec. leg.* 1.116, *de Josepho* 242, and especially *de vita Moysis* 1.84, which is dealing with communication by word of mouth, the word also expresses for Philo the idea of mediation. In this instance, accordingly, the angels are probably better seen as "intermediaries" than as "servitors." Philo's compounds in υπο- (one of them to enter Christian usage later as the designation of the subdeacon) are of incidental interest. They occur nine times, according to Leisegang's index,[2] against eight occurrences of uncompounded words (four of these in one passage of *de vita cont.*), without seeming to introduce a refinement of meaning. Even so, one could have expected rather more than these seventeen instances in literature of Philo's kind. Writing on the nature of prophecy, for example, a topic that elicited the use of διακον- by one of Philo's mentors, Plato, and by his near contemporary Paul, Philo seems almost to go out of his way to avoid the words (*de spec. leg.* 1.65). There is nothing to show, however, that Philo shares the Septaugint's apparent distaste for the words. For one thing, he uses them in writing of the patriarch Joseph (*de Josepho* 242) and of Aaron, spokesman of Moses (*de vita Moysis* 1.84).

Plato also writes of a human, if mythical, figure as the διάκονος of gods (*Lg.* 782b). This is Triptolemus, and Plato is referring to the belief that the Earth Mother and her daughter, Demeter and Ceres, appointed him to spread among men knowledge of agriculture, which had previously been possessed only by these other-worldly figures. The usage thus falls in with what Plato has had to say about the intermediary both in analytical discourse (*R.* 370e) and in treating specifically of communication with heaven (*Plt.* 290c).

When the same word occurs in parallel with the religious term "herald" in a fragment of Sophocles (fr. 137) we can be sure that despite the lack of context the meaning is "bird, herald and messenger [of God]." The phrase would then form part of a minor poetic motif of which several other expressions survive. Callimachus, as noted earlier, refers to the owl by the archaic but related term as διάκτορος of the goddess of Athene (*Etym. Magn.* 268.21). In Antipater of Sidon an eagle responds with "I announce" when addressed as Διὸς . . . διάκτορε (7.161), which shows that it is a "messenger of God." Sophocles himself calls the same bird "messenger of God" (*El.* 149), and the Greek term in this instance (ἄγγελος) occurs in parallel with the only other instance of διάκονος in this poet (*Ph.* 497). F. Ellendt long ago gave "messenger" as the meaning of both.[3] We shall see that the grammarian Pollux and his abstracters considered διάκονος, ἄγγελος, and κῆρυξ ("herald") to be terms that are equivalent or related in meaning within the area of religious message (8.137).

Constantine

The papyrus *P. Lond.* 878 (*SB* 9218) provides a particularly illuminating instance of διακονία in the area of religious message. Once the subject of a brief but learned debate, the papyrus has since been identified as part of a contemporary copy of the Emperor Constantine's proclamation in 324 concerning the Christian religion,[1] the text of which had already been preserved by Eusebius in *Life of Constantine* (2.24–42). The papyrus was fragmentary, however, and T. C. Skeat, who published it in 1950 and dated it early in the fourth century, hesitated to hazard a guess at the subject matter. In part its language suggested theological controversy but Skeat's interest lay in geographical details, which seemed to point to the author's past involvement in military service.[2] Various terms for "service" do appear, and working partly from these and from the mention of "pestilential disease," another papyrologist, W. Schubart, concluded that an eminent Christian, perhaps on the outbreak of plague, was issuing a warning to fellow Christians of God's punishment of infidelity. In particular he understood διακονία to refer to the author's "service to a humanity which was suffering from godlessness."[3]

In the light of our preceding discussions, however, this word is not likely to contribute to this line of interpretation. In a context that concerns fidelity to God in any way, the word is more likely to be designating what the author was doing on behalf of God, possibly covering the idea that he was addressing men in God's name. Eusebius's full text does in fact show that the word is designating the emperor's mission from God to extend the sway of Christian truth. If this settles the question, it is still instructive to see how the meaning is reached and to reflect that interpretation of the papyrus would have been facilitated had scholars not taken the narrow line of "beneficent service to someone in need" but had rather been able to view the word against the background we have been examining of communication of religious knowledge. At the same time it must be confessed that the meaning of this word is often elusive. Largely this is because the word is an abstract noun and, as such, shows different aspects of meaning according to context. No single English equivalent seems to express these suitably, and when a religious connotation is also present, translation becomes more awkward again, and yet the author may be relying on the distinctive connotation to establish his effect. To read the passage, which corresponds to 2.28–29 in Eusebius's *Life,* is also to observe the difference between the meaning of διακονία and of the other Greek words given in brackets which could each in its way be rightly translated into English as "service."

> And now, with such a mass of impiety oppressing the human race, and the commonwealth in danger of being utterly destroyed, as if by the agency of some pestilential disease, and therefore needing powerful and effectual aid [ἰατρεία]; what was the relief [βοήθεια], and what the remedy [θεραπεία] which the Divinity devised for these evils? . . . I myself, then, was the instrument whose services [ὑπηρεσία] He chose, and esteemed suited for the accomplishment of his will. Accordingly, beginning at the remote Britannic ocean, and the regions where, according to the law of nature, the sun sinks beneath the horizon, through the aid of divine power I banished and utterly removed every form of evil which prevailed, in the hope that the human race, enlightened through my instrumentality [ὑπουργία], might be recalled to a due observance [θεραπεία] of the holy laws of God, and at the same time our most blessed faith might prosper under the guidance of his almighty hand. I said, under the guidance of his hand; for I would desire never to be forgetful of the gratitude due to his grace. Believing, therefore, that this most excellent διακονίαν had been confided to me as a special gift [κεχαρισμένον . . . δῶρον], I proceeded as far as the regions of the East, which, being under the pressure of severe calamities, seemed to demand still more effectual remedies [θεραπεία] at my hands.[4]

Although the passage announces a great Christian initiative, its language is not that of a theologian but of an imperial notary, and—one or two phrases of a distinct Christian colouring aside ("special gift" is noteworthy)—it could be the proclamation of any ancient despot advancing what he believed to be his divine cause. Of the terms relating to the emperor's role, ὑπουργία means "service" as rendered by a subordinate; ὑπηρεσία, although by no means unknown as a designation for service under a divinity, is here taken from the language of the imperial civil service and designates the emperor's responsibility for the implementation of a divine decree; θεραπεία, in the phrase "observance of the holy laws of God," means in the normal way "service of God"; on its other occurrences it means "beneficent service to humanity," as do ἰατρεία and βοήθεια, all in the normal way within the context of a figurative "disease."

An orientation towards humanity is also implicit in διακονία, but here, instead of saying something about concern for the human race, as Schubart assumed, the word implies only that the human race is the object or end of the action under discussion. Its meaning is to be sought in the first place in the fact that it denotes something that has been received from God. In terms of "service" this would be an act of service originating in a divine command and effecting a divine intention in regard to men. The context then suggests that the service involves movement, not indeed from heaven to earth but from the west of the empire to the east. At the same time the word occurs at a point where the religious sentiment and conviction are at their peak; the word is enriched by this context and, as we may now say, also contributes to it. We note finally that the commission is exclusive to the emperor. It is a commission to go among the human race with a mandate from God to ensure that they are enlightened as to his holy laws and that they are faithful in observing the laws. Clearly the word "service" would be inadequate here; "commission" lacks religious overtones; "ministry" does not imply movement. A composite expression like "godly mission" is perhaps best, although we are then missing the idea of communicating in the name of God.

A Saviour and a Trickster

A similar instance of the abstract noun occurs in Epictetus, 3.22.69. The passage is now increasingly cited by theologians to explain the origin of Paul's usage in describing himself as a διάκονος engaged in a διακονία or apostolic mission, but because a large part of the interpretation, especially in regard to the common noun, is open to question, examination of the passage is best left to the chapter on the writings of this moralist. Another clear instance is provided, however, in one of the bizarre episodes of the novel *Leucippe and Clitophon*—"a somewhat foolish and improper novel," according to an eminent twentieth century novelist[1]— by Achilles Tatius (4.15.6).

After taking a draught administered to her by an attendant, Leucippe begins to show symptoms of madness, and her lover Clitophon is desperate. In a dream she mutters something about a certain Gorgias, but as Clitophon sets out to find this individual he is dumbfounded to be met by a young man who introduces himself as "saviour" (σωτήρ). Clitophon immediately takes him to be "one sent by God" (θεόπεμπτος), and asks if he happens to know Gorgias. The young man replies that he does not but that he is aware Gorgias is the cause of the misadventure. "I had to shiver at this," recounts Clitophon and, explaining that he had himself been given this name during the night by some "daemon" (δαίμων), asks how the disaster came about and who Gorgias might be. "You," he says to the young man, "are to be the interpreter of the heavenly message" (διηγητὴς γενοῦ τῶν θείων μεινυμάτων). Thereupon the young man tells what he had heard from Gorgias's servant. Gorgias, an Egyptian soldier, having fallen in love with Leucippe, had contrived to have a potion administered to her. The drug was too powerful, madness ensued, and now Gorgias, who was expert in drugs, would return to prepare an antidote. Clitophon is greatly relieved, and in thanking the young man for his part in the affair says, σοί . . . ἀγαθὰ γένοιτο τῆς διακονίας (4.15.6).

We could translate this last passage as "May you be rewarded for your errand" but "errand," which might be appropriate in a passage like Isaeus 1.23, would not indicate how Clitophon really looks upon the young man's unexpected intervention. From the moment the young man announces himself as "saviour" Clitophon thinks he is in the presence of "one sent by God" who is able to unravel "heavenly intimations," and he therefore sees the arrival of the young man and his disclosure of the course of events as the discharging of a mission on behalf of the gods. As a διακονία the young man's intervention is not a service to Clitophon but a message to him under divine mandate.

On another occasion in the same novel Clitophon is similarly convinced that he is dealing with a representative of the other world, an experience which this time elicits from him the word διάκονος. His friend Menelaus has persuaded Leucippe to take part in a cruel strategem at the expense of Clitophon. He places her in a coffin, arranges her to look like an eviscerated corpse, and leaves her for Clitophon to find. Clitophon duly does so and in desperation is about to take his own life when Menelaus arrives. Assuring Clitophon that he will restore Leucippe to life and wholeness, he begins to invoke the goddess of the underworld, Hecate. Clitophon hides his face in terror at the proximity of the praeternatural ("I really thought Hecate was there," he says later), Menelaus removes the disguises from Leucippe, and she emerges whole and entire. Clitophon is still overcome and implores Menelaus to explain the mystifying event. "If you are truly a διάκονος of the gods," he says, "I beg you to tell me where I am and what it is I have seen" (3.18.5). At this Leucippe intervenes to tell how it has all come about, and one suspects that it is as much the ludicrous irony as the distress of her lover which she can no longer bear.

The episode thus provides several indications that Clitophon's words have a special import. Leucippe's instant reaction to them probably shows that she feels it is wrong to toy with such deep events, and Clitophon's sense of awe certainly supposes that his appeal to Menelaus has been cast in highly reverential terms. In addition, as T. F. Carney has observed,[2] Clitophon had earlier criticised Menelaus and used the derogatory term μάγος ("sorcerer," 3.18.5–6), so that at this later desperate moment he would not wish his term διάκονος to be construed as further criticism. Rather he uses the term because he has heard Menelaus mouthing the powerful words of the dread goddess and assumes that Menelaus has yet further knowledge of the arcane. Thus Menelaus would be addressed less as a minister to a god than as a god's minister or appointee in dealings with human beings, "some duly sanctioned representative of the gods," as Carney puts it, and the term draws its religious connotation from the custom of designating heaven's spokesmen in this way.³

The Making of Aesop

In an anonymous life of Aesop composed around the time of the New Testament, the great storyteller, who had been active in the sixth century BCE and was widely venerated as something of a saint, is said to have received the power of speech and his artistic gift as a reward from Isis for kindnesses he had shown to a herald of hers who had lost her way in the course of an errand on earth. Isis is presented as the supreme god, the "precious bond of the universe," and it is she who calls the herald "my διάκονος" (*Vita Aesopi* 7).

The encounter of heaven's messenger with the dumb Aesop is touchingly described. Being lost and seeing Aesop working in a field nearby, the herald leaves the road to ask him the way to the city she wishes to reach. Aesop immediately recognises in her a heavenly being,¹ and duly reverences her, but as for giving directions can only make signs and point. He does however refresh her with his own food and water before setting her on her way. As she proceeds the herald thinks back on the piety and kindness of Aesop and raises her arms to Isis to pray that he may receive the gift of speech in return for the reverence shown to Isis in her person. Meanwhile, Aesop sleeps in the heat of the afternoon and Isis, hearing the prayer, approaches the Muses. She tells them that she will herself give this man the power to speak and they will endow him with the highest eloquence because out of piety he had overcome the impediment of his dumbness to show the way to "my straying messenger" (τὴν ἐμὴν διάκονον). Aesop's tongue is loosed, he wakes, and begins to name the things that he sees about him, his fork, his wallet, the ox, and ass, and the sheep.

At several points in the account the author moralises on piety and on the power of prayer but his main intent is to eulogise Aesop, whose kindness and religious spirit are carefully depicted against a strong mythical background. The herald is twice called the "form" of Isis, and her privileged role is further reflected in the term by which she is regularly designated, ἱεροφόφος, for this means "bearer of the sacred." Her gratitude to Aesop is emphasised and finds expression in a prayer of great fervour. This directs attention to Isis, who is grandly conceived as "precious bond of the universe" and who then formally proclaims to the Muses that her divine compassion will extend to the simple afflicted farm labourer. The occurrence on her lips at this point of the term διάκονος in place of "bearer of the sacred," which occurs in the narrative sections, is thus significant. It can have no less a religious value than the latter, and surely gains from the solemnity of the occasion and from being part of the language of a divinity. Perhaps it is part of the author's intention that Aesop, maker of fables, should be seen to have received the power to speak at the prayer of one whose own role under heaven was so clearly defined as speaking in the name of her god.

The Brother of a Reluctant Prophet

Endearing myth of this kind was anathema to a contemporary Jew like Philo. To him the voice of God had come in prophecy, which was enshrined in the scriptures. Writing on the evils of divination as cultivated among the Gentiles, he concludes with a statement on the rigour and purity of prophecy as exercised in the name of the true God. Extreme in its formulation, the statement nonetheless puts a picture that devotees of oracles and seers would have viewed with favour and which indeed has been reflected in many a Christian's view of biblical inspiration:

> Nothing of what [the prophet] says will be his own, for he that is truly under the control of divine inspiration has no power of apprehension when he speaks but serves as a channel for the insistent words of Another's prompting. For prophets are the interpreters of God, Who makes full use of their organs of speech to set forth what He will. (*de spec. leg.* 1.65)[1]

Immediacy of divine communication is what Philo requires. Accordingly, in writing of Moses, who hesitated to accept his brief from God, Philo writes:

> But, though he believed, he tried to refuse the mission, declaring that he was not eloquent, but feeble of voice and slow of tongue, especially ever since he heard God speaking to him; for he considered that human eloquence compared with God's was dumbness. . . .
> But God, though approving his modesty, answered: "Dost thou not know who it is that gave man a mouth, and formed his tongue and throat and all the organism of reasonable speech? It is I Myself: therefore, fear not, for at a sign from Me, all will become articulate and be brought over to method and order, so that none can hinder the stream of words from flowing easily and smoothly from a fountain undefiled." (*de vita Moysis* 83–84)[2]

Moses' words will thus be the words of God "undefiled."

Accounting then for the role of Aaron, who according to the book of Exodus (4:14) is merely one who can speak more volubly and articulately than Moses, Philo leaves us to suspect that Moses' words may not actually be intelligible, and he therefore introduces Aaron as an "interpreter," for God continues (in Colson's translation): "And if thou shouldst have need of an interpreter [ἑρμηνέως], thou wilt have in thy brother a mouth to assist thy service, to report to the people thy words, as thou reportest those of God to him." Authenticity is thus assured. Of interest to us is the phrase "a mouth to assist thy service." Philo wrote: ὑποδιακονικὸν στόμα (*de vita Moysis* 84). By virtue of the prefix ὑπο- (strictly "sub-") the translator envisages that Aaron is filling a subsidiary role within the "service" of Moses. As suggested in connection with *de Ab.* 115, however, this prefix seems to carry no weight with Philo, and it is part of a word that is in any case an adjective qualifying "mouth." Philo is saying that if Moses needs an interpreter, Aaron is well endowed for that role because he has "an intermediary's mouth," one that functions, in Philo's earlier phrase, "as a channel for the insistent words of Another's prompting."

Jeremiah

Another Jewish writer to recognise the aptness of these words for depicting the prophetic role is Flavius Josephus, who was writing at the end of the first century CE. During the Babylonian occupation of Jerusalem, Jewish leaders took counsel from Jeremiah as to whether they should remain in the city or escape with some of the people to Egypt. The prophet undertook, Josephus records, "to approach God on their behalf" (διακονήσειν αὐτοῖς πρὸς τὸν θεόν, AJ 10.177). Hudson translated here "to intercede for them," which expresses the notion of prayerful entreaty contained in both the Hebrew of Jeremiah and the Septuagint.[1] Josephus, however, chooses to avoid writing of the prophet at prayer and to represent him rather as a seer mediating between the people and God; according to Plato, priests exercised this kind of mediation, which he designated διάκονος (Plt. 290c). Josephus thus adopts the style of religious language familiar to his non-Jewish readers.

An Unlikely Prophet

Our last instance of the common noun as a designation of a messenger of God also occurs in Josephus. This is in two related passages of *The Jewish Wars* (*BJ* 3.354; 4.626). The passages concern a critical period in Jospheus's own career when in spite of being defeated at the head of a Jewish force and then surrendering to the Romans he succeeds in establishing himself as a friend and associate of Roman emperors. To justify his course of action, which would have seemed as opprobrious to Roman as to Jew, he attempts to show that he was impelled by a call from God. To this end he presents himself as διάκονος "of the voice of God."

H. St. John Thackeray was content to translate the term here "minister,"[1] but Josephus's meaning was more precise than this. R. H. Marcus, who prepared the entry in Thackeray's lexicon, noted the sense "instrument" at *BJ* 3.354,[2] and E. Schweizer gives "mediator of divine prophecy" at *BJ* 4.626.[3] The latter meaning would fall squarely within the area we have been examining, but neither this writer nor other New Testament scholars seem to have found these instances useful in determining what Paul, for example, intended by the same self-designation. This must be because the usage in Josephus and the usage in the New Testament are each seen in isolation instead of as parts of a usage common to Jew, Christian, and Gentile (if we may so divide writers of Greek in the first century). Thus Beyer, following Schlatter, even proposes that Josephus was the first to use this term in relation to God.[4]

By contrast our survey has shown that the religious connection is ancient so that the interest of the instances in Josephus lies in an entirely different quarter. Like Paul, who was only a generation older than himself, but unlike almost all other figures in the literature surveyed, Josephus was a historical figure whose career and character are well known, and in applying this term to himself in a context that is more complete and no less intensely religious than any in Paul he is seeking to take advantage of prestigious religious values which attached to it by reason of a long custom. We shall see in fact that it is the key term in his exposition of a theological conception that is staggering for a Jew in his personal and historical circumstances.

When Jotapata fell to Roman forces in Galilee under Vespasian, Josephus as defending general took refuge with other notable Jews in a nearby cave. He was discovered, and Vespasian sent to parley with him. While he was hesitating over an offer of security, some soldiers set fire to the cave and, with hostile crowds clamouring outside, he had to choose between surrender and a noble death. At this moment he had a vivid recollection of earlier religious experiences which told him that he still had a role to play in the changing fortunes of his nation. With this realisation the decision was taken out of his hands and the drama of his escape and of his eventual elevation to friendship with the emperors unfolds. He writes:

> Suddenly there came back into his mind those nightly dreams in which God had foretold to him the impending fate of the Jews and the destinies of the Roman sovereigns. He was an interpreter of dreams and skilled in divining the meaning of ambiguous utterances of the Deity; a priest himself and of priestly descent, he was not ignorant of the prophecies in the sacred books. At that hour he was inspired to read their meaning, and, recalling the dreadful images of his recent dreams, he offered up a silent prayer to God. (*BJ* 3.351–53)

From this carefully worded statement we can be sure what the tenor of the prayer will be. He will avow his readiness to live in order that he might announce a new revelation concerning Romans and Jews in response to a God-given call, in the interests of God's peeople, and as the consummation of ancient recorded prophecy. This last is a daring submission and makes his competency in divination complete, for in addition to having been personally informed by God and to being skilled in the reading of dreams he claims a unique insight into the meaning of the Jewish scriptures. We notice that his status as "priest" is cited only to indicate his familiarity with the scriptures and that he does not at this stage produce an entitlement like "prophet" (the word "interpreter" given in the earlier translation does not occur in the Greek) but describes his capacity in functional terms ($ἱκανός$, given in the translation as "skilled," perhaps better as "competent"; $ἔνθους$, "inspired", etc.)

In the prayer, similarly, he speaks of his election "to announce the things that are to come" and reserves to the last, in a neatly emphasised phrase, the indication of how he wishes himself to be acknowledged in the role that he assumes. This is as $διάκονος$:

> "Since it pleases thee," so it [the silent prayer] ran, "who didst create the Jewish nation, to break thy work, since fortune has wholly passed to the Romans, and since thou hast made choice of my spirit to announce the things that are to come, I willingly surrender to the Romans and consent to live; but I take thee to witness that I go, not as a traitor, but as thy minister [$ἀλλὰ\ σὸς\ ἄπειμι\ διάκονος$]." (*BJ* 3.354)

In phrasing his prayer in this manner Josephus is plainly hedging the expression of his commitment to God with an attestation of his integrity before men. In the moment of prayer itself, however, he should not have needed to reassure God that what he proposed to do was according to God's mind. In this formulation of the prayer then he is looking over his shoulder at fellow Jews and at Romans who might all well suspect that a general normally gives himself into the hands of his conquerors for something less than a high divine purpose, and he needed therefore to present himself in a way that commanded respect and, indeed, riveted attention on his role. The term διάκονος would be intended in other words as an accurate designation of a role of overriding importance, making irrelevant any considerations of personal honour or national self-interest. The prayer and its foreword indicate that the role consisted in being a man of prophecy or spokesman of God so that the term itself actually means, to use a phrase cited earlier, a "duly sanctioned representative" of the Jewish deity. The immediate context does not contain explicit indications of this meaning but Josephus knew that readers of his day could not mistake it, and his account of subsequent events develops from it.

Before recalling these events we note more exactly the nature and scale of Josephus's claim. From dreams it has been intimated to him that the end of the Jewish nation is at hand. His faith in God's purpose for Israel is not diminished, however, for he has been inspired to perceive in the scriptures that Israel's destruction was necessary if God's rule was to expand through the instrumentality of the Roman empire. What the Jewish deity now requires is for this knowledge to be authoritatively conveyed to the Romans. To this task Josephus has been elected, and in assuming the title διάκονος he states his credentials. He goes to the Romans under a heavenly compulsion to announce in the name of God that the providence that had formed his nation has culminated in their exercise of power.

Josephus's decision to surrender is bitterly met by his forty companions. They judge that death by suicide is the only honourable course and threaten to kill him. Josephus argues the immorality of suicide and counters their threat by proposing that they should all enter a pact to die at one another's hand in an order to be determined by lot. In this he professes to have been trusting in God's protection, and providence duly sees to it that as the massacre draws to a close he is left alone with one other, whom he persuades to live. He surrenders, is taken to the Roman commander, by his own account impresses all, and on hearing in his prison that he is to be sent to the emperor Nero requests an audience with Vespasian; Vespasian's son Titus and two of his friends are present to hear Josephus address the Roman: "You imagine, Vespasian, that in the person of Josephus you have taken a mere captive; but I come to you as a messenger of greater destinies [ἐγὼ δ'ἄγγελος ἥκω σοι μειζόνων]. Had I not been sent on this errand by God, I knew the law of the Jews and how it becomes a general to die" (*BJ* 3.400). This is a frank assertion of a claim to be God's special messenger (ἄγγελος instead of διάκονος), the leading point of honour being boldly advanced in its support. Josephus proceeds to state more plainly that it is with Vespasian and not with reigning emperor or his successors that God would communicate (401–2), but the message itself, that Vespasian will be emperor, loses in directness and authority in being also the vehicle of an appeal in Josephus's own favour:

"To Nero do you send me? Why then? Think you that [Nero and] those who before your accession succeed him will continue? You will be Caesar, Vespasian, you will be emperor, you and your son here. Bind me then yet more securely in chains and keep me for yourself; for you, Caesar, are master not of me only, but of land and sea and the whole human race. For myself, I ask to be punished by stricter custody, if I have dared to trifle with the words of God."

Vespasian is rightly suspicious of what are later called "fabrications fashioned by fear," and Josephus is left in chains, but as events unfold and Nero dies and the Roman succession is in difficulties, his own ambition rises and quickly enough the troops proclaim him emperor in Caesarea. At this point, which is much further on in Josephus's history (*BJ* 4.622), attention shifts to Vespasian's religious disposition and to his reflections on the course events have taken. He judges that "divine providence had placed the empire within his grasp and that some destiny had brought the sovereignty of the world around to him." He recalls the heavenly signs that had pointed this way, and in particular "the words of Josephus, who had ventured, even in Nero's lifetime, to address him as emperor" (623), and is shocked to realise that the man is still his prisoner. Accordingly he convenes officers and friends, recounts the military prowess of Josephus, and reminds them of "the prophecies which time and event had shown to be divine." He then declares: "It is disgraceful that the one who foretold my attainment of power and was the minister of God's voice [τὸν . . . διάκονον τῆς τοῦ θεοῦ φωνῆς] should still rank as a prisoner of war or endure the lot of a bound criminal" (*BJ* 4.626). Calling for Josephus he orders his release, Titus successfully urges full citizen's rights, and Josephus closes the account with an observation that seems all too inadequate: "Thus Josephus won his enfranchisement as the reward of his divination, and his power of insight into the future was no longer discredited" (*BJ* 4.629).

And so Josephus enters history on the ground of having presented himself and of having been acknowledged by the highest authority on earth as "the διάκονος of the voice of God." The events have been truly momentous because Vespasian, the destroyer of the Jewish nation and master of the world, is brought under the providence of the Jewish God, and Josephus proposes that this is the part of the Jewish prophecies that he had been given to understand and to pass on. The other picture of himself as the diviner of dreams had suggested to H. St. John Thackeray that Josephus may have intended to present himself as an antitype of the patriarch Joseph, interpreter of dreams,[5] but while he does invest himself with the mystery of the seer he is also the spokesman of the Jewish deity designated by a term from the Gentile religious tradition. As befits the tradition and nature of such a διάκονος, he protests that he will not trifle with the word he has been entrusted to deliver, and his role is eventually recognised by Vespasian, who in recalling "the words of Josephus" acknowledges them to be "the word of God" (φωνή in both expressions).

The discordant note struck by Josephus's self-interest may raise doubts as to how genuinely he was in fact acting under inspiration[6] but ironically enough it serves to set off the theme of διάκονος the more clearly, because that word emerges at two crucial moments. In the first it has every appearance of being the climax of a carefully organised statement of his credentials and mission, and in the second it is tied into an equally portentous context where the notion of speaking in the name of God is explicit. Thus it appears on his own lips only at the moment of his dedication and on those of the emperor at the moment of his acclaim. This deployment of the term within a narrative charged with ideas of divine inspiration, election, and mission makes it the centrepiece of his apologia.

The import of this word, then, may be gauged from the context itself, from the religious language and professedly religious attitude of the only two men to use it, and from the fact that the history Josephus was to write and the final stages of which he was to observe was the creation by God and destruction by Romans of the chosen people. As a man devoted to the people's sacred books he had been given to understand the people's destiny; as a man who read the times he was given to perceive how the destiny had to work itself out; and as διάκονος he was designated as one empowered and constrained to declare in the name of God that the Roman supremacy was its culmination.

"How Sweetly You Do Minister to Love"

It may seem to be going wide afield to cite Shakespeare in illustration of our next point in Greek usage but a phrase of his about the ministry of love puts a thought which, elusive though it remains for the modern reader until clarified in subsequent statements, the ancient Greek would have recognised at once as pertaining to the activity of the go-between or matchmaker in affairs of the heart. Because there are also some indications that Shakespeare may have been indebted to Greek romance for the way he expressed himself, an observation on his language will form a helpful introduction to a number of instances where the Greek word group expresses a notion simple in itself but unexpectedly rich in its implications. The usage is basically the same as that which has been examined in relation to message for the gods and is argued in the following pages to have established itself in the field of romance by reason of an attitude of the ancients that love was itself a divine visitation and that the intense or merely tragicomic interchanges by which love is brought to its consummation were effected under divine influence.

Early in *Much Ado about Nothing* the honourable Don Pedro, realising that Claudio is very much in love with Hero, offers to advance his cause with her. "I will break with her," he says, "and with her father, And thou shalt have her." A grateful Claudio replies, "How sweetly you do minister to love" (act 1, sc. 1, line 295). Don Pedro proceeds to declare love to Hero in Claudio's name but the conversation is overheard and Claudio is maliciously informed that Don Pedro is himself enamoured of Hero. Sensing that he has been betrayed, Claudio reflects:

> 'Tis certain so—the prince wooes for himself.
> Friendship is constant in all other things

> Save in the office and affairs of love:
> Therefore all hearts in love use their own tongues. . . .
> Let every eye negotiate for itself,
> And trust no agent.

Expressions here like "agent," "negotiate for someone," "use another's tongue," which refer to the ministry performed by Don Pedro, are such as some passages from Greek romance might require in the translation of διακονία and its cognates.

Another Elizabethan, John Lyly in *Euphues and His England,* similarly wrote of a lover's letter as "the minister of his love."[1] A student of sixteenth-century language could probably inform us of the provenance of "minister" in the sense of "go-between" but it would seem not to come from the Latin where, for example, Cicero writes of "ministers of pleasure" and means the loved as instruments of the lover.[2] In fact the plot of *Much Ado about Nothing* is a variation on what was one of the most admired and frequently handled stories of Shakespeare's day and has its origins in Chariton's tale of Chaereas and Callirhoe, which is perhaps the earliest of the Greek novels. The play has also been said to be closest in spirit to the Renaissance, and its men characters to be "choke-full of the classic lore of the new time,"[3] so that it would be fitting for Claudio, notable among them for his eloquence, to speak in the idiom of ancient romance. Shakespeare's Italian source, Matteo Bandello's *Timbreo and Fenicia,* to which he is indebted especially for the role attributed to Don Pedro, speaks of the "commission" of the matchmaker and of his "mediation," and these again are ideas which in like situations the Greeks expressed through διακονία.

In Don Pedro's "sweet ministry," then, Shakespeare is referring to the commission laid upon a party to transmit a lover's intention to the beloved. Terms like "ministry" do not occur in anything like this sense elsewhere in his plays, and his dependence on sources at this point is such that a classical allusion could well be present. In any event for our purposes this instance of the word is an apt illustration of what his forbears in the telling of romance intended in similar circumstances by διακονία. At the same time Shakespeare's word is not so nuanced as we shall see the Greek to be, nor did it long remain in use to signify the conveying of information. On this point it is interesting to observe that in a popular handbook of the time a near contemporary unwittingly came closer to the Greek in depicting the role of the go-between as an "ambassade on the lovers behalfe," this being a bastardised form of an Italian original and representing an attempt to write appropriately in English of a delicate and elevated role in the office and affairs of love.[4]

To the Greeks falling in love was, of course, no less unpredictable than to us, but they liked to attribute the experience to a divine initiative. "The lovers' frenzy is divinely inspired," writes Plutarch (*Mor.* 759d), and he is referring not only to the delights of the senses, which are the gift of Aphrodite, but to the union in affection, which is the work of Eros (*Mor.* 756e). Thus when Heliodorus's heroine Charicleia is forcibly betrothed to the robber chief Thyamis, and wishes to reassure her hero Theagenes that the outcome of their own true love will be a happy one, she says she has entrusted that future "to the deity appointed to preside over our love from its beginning" (*Hld.* 1.26).[5] A special providence for lovers is indeed what still helps us to tolerate the extravagances of the Greek novel—perilous changes in fortune, unlikely escapes, cruel separations—until the chaste union of lover and beloved.

Just as mythology showed Zeus himself turning to Hermes to arrange assignations with earthly consorts (*DDeor.* 24.1–2), so frustrated human lovers learnt to look to a divine intermediary of their choice. A prayer of the third century CE apostrophises a magic lamp which is credited with the power of bringing lovers together because it had once brought the god Osornophoris into an incestuous union. We notice that the invocation is designed to be used by any number of suppliants: "You were the medium for him [Διακόνησας αὐτῷ] when he was in love with his own sister. . . . So now be a medium for me [διακόνησον κἀμοί], so and so, in regard to so and so [πρὸς τὴν δῖνα]" (*P. Warren* 21.4,8). More eloquent is another invocation from the same century composed in his own ardent interests by one Ptolemaios, who was something of a necromancer. Swearing by the ten magical names of "the Lord God who sustains the universe," he expects a perfect requital of his love for Aplonous through the operation of what is apparently the spirit of a dead man (νεκυδαῖμον). "I adjure you, spirit of the dead one, and I summon up your spirit: be a medium for me unto Aplonous [διακόνησόν μοι εἰς 'Α.]" (*SB* 4947.2). What the suppliant has in mind, as the rest of the spell shows, is that the daemon will stir such transports of desire in Aplonous that her soul will actually fuse with his own. He also says more simply, "Make her love me for ever," something that he has not been able to do himself but which he thinks is within the gift of an other-worldly go-between.

The prayer illustrates in what real sense the ancients could view love as the meeting of two spirits. Xenophon of Ephesus, in a passage that follows, relies on this conception when he makes the lovers' eyes the gates along love's pathway. He sees more here than the magic of answering glances which Plutarch describes (*Mor.* 681b,c) and to which Shakespeare refers—"sometimes from her eyes I did receive fair speechless messages" (*Merchant,* act 1, sc. 1, line 163)—and even more than a modern writer makes of an intense, wordless, and much discussed encounter of a young man with a girl who felt "the worship of his eyes," whose own eyes "had called to him and his soul had leaped out at the call."[6] In the latter case, of course, it is not a release of love but of creativity, and the young man is in no danger of letting go his hold on a leaping soul. By contrast, the wise Calisiris in Heliodorus's novel relates such familiar experiences to a physical theory of the day that saw the air as a medium for emotional interaction; the envy, anger, or benevolence of one person actually transfuses the air about him and then, by contact of the air with the eyes of another, passes into the other's psyche. "One proof of this," says Calisiris, "is the genesis of love, in which mere sight provides a beginning and a leverage; it is through the eye that the passion penetrates to the soul" (*Hld.* 3.7).[7]

Xenophon of Ephesus is writing of Anthia as she keeps watch by her sleeping lover, kisses his eyes, and addresses them in these words:

> "These eyes, at times so disturbing, and next so full of love—in bringing me grief as, oh, you have often done, in stirring my heart with, oh, the first thrill of love, you have kept me in perfect touch [καλῶς μοι διηκονήσατε] and have

unerringly guided my love into the soul of Habrocomos. So now I kiss you again and again. And I press my own eyes against you, for to Habrocomos they have told the story of my love [τοὺς Ἁβροκόμου διακόνους]. May you always see what you see today: never let any woman but me look attractive to Habrocomos, and among men may he alone look attractive to me. You take hold of souls whose burning passion has come from you: keep them in that grasp." (*X. Eph.* 1.9.7–8)

Here the phrase "they have told the story of my love" is admittedly a paraphrase of the Greek at *X. Eph.* 1.9.8, but to keep to the letter ("my eyes, the ministers of Habrocomos") is to give the impression that Habrocomos is somehow their master.[8] That is not Anthia's thought. The eyes are apart from both persons, and are "ministers" or "ministering" in the rare English sense alluded to in words of Shakespeare's Claudio and exemplified clearly in the words of a modern novelist: a character complains that the reading of novels "excites mere feelings without at the same time ministering an impulse to action."[9] In the same way the eyes of the lovers are ministering the movements of two hearts. They have inserted a pang ("sharp pang") in her soul, and have led her passion into the soul of her lover. They are nature's bawds.

The word is Thomas Underdowne's, the Elizabethan translator of the novel by Heliodorus already referred to, although at the point where he uses it (*Hld.* 7.9.4) the Greek is much less succinct and has no derogatory overtone.[10] Heliodorus, who may have been writing as early as the first half of the second century CE,[11] has no sooner relieved his heroine Charicleia of the threat to her virginity at the hands of the robber chieftain Thyamis than he brings her true lover Theagenes to Memphis where his virtue in turn is at once under threat from the powerful and love-crazed Arsace. Arsace is the sister of the great king of Persia and wife of the satrap Oroondates, who is absent for the time on an expedition. Arsace has been stirred at the sight of the handsome Theagenes only to see him then in a chaste embrace with Charicleia. Further inflamed, she retires to her chamber to toss despairingly on her bed.

> She would summon her maid without cause, and then dismiss her without orders. In a word, her passion would have insensibly proceeded to madness but for the intervention of an old crone named Cybele, who was her chamber-woman and used to *serving her amours*. Cybele hurried into Arsace's room, for nothing escaped her notice: she was like a lighted lamp and like a lighted lamp she kindled Arsace's passion. "What is this, mistress?" she said. "Is some new or strange pain hurting you? Whose looks have disturbed my pet? Who is so presumptuous and insensible as not to succumb to beauty like yours, as not to account the love you offer as bliss, as to disdain your complaisant nod? Just tell me, my sweetest baby. No one is so adamant as not to be susceptible to my philters. Just tell me, and your desires will be fulfilled at once. You have had enough experience of my talents, I think."[12]

Here, "serving her amours," the phrase I have italicised, is Underdowne's "bawd." The Greek is τὰ ἐρωτικὰ τῇ Ἀρσάκῃ διακονουμένων (*Hld.* 7.9.4), and more closely in English is "ministering experiences of love for Arsace."[13] The syntax is the same as in expressions for delivering a message (for example, Lucian, *Cont.*

1). We cannot help noticing also the comparison Heliodorus draws between this old bawd and a lamp of the kind that was to kindle passion by "ministering" in the name of the suppliant in *P. Warren* 21. If the instances of the usage are infrequent, such correspondences nonetheless say much for the ease with which the Greek expressed himself in these terms.

Underlying the expressions is the notion of mediation carried out by a go-between, a notion more subtly conveyed by Heliodorus as the nasty intrigues at the court of Arsace develop. Achaemenes is Cybele's son and the member of the satrap's guard who had earlier been commissioned to take Theagenes in charge as a prisoner of war for later despatch to the Persian king. In a rout of the guards the prisoner had escaped to arrive incognito with Charicleia and his rescuers at Memphis. Here Achaemenes, who is also smitten with love for Charicleia, recognised him and suspects that the two close foreigners, who are being so honourably received by Arsace, are more than the brother and sister that they pretend. He takes good note of Arsace's passion for the young man and perceives that by collusion with his mother he can expose Theagenes as a slave and prisoner of war, thus prising the two young lovers apart and leaving Charicleia free to be taken in marriage by himself. As he first observes Theagenes' favour at the court, and begins to ponder on his own prospects, he suspects that Theagenes is for Arsace just another—what Heliodorus calls—ἀφροδίσιον διακόνημα (*Hld.* 7.16.1). The first word is simple and means "pertaining to the goddess of love, Aphrodite," her kind of love being sexual pleasure itself, as our word "aphrodisiac" still reminds us. The second word is a rare cognate within our group. Translators can only paraphrase (I might venture "sexual plaything"),[14] but miss what the second word is meant in the first place to denote, namely, that Theagenes is a vessel through which sexual pleasure is to come to Arsace. This is more than to say that he is a sex object—as today's parlance might be; he is a sex medium, and not a person at all. The word has occurred because διακον- lends itself to statements about mediation and has a customary place, in particular, in language about experiences of a heavenly origin, which is where talk about Aphrodite belongs.

In the interests of Arsace, Cybele pursues a resolute Theagenes, and among her first questions is "Why this rejection of Aphrodite?" Arsace's husband is away, she explains, "and you have me who nurtured her, and the keeper of all her secrets, to procure this union [τὴν ὁμιλίαν διακονούσης]" (*Hld.* 7.20.2).[15]

If Cybele is a sordid old retainer, and Arsace a decadent aristocrat, we are to bear in mind that Heliodorus builds so large a part of his story around the lascivious designs of such people because he intends to embroil his readers in the dangers attending the pursuit of chaste love. Theagenes and Charicleia are virgins to the end. The following passage from the midst of the lovers' trials is eloquent of the author's quaint intent:

> Then were Theagenes and Chariclea left in the den alone, who accounted even the extremity of their present dangers to be the height of joy. For this was the first time that ever they were by themselves, delivered from all who might trouble them. Wherefore they took their fill without hindrance of kisses and close embracings, forgetful of all else beside as they clung to one another as if they had been one body, but content still to satisfy themselves with chaste love, temper-

ating their affection with tears and cleanly kisses. For Chariclea, if at any time she perceived Theagenes to pass the bounds of seemliness and deal with her over wantonly, would rebuke him by telling him of his oath; and he would suffer himself to be reformed with little labour and brought again to temperateness, in as much as he could master his desires although he could not master his love.[16]

In desperation Arsace eventually commits Theagenes to torture at the hands of a jealous eunuch only for the young man to rejoice in the opportunity of demonstrating his fidelity, and "the chastity of his spirit gathered strength"; when, next, Charicleia is unjustly condemned to die by fire, the flames will not touch her; and the novel concludes in a spectacular scene of human sacrifice at the court of Ethiopia at which the lovers are to be chief victims and again demonstrate chastity through ordeal by fire. Charicleia is recognised as the lost princess of the kingdom, the two are declared man and wife before "Lord Sun and Lady Moon," are invested with the priesthood, and are grandly escorted to Meroë "where the more august marriage ceremonies would be celebrated with greater brilliance."[17]

The point of filling out the story a little is to illustrate that Heliodorus's ideal of love is never lost to view. The fact that talk about "ministering" in love comes from the lips of the foul Cybele does not cheapen its currency. Heliodorus writes uniformly in an especially ornate style, and his Cybele falls easily into the formal mode of the courtier, lending indeed an air of respectability to sexual misdemeanours.

Other writers too may be judged to adhere to this minor convention in the telling of salacious tales. One of the letters of Aristaenetus, a late writer, tells of the comeuppance of an adulterer. The maid of the house has been acting as go-between in his affair. A virgin herself but passionately aroused by what she comes to hear and see, she confronts her master with the choice of making love to her or losing his liaison. He willingly complies and is, of course, discovered in the act by his wife. The story is really a vehicle for the wife's amusing lecture to the girl on the joys of mature sexual experience. For this reason Aristaenetus writes ponderously of the girl's arousal as a visitation of Eros, who enters her—after the manner alluded to by Heliodorus and Xenophon of Ephesus—"through her ears and eyes." Thus possessed, she takes the god's counsel to address her master: "If you want me to cooperate and keep acting as your willing go-between [διακονεῖσθαί σοι]—well, what do I say next? You already perceive that I have passionate desires" (*Aristaenet.* 2.7.15). The account opens by informing us that the maid "first learnt of love in the course of taking plans for assignations between two lovers" (διακονουμένη . . . αμφοῖν τὰ δοκοῦντα, *Aristaenet.* 2.7.2).

Josephus enjoys telling the story of how a prominent Jew of the third century BCE came to have a son by his own brother's daughter. On a visit to Alexandria, and during entertainment as a guest at dinner of Ptolemy III Euergetes, Joseph fell in love with a dancing girl. Because Jewish law made it impossible for him to be known to be consorting with a foreign woman, he put it to his brother that an affair could be arranged, and his transgression concealed, if his brother would consent to act as "a reliable go-between" (διάκονον ἀγαθόν, *AJ* 12.187). His brother was more than willing to oblige—"to take up the liaising commission" (δεξάμενος τὴν διακονίαν, *AJ* 12.188)—because he saw the chance of achieving the object of his own visit to Alexandria. This was to marry off his young daughter, who was accompanying him, to a prominent Jew. Accordingly he decked out his daughter appropriately and brought her to Joseph at night. The arrangement proved agreeable to Joseph, even after explanations from his brother, and issued in marriage and a son.

It is worth noticing the difference in the translation provided by R. Marcus at *AJ* 12.187, 188 in the edition begun by Thackeray and that provided by his successor in the series, H. Feldman, for the only other instance of this use in Josephus (*AJ* 18.70). For the first of the three, following closely Hudson's eighteenth-century Latin version,[18] Marcus has "do him a good service," and for the second, "undertaking to be of service." Feldman, by contrast, correctly avoids this note of benevolence in rendering the word προδεδιακονημένοις (*AJ* 18.70) by "previous agents." The word is designating those who had been attempting to act for Decius Mundus in the seduction of Paulina, a narrative we shall be examining at a later stage for its religious overtone.

These few instances in Josephus would seem to exemplify further a stylistic tendency already noted in connection with *BJ* 3.354; 4.626. This is that in writing Jewish history for Gentile consumption Josephus sees some virtue in adopting an idiom whose special connotations are likely to be appreciated by his non-Jewish readers. Minor as the indications are from his references to meddling in love affairs, the observation is probably supported by the different twist given to the same idiom by two other Jewish writers.

In the *Testament of Juda* the role of this kind of "minister" is metaphorically adapted to an injunction against illicit sexual activity. Jacob addresses his offspring:

> "My children, do not get drunk on wine, for wine takes the mind off truth, sets up a sexual urge, and starts the eyes roving. In fact the spirit of lust keeps wine as an emissary for pleasurably stimulating the mind [ὡς διάκονον ἔχει πρὸς τὴν ἡδονὴν τοῦ νοός]. These two things, lustfulness and drunkenness, take away a man's discernment." (*Test. Juda* 14.1–2)

In introducing the thought of a minister or emissary here, the writer has in mind the kind of lustful prospect by which the bawd Cybele tried to entice Theagenes into the arms of Arsace (*Hld.* 7.20.2). He is saying that lust sends wine to lure the mind to its own embrace. Whatever the original language of this testament,[19] such a use of the Greek word reflects an awareness on the part of the writer or translator of the special place held by it and its cognates in talk of love. We can also probably see him assuming, from familiarity with the general run of stories about ministers of the wayward favours of Aphrodite, that when a person in love needs to communicate through a minister his purposes are not likely to be in accord with Jewish morality. The term itself, in other words, is here connoting the clandestine and illicit. Unless this connotation is admitted, Jacob's line of thinking in 14.1, where wine is said to be sexually stimulating, does not carry through to 14.2, where, by reading for example "the spirit of fornication hath wine as a minister to give pleasure to the mind,"[20] wine is seen as leading only to drunkenness, and the connection with lust is broken.

With *Joseph and Asenath* 15.7 we are in the field of allegory. This is a work of Hellenistic Judaism, and is likely to date from the turn of the first century CE.[21] Its precise genre has been a matter of some discussion—Denis's "midrashic novel" is a fitting description[22]—but it must be judged to stand in a definite relationship to the romances of its era. Asenath is the daughter of an Egyptian priest and was to become the wife of the patriarch Joseph and mother of Menasseh and Ephraim (Gen. 41:45, 50; 46:20). The first part of the book recounts how the woman and the patriarch came together. Asenath is introduced as a girl living apart in separate chambers, surrounded by her gods and treasures. Her piety and especially her consummate chastity are emphasised. In seven of her ten chambers live the seven virgins who "attend on" her (2.11: διακονοῦσαι), and "never did a man converse with them," so that the reader has no occasion to anticipate that in this romance one or other of them might be used by the heroine to liaise with a secret lover. She has already rejected many suitors by the time Joseph arrives on official business in her home region of Heliopolis, but is overcome at the particular beauty of the man who "has come in his chariot as the sun from heaven" and "son of God." Presented to him eventually, she is gently repudiated by him because she does not bless the living God or eat the bread of life.

Disconsolate at this rejection, but drawing comfort from Joseph's prayer for her, she withdraws for seven days to her chamber where she discards her luxuries and false images and abases herself in mourning, fasting, and penitence before the God of the Hebrews. On the eighth day she is able to pray for forgiveness of what she sees as her most grievous sin, her blasphemous ignorance that Joseph is God's son. She commits him to God's keeping, and for herself prays only that she might be his slave girl, "that I might wash his feet, attend on him [διακονήσω], and be his slave [δουλεύσω]" (13.12). In a protestation of this kind Kerenyi has seen a direct borrowing from the language of Greek romance,[23] but no sooner is it made here than the course of Asenath's love for Joseph merges for a time into allegory about God's love for the soul. S. West contests this line of interpretation, despite adducing the sixth-century Jewish opinion of Moses of Aggel that the truth (θεωρία) underlying the narrative (ἰστορία) is the union of God with the soul; further developments on this eighth day, however, make hers a difficult position, "the author gives no hint that he is speaking in parables, that there is a deeper level of meaning."[24]

What we have in chapter 14 is a glorious visitation from God's angel Michael with a message of comfort. He instructs her to lay aside her mourning and to return in robes of the virgin. In the next chapter she re-presents herself for Michael's proclamation. From this day, he tells her, she will be renewed, eating the blessed bread of life and drinking the cup of immortality, and the Lord God has given her as a bride to Joseph. In recognition of her new status before God, she is to lose her pagan name, which meant "belonging to the goddess Neit," and to become "City of Refuge," because those who adhere to God through penitence shall be protected in that fortress. At this stage Asenath is surely a figure larger than life, and in what follows penitence itself is personified as the virgin daughter of the Most High who has brought Asenath to God. "Through every hour," says Michael, "she entreats him on behalf of those who repent, because for those who love her [or him—the Greek is uncertain] he [she?] has prepared a heavenly bridechamber.'' There seem to be sound reasons to see here, where attention shifts for the moment from Asenath's own happy prospect of union with Joseph, a picture of Penitence filling the role that Greek mythology and romance attributed to Hermes, Eros, and the likes of Heliodorus's Cybele in bringing lovers to their great masters and mistresses. In what Michael goes on to say, there would indeed seem to be a direct allusion to the idiom we have been observing: "and she will herself minister to them [διακονήσει αὐτοῖς] for ever" (*Asenath* 15.7). On meeting this word the modern reader might think in the first place of Penitence attending upon brides in the eternal bridechamber, but because the role of Penitence is actually continual intercession before God for the purpose of maintaining converted Gentiles in God's love, the author might rather have been suggesting the kind of ministry that is mediation between lovers. By translating "she shall minister for them" (instead of "to them") such an allusion is not difficult to recognise. We have already met several instances where the construction used here (διακονεῖν τινι) expresses the idea of doing messages or acting as a go-between;[25] given a suitable context, such is in fact more likely to be the meaning of the Greek than ideas to do with menial attendance.

The author of *Joseph and Aseneth*, then, may well have felt that the moral of his romance was not fully told until an allegory turned also on the familiar figure of the go-between who mediated to heroes and heroines the divine mania of love. One part of the didacticism, shrouded in the beauty and graciousness of Penitence, would be the comfort that conversion brings. Another would be that her task is not the arousal of passion in this god's clients but their spiritual edification, for this section moves towards the transformation of carnal love into the heavenly love of God. The account also makes it clear that such delight, an actual participation in the love that is at the centre of the story, is available to all those who repent, a gift that it was beyond the powers of profane romance to bestow, indeed beyond the intentions of its creators because an attainable love would destroy their art form. Readers of this novel, however, are left in no doubt that they can be transported to a heaven on earth where love abides and where the virgin Penitence is a go-between assuring love's permanence.

The foregoing analysis of passages touching on what Shakespeare called ministering to love has sought to reveal a little more of the value that interpreters might need to look for in our cognate words. The idea, as always thus far, has been of action in a middle ground and, as in the earlier sections of this chapter, of some form of communication. Just as significant are some signs that the usage has its roots in religious language. Even where these signs are not clear, there are grounds for suggesting that writers were at least adhering to a stylistic convention of romantic literature for reasons not unlike those that led medieval troubadours to fall back on stock expressions. At the same time the instances are few, only fifteen from eight sources and in only nine contexts. This is perhaps surprising, given the extent of the ancient literature on love, but fits with the comparative rarity of the words. Writers could always use other terms (thus ὑπηρετ- in Plutarch, *Mor.* 64f; Heliodorus 3.16) or their plots might not call for the office of go-between. Longus's *Daphnis and Chloe* would be an example. Of the thirty-six plots sketched out by Parthenius only four or five make mention of such a role. If on the other hand the interpretation offered previously of *Test. Juda* 14.2 and *Asenath* 15.7 is correct, there was a considerable sensitivity on the part of Hellenistic Jewish writers to the connotations that the words possessed in well-defined contexts of erotic love and which even inspired some moralising.

That the literate Greek was expected to discern these connotations is illustrated finally from a casual phrase of Lucian. He writes of entertainers rendering love ballads at dinner parties as τοῖς . . . τὰ ἐρωτικὰ ταῦτα διακονουμένοις (*Merc. Cond.* 27), which Harmon gives as "those who render these services to passion"[26] but Fowler more accurately and sensitively as "the minister of Love's pleasures."[27] The better translation is saying that the performers stimulate sensual passion by drawing it, as it were, from its source in their songs, even—since the songs are about Aphrodite and Eros—from its source in gods of love. Aspects of this conceptualisation are what we have been examining.

Errands

At the beginning of the chapter we examined Lysias 1.16, deciding that the passage made best sense when the slave girl described there as διακονοῦσαν ὑμῖν was understood to be doing errands to the market rather than chores about the house. Usage examined in the intervening sections will now be seen to support that interpretation, but other passages also, which like Lysias 1.16 usually have no religious significance, exhibit the word group in senses of doing errands, conveying a message, or delivering something. Thus, according to Theophrastus, a telltale sign of a flatterer, who was a character detested among the ancients, is that "without drawing breath he is capable of fetching things from the women's market" (τὰ ἐκ . . . ἀγορᾶς διακονῆσαι, Thphr., *Char* 2.9). We notice the direct object. What Theophrastus is ridiculing is not the slavish behaviour[1] but the man's lack of self-respect in plunging with such alacrity into woman's business in the hope merely of a little praise.

Similar instances of the verb occur in the comic poets Archedicus and Menander, in the novelists Achilles Tatius and Chariton, in the papyrus *UPZ* 18, and in the *Anacreontea*.[2] Leaving aside for the moment *UPZ* 18 and Chariton, which both have special points of interest, we take up briefly with the last mentioned. This is a collection of lyrics in the style of Anacreon, a poet of the sixth century BCE. In one ode, which seems to have been composed around the beginning of the Christian era, a dove is reflecting on the comparative advantages of life in the wild and life as a slave of Anacreon, who has bought her from Venus and sent her to Bathyllus. "I do so many messages for Anacreon ['Ανακρέοντι / διακονῶ τοσαῦτα]," she complains, "and now look what letters of his I bear" (*Anacreontea* 15.14).[3]

Other parts of speech show related meanings. Thus an adjective appears in Demosthenes to denote that a slave carrying a bronze pitcher is a porter (παῖδα διάκονον, Dem. 47.52). Twice in dramatic poetry the same word functions as a noun meaning "messenger." In the instance in Sophocles parallelism has long led translators to give this meaning when Philoctetes, abandoned on Lemnos, begs Neoptolemus to become his messenger (ἄγγελος) because he fears that messages committed to other couriers (τὰ τῶν διακόνων, Sophocles, *Ph.* 497, cf. 500) have gone astray. In the second instance a servant girl issuing a summons to a feast calls herself ἡ διάκονος (Aristophanes, *Ec.* 1116); she is probably identical with the herald mentioned earlier (κηρύκαινα, *Ec.* 834),[4] and is recognised by the chorus as a messenger (*Ec.* 1137, 1175). The noun διακονία refers to the delivery of a new pair of shoes in one of the humourous letters of Alciphro (21.2), to the fetching of a magistrate to the deathbed in Isaeus (1.23),[5] to the bearing of libations in a religious procession in Plutarch (*Arist.* 21.4) and to the bearing of letters in Thucydides (1.133), a passage we shall return to.

All of these uses connote local movement. We see the difference between διακονεῖν and ὑπηρετεῖν in this respect in a passage of Lucian. The pirate Sostratus is arguing before Minos that guilt should not be imputed to him for murder because all men are but instruments (ὑπηρετοῦντες) of fate, and he asks if the public executioner is a murderer for killing at the behest of a judge. Minos agrees that no culpability attaches to the executioner; as well, he says, blame the sword, which functions only as an instrument (ὑπηρετεῖ). Thereupon Sostratus takes a different example: suppose a master sent someone along with a precious gift, to whom would thanks be due? Minos must answer, "To the sender"; and he explains, "the one who delivers it is, of course, a courier [διάκονος]" (*DMort.* 30.2). Sostratus is immediately able to retort: "See how unjust you are, then. You punish us, who are agents [ὑπηρέτας] of what fate has ordained, and honour those who *act as couriers* [τοὺς διακονησαμένους] in respect of some other person's good deeds" (*DMort.* 30.3).[6] Of some moment, but of incidental interest for our present discussion, is that Sostratus's point of honour holds only if the term for courier (30.2) carries distinction; it is in fact noticeably emphasised at the beginning of the Greek sentence. What we are stressing, rather, is that in these ripostes of Sostratus on the moral responsibility of agents Lucian changes from ὑπηρετ- to διακον- when the agent is on the move.

It is this aspect of meaning that appeared so clearly at Plato, *R.* 370e, in a passage that defined the function of the merchant in terms of courier. Cognates pervade the discussion of commercial activity there principally because they express so well in a nontechnical way the idea of an agent at the centre of a process of exchange. The notion of movement is not always present, as in the cases of retailers and labourers (*R.* 371c,d,e), and in two passages in *Laws*. In the first of these the abstract noun recurs in connection with trade in the statement that love of wealth lures citizens from the pursuit of the martial and noble arts, turning the valiant to crime as a way to riches and turning the temperate into "merchants, sea-captains, and middlemen of all sorts [διακόνους πάντως]" (Plato, *Lg.* 831e).[7] The connotation of movement is even more remote in the law proscribing trade, although in its second part the law is so framed as to cover transactions that could require movement: "No one shall willingly or unwillingly be a retailer or a merchant [ἔμπορος], nor shall any one even hold a commission of any kind from private individuals [διακονίαν . . . κεκτημένος ἰδιώταις] unless . . ." (Plato, *Lg.* 919d).

These few statements in two works of a writer as important as Plato exhibit an affinity between notions expressed by διακον- and the notion of trade as traffic. The affinity rests on or came to nourish the belief that Hermes, who was διάκονος par excellence (cf. Epictetus 3.1.37), was the god of commerce (Aristophanes, *Pl* 115: "trafficker") and "patron of marketing and commercial travel [ἐμπορικῷ]" (Heliodorus 5.13), as we have had occasion to observe. The affinity is noted again here in the context of errand because it seems to elucidate an instance of the abstract noun in the *Testament of Job*. Recounting that many people had joined him in succouring the poor, Job states that even those without cash in hand to contribute to his undertaking would borrow from him, go off to far cities, and by "trading" (ἐμπορευσάμενοι) enable themselves "to perform for the poor a sacred work" (τοῖς πένησιν . . . ποιήσασθαι διακονίαν, *Test. Job* 11.3b). In this phrase I read the dative with the infinitive, and then translate the noun by "sacred work" instead of by "mission" or "expedition"—despite the reference to travel—because the meaning of the word is coloured by its two prior occurrences in this short passage (11.1,3a). At the same time the author has marked a transition from "this" work mentioned in 11.3a to another and different kind of work in 11.3b by calling the latter simply "a" work, and even in his simple Greek possibly intends the same correlation between διακονία and trading or trafficking as exists in Plato. We will return to the passage in a later chapter. Meanwhile we shall take up the two passages singled out earlier in this section, the one in Chariton because it contains an expression parallel to an expression used by Paul, and the other in the papyrus *UPZ* 18, which has seemed obscure but which falls into place when we allow for the idea of errand.

A Letter from Chaerea

In his second letter to the Corinthians Paul refers to a letter in a phrase that the Authorised Version translated as "ministered by us" (2 Cor. 3:3). What the translators may have intended by this is no longer clear from their English but their phrase was retained by the revisers of 1881 *(RV)*; something as indeterminate— "the result of our ministry"—was preferred nearly a century later in the *New International Version*, while other more recent renderings divide around two different interpretations of the Greek, with the letter being either "written" by Paul (Moffatt, Phillips, *JB*, *LB*) or "delivered" by him *(RSV, NEB, GN);* curiously the *New Jerusalem Bible (NJB)* of 1985 picks up the phrase of Weymouth's *Modern Speech* version of 1902, "entrusted to our care," which fudges meaning.[1] Because whether Paul saw himself as having written or delivered the letter is not without implications for his understanding of the role of an apostle, a passage of interest is Chariton 8.8.5 where a similar phrase occurs. Of added interest is that modern editors of this novel have emended the text in order to accommodate it to a meaning they think it should bear. We have seen one other emendation conjectured for such reasons in connection with Aristophanes, *Av.* 1253.

Chariton's tale of Chaerea and Callirhoe is possibly the earliest of the extant novels,[2] and in a situation that characterises these works the lovers are separated by misfortune. With a sudden turn of heart the hero's tormentor urges him to contact Callirhoe by letter. This Chaerea attempts to do, but the letter falls into the wrong hands. Explaining how this came about, Chariton wrote (and the punctuation of the Greek is that of Reiske's eighteenth-century edition):[3] ἀμελείᾳ δὲ τοῦ διακονουμένου, τὴν ἐπιστολὴν ἔλαβεν αὐτὴν Διονύσιος. This punctuation invites the translation: "But through the negligence of the servant, Dionysius received the letter itself." As Reiske noticed, however, and as even appears in translation, this is not a happy sentence. The editor of the Oxford edition, W. E. Blake, who omits the comma, adopted the nineteenth-century conjecture of R. Hercher, αὐτός for αὐτήν,[4] for the purpose, it would seem, of obtaining the sense: "But through the negligence of the servant, Dionysius himself received the letter." Reiske had already raised a question of usage concerning the participle translated here by "servant"; he noted a few instances of direct objects after this verb, including "the erotic songs" we met at Lucian, *Merc. Cond.* 27, and judged that the word for "letter" is grammatically the object of the participle. Accordingly he shifted the comma from after the participle to after the word for "letter," and produced a balanced sentence which we are to translate: "But through the negligence of the person who delivered the letter, Dionysius received it."[5]

The Twins' Petition

Less clear than this instance from Chariton as a statement about errand, but more instructive, are two statements from successive drafts of a petition to the Egyptian king that was drawn up in the second century BCE on behalf of twin sisters in a religious sanctuary in Memphis (*UPZ* 18.23; 19.25). Ulrich Wilcken's celebrated reconstruction of the circumstances occasioning the petition, and his ordering of the drafts do not entirely account for what the petition meant in stating that the twins' brother was to "minister" for them.[1]

The Serapeium at Memphis had been the mausoleum of the Apis bulls from the thirteenth century, but by the Ptolemaic period a thousand years later many other cults had also been attracted to the site.[2] The native cult of the Apis traditionally involved the services of twin girls, who were co-opted in the first instance for the purpose of participating in the mourning at Memphis on the death of an Apis, and thereafter were to reside throughout the life of the new Apis within the Serapeium on the heights outside the city. Although not technically priestesses, they were cultic personnel, offering libations to Osarapis as the incarnation of the dead Apis and to Asclepios.[3] The forces of syncretism were such, however, that by their time in the second century BCE the twins looked upon themselves as dedicated to Sarapis, a god amalgamated from several mythologies and for political purposes by Ptolemy I.[4] Their cultic status, in any event, drew them a royal allowance of bread and oil. Wilcken makes the point that whereas the bread was distributed within the temple precincts on a daily basis, the oil was issued only as an annual ration on the presentation of a chit to the royal depository in Memphis.[5] The final draft of the petition, accordingly, requests that a certain Demetrios be accredited to act in this matter on behalf of the twins (*UPZ* 20.48–55).

The twins, whose names were Thages and Thaous, were complaining that among other things they had not been receiving these allowances. The petitions were drawn up by the temple recluse Ptolemaios, and began in 164 BCE. The problem arose through the straitened circumstances in which the twins had entered service. Rejected by their mother on her taking up with a new husband, and then deprived by her of their inheritance on the untimely death of their natural father, they were without other security than their emoluments from their position in the temple. Even within the Serapeium, however, they were not beyond the reach of their mother's ill-will, for when family friends had prevailed on them to take on her son by second marriage, Pachrates, for the purpose εἵνα δειακονεῖ ἡμῖν (*UPZ* 18.23)—or, as the second draft puts it with the infinitive, διακονεῖν (*UPZ* 19.25)— he not only rifled their possessions but made off with the oil that he had collected from Memphis, and returned to his mother.

Wilcken is undoubtedly correct in thinking that this verb does not point to Pachrates as a house boy or daily menial of his stepsisters. Such a role would be out of keeping with the context. In his judgement the context points rather to a meaning "so general that we cannot exclude the idea of powers of attorney."[6] This is a judicious comment, but if the verb really meant "to minister" in this sense Wilcken might need to explain why, in the fair copy of the petition (*UPZ* 20.48–50), such a useful expression of an apparently sound legal point was dropped. There is no reason at all, on the other hand, to suspect that the ministry was cultic in character, that is, that Pachrates was to assist or stand in for his stepsisters in their rituals. The cult rested on the ministrations of twin girls, and the correspondence is charged with the twins' awareness of this privilege; on one occasion (*UPZ* 57), it is true, an inmate of the temple is recorded as having made libations in their stead, but throughout the letters the terms for ritual ministry are θεραπεύειν and especially λειτουργεῖν. The nature of the ministry is to be gathered rather, as Wilcken has shown, from the close association of the verb in both drafts with the task (κομίζειν, "to carry") of getting the year's supply of oil, which was some nine gallons, into the Serapeium. For the twins, given their regular hours of duty and their distance from the depot in Memphis, this would have presented practical difficulties. In recording that they agreed to take on Pachrates "in order that he might minister for us," they are referring to this laborious errand. Their record goes on immediately to say, "We sent him to bring our ration here from the royal depository" (*UPZ* 18.25–26). Pachrates, in fact, took the chit surreptitiously, collected the oil, and absconded. Because their plan misfired in this way, the incident of the errand is central to the twins' complaint at this stage. By the time *UPZ* 20 was drawn up, Ptolemaios had arranged for his friend Demetrios to represent the girls, and the incident involving Pachrates is clouded over.

In High Places

Our words occur in a number of passages recounting messages and missions on behalf of persons in high standing. When David fled Jerusalem on the revolt of Absalom (2 Sam. 15), he arranged for news of developments in the city to be brought to him by the priests' sons Ahimaaz and Jonathan. Josephus twice calls them "faithful messengers" (πιστοὺς . . . διακόνους, AJ 7.201, 224); no such designation appears in the biblical narrative. The same word recurs to designate the eunuch Achratheus (RSV: Hathach) as Esther's go-between with Mordecai. Here, where the Hebrew reads simply, "Esther told them to reply to Mordecai" (Esth. 4:15), the Septuagint has the prolix phrase, "Esther sent back to Mordecai the same one who had come to her," which Josephus simplifies with "sending the messenger [διάκονον]" (AJ 11.228).[1] Again, when David sent some of his men from the desert to raise provision from Nabal of Carmel, instructing them in the way they should address the wealthy man, the biblical account reports their mission in the following terms: "When David's young men came, they said all this to Nabal in the name of David" (1 Sam. 25:9); for Josephus, for whom both David and Nabal are eminent personages, this statement becomes: ταῦτα δὲ τῶν πεμφθέντων διακονησάντων πρὸς τὸν Νάβαλον (AJ 6.298, "When those who had been sent conveyed all this to Nabal . . ."). Here the Greek participle διακονησάντων implies—what the English translation "conveyed" fails to do and what the Hebrew and the Septuagint state in so many words—that the message is delivered "in the name of" David.

The identity of the person who arranges for a message to be delivered in his name can be indicated in Greek by appending a personal dative. We see this illustrated in another passage where Josephus is writing an account parallel to the Bible. In the first Book of Kings we read that Adonijah attempted to gain Abishag in marriage through Bathsheba's intervention with Solomon. According to the Septuagint, Adonijah said to the queen, "Speak [εἰπόν] to King Solomon . . ." (3 Kings 2:17), which Josephus records more elegantly and in terms more suited to courtly intrigue as he requested her "to intervene for him with his brother" (διακονῆσαι πρὸς τὸν ἀδελφὸν αὐτῷ, AJ 8.5). Where the Septuagint reports Bathsheba going on to say, "I will speak to the king about you" (λαλήσω περὶ σοῦ), Josephus writes simply that she promised "to intervene" (διακονῆσαι, AJ 8.6).[2]

The notion of acting in another's behalf is less apparent to the modern reader when the verb does not have a direct object. Demosthenes records that the Athenians outlawed Arthmius because "while on a mission for his king [τῷ δεσπότῃ διακονῶν] he brought gold into Peloponnesus" (Dem. 9.43).[3] Again, mounting an attack on Aeschines' embassy to Phillip of Macedon, Demosthenes contrasts the performance of the Athenian envoys (πρέσβεις, 19.69) on that occasion with the performance of Antipater and Parmenio, who were Philip's envoys (πρεσβευτής, 19.68) to Athens. These are referred to as δεσπότῃ διακονοῦντες (19.69), a phrase usually translated in the style of "under that despotic master"[4] but meaning, like the phrase in the preceding passage, "on a mission for a master." It is instructive to see the place this sense holds in Demosthenes' exposition. He asks his audience to consider the reports made to the Athenians by both these parties in the light of two factors. The first is that the Athenians enjoy political freedom whereas the Macedonians are subject to a despot, and the second is that the Athenian envoys are permanent residents of Athens whereas the Macedonians were on a fleeting visit there. Demosthenes proposes that these factors could have led the Athenians to expect lies from the foreign envoys, whereas they had had the facts put truly before them, and to expect the truth from their own, whereas they had been deceived. In using the phrase "on a mission for a master," he is referring at one point ("master," "despot") to Macedonian despotism, and at the other ("on a mission," διακονοῦντες) to the visitors, who, as he adds, were not likely to meet the Athenians again. This additional comment was more than he needed to say because as διακονοῦντες from Macedon the envoys would be returning there any way.

Opening his *History of the Goths* the late historian Priscus introduces Esla, a Hun who is to take a prominent part in dealings between King Ruga of the Huns and the Romans. We read that he was "accustomed to mediating on behalf of both Ruga and the Romans in cases where there was disagreement" (εἰωθότα τοῖς διαφόροις αὐτῷ τε καὶ 'Ρωμαίοις διακονεῖσθαι, *HGM* 276.7). There are two points of interest here. One is that when Esla is acting on behalf of both warring parties the verb is not likely to be a technical designation of an ambassadorial function; an embassy represents one party only, and the normal Greek term is πρεσβεία, which Priscus in fact uses a few lines later in referring to an approach to the Huns by the Romans. The second point is the dative "disagreements." The normal dative after this verb nominates the person in whose behalf or on whose authority an action is carried out, as here also in the other dative phrase "for him and the Romans." Any other object is normally direct, as in the passage recently seen, "they conveyed *these things* to Nabal" (Josephus, *AJ* 6.298). There is also the infrequent use of the dative exemplified in Lucian's ἀγαθοῖς, which we translated as "ministering in the discharge *of someone else's good deeds*" (*DMort.* 30.3) but in such a use the good deeds are something that the minister puts into effect. The same cannot be said of "disagreements"; that dative indicates a situation on the occasion of which Esla ministers. Perhaps the only other instances of this use are Heliodorus 5.8 and Acts 6:2, both in reference to ministry at table.[5]

Couriers of the type appointed by King David in Josephus's account (*AJ* 7.201) occur in Thucydides. This and a cognate in the same passage are the only instances of the word group in the historian, a fact that would seem to be telling us something. The passage (1.128–38) narrates the death that the Spartan ephors eventually contrived to bring upon their regent, Pausanias, whom they suspected of collaborating with the Persians. Pausanias is reported as having been communicating with Xerxes through letters which included the instruction that their bearers should be put to death. The last of the couriers reads the letter he is to carry, informs the ephors, and agrees to act as decoy for them in a plot to expose Pausanias' supposedly treacherous dealings with the enemy. With the ephors hiding within earshot, he accordingly confesses to Pausanias that he has read the letter, pleads his past fidelity in "missions [διακονίαις] to the king," and pleads to be spared the death meted out to the rest of the "messengers" (διακόνων, Thucydides 1.133). Pausanias discovers his danger, flees to a sanctuary, and is there walled in to die. The incident is narrated in a manner widely recognised as uncharacteristic of Thucydides, and his source was possibly the official story put out by the ephors to justify their summary execution of a man who was an embarrassment and possibly a threat to them.[6] In such a case the occurrences would be due either to the formal language of the public record or to the conservative style of the Spartans themselves.

Communicating

If this concludes the series of instances relating to the conveying of messages from one place to another and to the performance of errands, a few instances remain that do not imply local movement. These illustrate the intimate connection between word and διακον- all the more effectively in that the usage ties in with the kind to be met in the following chapter where the notion of mediation arises in connection with agents and instruments. There we will be dealing with deeds and actions, whereas these few passages are about the spoken or written word.

Two in fact have already been seen, one in Aristides' phrase about interpreting the writings of past scholars for scholars of a later generation (Aristides 1.82), the other in Lucian's phrase about singers stirring emotion in their audience (Lucian, *Merc. Cond.* 27). An expression as succinct in the Greek as either of these occurs in Euripides. When Creusa confides to Ion the sorry tale of how her only son, fathered by Apollo, had been lost to her—Ion was in fact that son—she describes herself as διακονοῦσα κρυπτά (Euripides, *Ion* 396), in A. S. Way's admirable rendering, "handling secrets."[1] In Heliodorus's novel much turns on the true parentage of the heroine Charicleia. Exposed at birth lest her fair complexion should lead to a wrongful charge of adultery against her dark-skinned mother, queen of Ethiopia, she was wrapped in fabric on which her mother had inscribed a heart-rending statement of the infant's lineage. In royal script, which only the learned Calisiris much later was able to decipher, the queen's letter to her daughter concludes with the explanation, "I speak to you now through the medium of writing [τὸ γράμμα διάκονον εὑραμένη] because some divinity has deprived me of living conversation with you" (Heliodorus 4.8).[2]

Equally expressive of the notion of mediation is Achilles Tatius's use of the verb in an elegant statement about begging for mercy. The novelist relates that when Clitophon fell into the hands of Egyptian bandits he considered his misfortune to be all the more severe in that Egyptians would be unmoved by the kind of plea that would arouse compassion for the captive in the sensitive heart of a Greek. Reflecting on the efficacy of entreaty, Clitophon draws on the image of a suppliant at a religious shrine or festival who carries an olive branch to signify the prayer of his heart. He says: "The tongue subdues the rage in the listeners' breast by bringing out [διακονουμένη], in place of an olive branch, the anguish in his own" (Achilles Tatius 3.10.2).[3] Some comment on this translation will be helpful because "bringing out" is a poor substitute for "mediating"[4] and because other commentators have had some trouble accounting for all the elements in the Greek. S. Gaselee, who read a dative (τῷ . . . πονοῦντι) instead of the accusative (τὸ . . . πονοῦν) after διακονουμένη, translated "the tongue, ministering to him that is in anguish of soul by helping him to express supplication,"[5] but this is to obscure the function of such a dative (implying help for the person designated by the dative instead of nominating the person at whose direction an action is performed) and to make a difficulty of the Greek genitive ("of soul"). E. Vilborg, who makes the latter point,⁶ has in any case found the better reading to be the accusative,⁷ which T. F. Carney also accepts as "a difficult reading" and translates, "As far as concerns the suffering undergone by the soul, the tongue, serving for a suppliant's emblem . . .";⁸ on Carney's own admission, however, this accusative of respect is something of a guess, it is unlikely in such a nicely balanced sentence as well as being unnecessary from the point of view of grammar, and "serving for" is a meaning without precedent.⁹ Vilborg himself, imputing an intensive value to the middle participle for no sound reason, arrives at a translation by supposing it to have also a pregnant sense, "the tongue willingly transforms the suffering of the soul into a prayer in words."¹⁰ The idea of a transformation of suffering is an even higher conceit than that attempted by Achilles Tatius, who is simply employing διακονουμένη in a customary way to follow through the notion of mediation introduced in the sentence immediately preceding the one in question; "speech," observes Clitophon there, "is often the go-between of compassion," as Gaselee translates it.

6

Deed

In a much more confined ambit—and with fewer stories to tell, although with one or two longish ones, beginning with an alluring tale about Hermes—we now move into the field of "deed," where we encounter usage about agents and their activities. These are wide-ranging, reaching from some high abstractions of Aristotelian philosophers to a meaner level of political manoeuvrings and social toadyings, but taking in significant nationalistic and civic undertakings and even implementations of imperial policy; we shall see the words designating duties as varied as contract killings and priestly and prophetic callings, concluding with a vision of the cosmic daemons ordering the world for their God.

A Virgin's Ransom from Hermes

Star-crossed lovers are the stuff of the ancient Greek novel. Heliodorus was only half-way through his Ethiopian romance when Theagenes, his clean Greek boy, concluded from a truly bitter experience that the gods would never allow him to be united with his chaste Charicleia. As a marauding party was rowing across the lake towards them, and with delicious complexities of plot still up Heliodorus's sleeve, Theagenes is already ruefully reflecting on the ways of the gods with young men and women in love.

> How long shall we flee the fate that followeth us everywhere? Let us yield to fortune and withstand no longer the violence that is ready to assault us. What else shall we gain but fruitless travel, and a life of banishment, and the continued mocking of the gods? Do you not see how to our exile they join the robberies of pirates, and go about with great effort and diligence to bring us into greater dangers by land than erst we have found by sea? Not long ago they made a battle about us: then they brought down thieves: afterwards they made us prisoners: then they left us alone but at liberty making us believe that we might go whither

we would: and now they have brought us into the hands of such as shall kill us. This war for their sport have they made against us, devising, as it were, a comedy out of our affairs. Why then should we not cut short this tragical poem of theirs and yield us to those who wish to slay us; lest perchance, if they mean to make an intolerable end to our tragedy, we be forced to kill ourselves.[1]

In the face of Theagenes' fatalism, Characleia gives the impetus to true love that is expected of all good girls in novels—especially of white girls born to black Ethiopian queens. While she concedes that Theagenes has probably taken the only view possible of heaven's dealings with them both, she sees no sense in limply yielding to fate, and urges that they run for their lives. Mitranes' men take them, nonetheless, Theagenes is spared to enhance with his beauty the domestic service of the god king in distant Babylon, and Charicleia becomes of course the prize of the marauding captain.

Mitranes is not the only man in the area, however, who is in a position to take advantage of a virgin in distress. His business associate, the merchant Nausicles, also has an eye for a pretty girl, and he at once puts it to Mitranes that Charicleia is in fact Thisbe, his own lost love, whom Mitranes had contracted to find and deliver to Nausicles. And as Charicleia now passes from the hands of an officer to those of a merchant, only Calisiris, the wise Egyptian priest who is the mentor and guardian of Charicleia and has presented himself as her father, knows that even more than by the love of women Nausicles can be snared by the lust for gold. To redeem the girl from her wretched fate, therefore, the priest puts the unlikely proposition to the merchant that upon due prayer untold riches might be forthcoming from the heavens. Nausicles is both credulous enough to accept the priest's assurance and hard headed enough to offer to accept such wealth as a ransom for the girl. Calisiris has already heard in a whisper from Charicleia that she bears on her person a precious ring from her father, so that when Nausicles puts the priest to prayer while he himself sets about a public sacrifice, Calisiris is able to hatch a plan to hoodwink Nausicles. By sleight of hand he will produce Charicleia's ring from the sacred fire. Nausicles will surmise that in return for his sumptuous sacrifice the gods are endowing the benefactor of the gathering and their own chief devotee, the merchant Nausicles, with wealth beyond his imagining.

> After they came to the temple of Mercury [Hermes]—for Nausicles made his sacrifice to him and honoured him more than the rest, as being the god who has most care of merchants—and the offering was begun, Calisiris looked a little upon the entrails, and by the diverse changes of his countenance declared the pleasure and pains of that which was to come. While the fire yet burned upon the altar, he thrust in his hand and made as though he pulled out of the fire that which he already held in his hand, and said: "This price of Charicleia's redemption the gods proffer thee, Nausicles, by me." And therewith he delivered him a princely ring, a passing heavenly thing. As touching the hoop, it was of electrum, wherein was set a bright amethyst of Ethiopia, as great as a maiden's eye, in beauty far better than those of Iberia or Britain.[2]

The stone in this royal ring with its lights and skilful workings Heliodorus describes over the next page and more. "Such was the ring," he writes at length, and turns to Nausicles, who is reacting as Calisiris had predicted, for he is immediately ready to discard the beautiful girl if by so doing he can gain the ring. To veil his grasping heart, however, he turns a speech of the most pious sentiment and mealymouthed self-abnegation.

> Nausicles astonished at the strangeness of its coming and delighted even more at the value of the stone, esteeming the ring of more price than all the goods he had beside, spake thus: "Good Calisiris, I did but jest, and when I asked you somewhat for the ransom of your daughter, it was but words: for I had determined to let you have her for nothing. But since, as you say, the glorious gifts of heaven are not to be refused, I take this stone sent by the gods, persuading myself that it comes to me from Mercury, according to his wont the fairest and kindest of all gods, who hath given this gift to you through fire, as may be seen still by the flaming thereof. And besides I deem that gain to be best, which without damage to the giver doth enrich him that receiveth it."[3]

This said, Nausicles withdraws with the men to the outer court of the temple to enjoy with them the sacrificial banquet, thereafter, over his cups, to enter into the kind of discourse appropriate to sacrifices in honour of Hermes,[4] while the women remain within to revel in dances honouring the earth goddess. Charicleia retires alone to pray for the safekeeping of her beloved Theagenes.

This incident in the fifth book of the novel is as adroitly manipulated as any in Heliodorus. In the telling it is also quaintly evocative of the bookish culture to which Heliodorus and his clientele belonged. A signal instance of our verb occurs at the almost cathartic moment—for it marks the redemption of Charicleia—when Nausicles confesses to seeing the work of Hermes in the appearance of the ring. Wright's revision of Underdowne's sixteenth-century translation of this passage (Heliodorus 5.15) has been used here because its old-worldly air helps keep us in touch with the ornateness of Heliodorus, and according to Wright *Hermes has given* a gift to Calasiris through the fire. Other translations put the matter variously: Hermes sent a stone through the fire, using *the ministry of Calisiris;*[5] *Hermes conveyed* a treasure trove to Calisiris through the fire;[6] Hermes made the gift reach Nausicles which Calisiris had found in the midst of the fire;[7] *Hermes made the gift available* to Calisiris through the fire.[8] These italicised phrases seem to be what the translations make of the participle διακονήσαντος (and one translation seems not to render it at all). Among several noteworthy discrepancies to be observed in the translations,[9] one would concern whether Hermes was dealing with Calisiris or with Nausicles, and another, whether Hermes was acting personally through the fire or using Calisiris's ministrations. Differences of opinion at these points arise from different understandings of the genitive participle, which comes at the end of a longish phrase containing two other genitives ("Hermes" and "fire"). To determine whether the participle is saying something about Hermes or something about the fire, and what in any case that might be, we need to go back into the context once more.

In his speech Nausicles is referring summarily to events that have already been recounted by the novelist. Readers are informed about just what Calisiris had in mind in declaring that the gods provide riches in response to the prayer of the wise man. They also recall how explicit Calisiris was on the nature of his own role, for as he produced the ring he had announced that the gods were offering the ring to Nausicles through Calisiris as a ransom for Charicleia.[10] At this point too we had the description of Calisiris thrusting his hand into the flames while the sacrificial fire still burned. In his own speech Nausicles is intending to refer to this clearly recorded incident. He understands that he is the recipient of a valuable ring that heaven has made available as a ransom for Charicleia through a process involving a sacrificial fire burning in honour of Hermes and a holy man praying for this happy outcome. The only elements additional to what the novelist has already related are the explicit—and implicit—references to Hermes. Nausicles is careful to include this god because Hermes is patron of his trade and is the object of the present cult. We are to assume then that Nausicles understands Hermes to be dealing personally with himself—a highly flattering idea that had been cultivated by Calisiris—but mediately, through the role of Calisiris at the fire. All these elements, we now note, the novelist has contrived to include in the terms of Nausicles' speech:

> I accept this jewel, sent as it is from heaven, confident that this godsend has come to me—altogether appropriately—from Hermes, the noblest and kindest of the gods, and through the fire which transmitted the gift at your behest: for, as you can see, fire still flashes from it.[11]

Understood in this way, Nausicles' statement of the events corresponds exactly to what the narrator has already told us. Calisiris had claimed he could win wealth through sincere prayer, had been invited by the merchant to undertake this task, had then worked his sleight of hand in the flames, produced the ring, and declared, "For you, Nausicles, and through me the gods have brought forth this ransom for Charicleia." From his own later statement, with its use of διακονήσαντος, we can see that Nausicles has understood that the fire was in fact a medium which, at the prayer of Calisiris, transferred the stone from Hermes—indeed, this is the point of Nausicles' observation that by its flaming brilliance the stone betrayed the path it had traversed. We note also the highly religious character of the occasion: a gift beyond price has come from heaven in answer to the prayer of saintly Calisiris and in the midst of sacrifice.

A Little Philosophy

We have had occasion to emphasise that the word group sits easily in talk about the kind of communication that ancients conceived of as going on with the other world whether by way of message, entreaty, or the transient thrill of love. In such contexts, as in the passage from Heliodorus discussed in the preceding section, the notion of mediation at times suggests itself. In line with this, several passages from philosophers, mainly the commentators on Aristotle of the second century CE and later, show that words of the group are invoked for the express purpose of conveying the idea of mediation. The usage reflects that in Plato's analysis of various social and political activities but even more than there illustrates the philosopher's appreciation of a set of words capable of expressing the abstract notion of mediation without connotations that might distort the course of his exposition.

Thus, in considering the nature of translucent objects, Alexander of Aphrodisias observes that an object like air or water, which has no firm boundaries or inherent colour, is able both to take on and to transmit the colours of other things. To denote the function of transmitting colour the philosopher uses a form of διακον- (marked by italicised words in the following translation of the passage in *de Mixtione*):

> A formless translucent body like air or water has no predetermined inherent shape, nor does it have a colour of its own; in respect of the colours of other things, however, it is both *conductive* and receptive [τῶν ἀλλοτρίων χρωμάτων διάκονόν τε καὶ δεκτικόν, Mixt. 5.14]—just as it is receptive of shapes. With good reason has nature made colourless the translucent body which is to *transmit* the colours of other things [τὸ διακονησόμενον τοῖς ἀλλοτρίοις χρώμασιν διαφανές, Mixt. 5.16].[1]

In his commentary on Aristotle's *Meteorologica* Alexander writes of objects that, though not themselves having to endure the influence of other objects, are "conductive" (διάκονα, *in Mete.* 18.25) in regard to the next objects in line, and asks, in the light of that, if it is remarkable that the moon should pass on (διαδίδωσι) the sun's influence to objects that are naturally susceptible to it without itself coming under that influence: thus, the moon is not set alight by its proximity to the sun but "passes on" (διακονεῖσθαι, 19.6) the sun's influence to other objects.[2] William of Moerbeke, who introduced this tract to the scholastics about 1260 CE, translated the adjective at 18.25 by "delativa" and the verb at 19.6 by "deferret," words virtually synonymous with the English used here.[3]

According to Ammonius a similar process takes place in the chain of communication by which ideas are kept in circulation in our world. When souls are united with bodies, he states in his commentary on Aristotle's *Categoriae,* they are not naturally endowed with ideas but come by them through sharing in a society of souls where ideas are spread by "the human voice *acting as a medium* for the souls" (διακονούσης αὐταῖς τῆς φωνῆς, *in Cat.* 15.9). In another area of ancient psychology Themistius explains that the faculties of sight, hearing, and smell, unlike touch, do not come into contact with their natural objects directly but only indirectly "with some other matter in between *acting as a medium*" (ἑτέρου σώματος μεταξὺ διακονοῦντος, *in de An.* 125.9). Commenting on the same distinction between the senses, Alexander Aphrodisiensis makes the more detailed observation that the faculties of sight, hearing, and smell are not themselves informed by the impressions they receive—the eye's pupil, for example, does not take on any colour—but the faculty of touch appears to be more completely a mediating faculty (μεσότητος) in that it receives unto itself impressions that are indeed "to be passed on" but which in the course of that process are "experienced" (διακονητικά τὲ καὶ μηνυτικά, *de An.* 59.14).[4]

When Iamblichus contrasts the body's automatic responses to stimuli with actions deliberately performed, he sees the difference in the way the body operates in the latter case only under the direction of the mind, not of itself as in the former case but by becoming the means or instrument "effecting" the mind's intention [ἡ διάνοια] δεῖται . . . τοῦ διακονήσοντος σώματος, Protr. 6). This takes us away from the abstract concept of mediation illustrated in the preceding citations from philosophers into an area we shall see much of and for convenience shall call agency. Ears, hands, teeth, and the sense faculty as a whole are each seen as instruments of the mind by Philo,[5] and Themistius states that every living creature requires organs "working as a means" for the end of nourishment (ὀργάνων . . . τῶν πρὸς τοῦτο διακονούντων, in de An. 42.6). In demonstrating the inferiority of the sense faculties to what he calls the faculty of discernment, the Stoic Epictetus uses the word group to denote what characterises the subsidiary role of the senses. Because in exercising discernment we are assessing sense impressions and selecting from them, the faculty by which we discern will be superior to those by which we sense. Sense faculties, we read, "are appointed to act like butlers and maids [ὡς διάκονοι καὶ δοῦλαι] for the faculty which uses sense impressions" (2.23.7). This, of course, is figurative language drawing on an image of domestic servants as does Plato's language at Grg. 517b and related passages; the inferiority of the sense faculties is established, however, not by the mere comparison with a menial status but rather, as in Plato, by a consideration of the circumscribed nature of activities of which menials are capable. As Epictetus proceeds to argue, "How can any other faculty be greater than this one which uses other faculties as agents [διακόνοις]?" (2.23.8).[6] Sense faculties, he then says, are designed "to carry out and effect" their roles under the direction of the higher faculty of discernment (διακονεῖν ταύτῃ [δυνάμει] καὶ ὑπηρετεῖν, 2.23.11). It is not possible, as he proposes shortly, that "the agent" (τὸ διακονοῦν)—a generic use—should be superior to the one it "is acting for" (ᾧ διακονεῖ, 2.23.16), just as the horse is inferior to the rider, the dog to the hunter, the harp to the musician, and officials (ὑπηρέται)[7] to a king. Elsewhere, in figurative terms again, Epictetus puts it to the Epicurean that it is natural for man to subordinate "pleasure" to duty "as an agent"—ὡς διάκονον, ὡς ὑπηρέτιν (3.7.28: feminine nouns in agreement with "pleasure")—for the purpose both of stimulating desire and of keeping us acting in accord with nature. Philostratus maior designates Aesop's fox διακόνῳ τῶν πλείστων ὑποθέσεων (Im. 1.3.2), and we are to understand that he is speaking of the animal as the "vehicle" of most of Aesop's themes.[8] Similar usage occurs in less philosophical writing, as we shall now see.

Double-Crossing Agents and Others

In Heliodorus's tale Cnemon tells how in his young days in Athens his high-placed father, on being widowed, married a young woman who developed a passionate attraction for her stepson. Repulsed by the god-fearing Cnemon, she devised a cruel revenge. She arranged that her maid Thisbe should arouse his suspicions about his stepmother's virtue. One night, duly advised by Thisbe that his father was away and that the adulterer was visiting, Cnemon seized a knife, kicked in the door of the bedroom, and was about to begin the righteous slaughter when he realised that the man who had leapt from the bed was his father and he was now on his knees begging for his life. Aghast, Cnemon looked around for Thisbe, but the wicked thing had fled. Cnemon was bound, dragged before the people, and exiled. Relating Thisbe's part in his painful experience, Cnemon says his stepmother "made her the agent of the scheme against him."[1]

In the middle of the first Jewish revolt, with the Romans already in command of Galilee and Vespasian poised to advance on Jerusalem, Josephus describes the city and temple in the murderous grip of the Zealots. Drawing on a saying that city and temple would fall "if ever the citizens strove with each other and Jewish hands were the first to pollute the house of God," he comments, "The truth of this the Zealots did not question; but they made themselves the means [διακόνους] of its fulfilment" (*BJ* 4.388).[2] At an earlier stage of this gruesome period in Jerusalem, John of Geschala persuaded the volatile Idumaeans that the only way to prevent the priestly party from betraying the city and nation to the Romans was for them to bring a strong force to the relief of the beleagured Zealots in the temple compound. The eager force arrived only to find the city gates closed to them. The high priest's spokesman addressed them, nonetheless, anxious to allay the slander that his party was seeking a truce with Vespasian. If there had been dealings with the enemy, he argued, "let the Zealots name the friends we sent, the creatures who *perpetrated* the act of treason" (*BJ* 4.252).[3]

Two other instances in Josephus are worth attention here because they occur in his retelling of incidents from the Greek Bible, a vast literature that virtually eschewed any use of the word group. Gen. 29:12 relates that after Jacob had watered Rachel's sheep, kissed her, and confided to her his identity, "she ran and told her father." In Josephus's account Jacob looks back on this moment when he is asking Laban for Rachel's hand; he diplomatically informs Laban that not least among his daughter's qualities is the fact that she had been the one who introduced him to Laban, "became for him an *agent* in reaching Laban" (*AJ* 1.298).[4] Similarly, at Haman's downfall in the story of Esther, when Artaxerxes despatches him to convey to Mordecai the honours that Haman had anticipated were his own, Josephus adds to the biblical version of the king's words (Esth. 6:10) the succinct phrase translated so well by Marcus, "you shall be *the one to carry out* those things about which you have given good counsel" (*AJ* 11.255).[5]

In more formal language, court proceedings from Memphis record the following interchange between the advocate and the prefect (*SB* 7696.31):

> *Advocate:* The prytanis convenes the senate in name and appearance only, for the real convener is the law.
> *Prefect:* The law using some *instrument,* some prytanis or other was the convener.[6]

Again, a funerary inscription from Memphis in the first century CE records that a boy died by "the common law of death," which used coughing as its *instrument* (*SB* 7871.15).[7]

In a political essay Lucian criticises Hyperides for making himself available as an *agent* of the people's passing fancies (*Dem. Enc.* 31),[8] just as in a passage where menial roles are distinguished from nonmenial roles Plato uses the terms οἰκέτας τε καὶ δούλους in instructing the wardens of the republic to do without "servants and slaves" to meet their domestic needs but uses the term διακόνοις in warning them off drawing on those of local farmers and villagers to act as *agents* in the wardens' personal business affairs; such people might be called upon only in matters affecting the community (*Lg.* 763a).[9] The term occurs to the same effect in Socrates' description of the "affluent community"; after he has agreed to ease the rigours of the "authentic community" by allowing within it a range of callings that are not strictly necessary, including huntsmen, the artistic profession,[10] and those who do work that would normally be done by women, he adds a further class whom he designates by the phrase καὶ δὴ διακόνων πλειόνων δεησόμεθα (*R.* 373c), and instances as belonging to this class "tutors, nurses, nannies, beauticians, hairdressers, cooks and butchers." The word διακόνων, as at *R.* 370e, 371a, e, is the generic designation of those functionaries within the social system, and we are to understand Plato as writing, "and so we will need still more functionaries."[11]

Our reading of Plato, *Grg.* 521a, has already shown that "to perform a task in order to curry favour" (διακονεῖν πρὸς χάριν) is the essence of "flattery" (κολακεία), and characterises the actions of the domestic attendant, who typically does whatever he thinks will win approbation. This kind of thinking resurfaces when Aristides rebuts Plato's criticism in the *Gorgias* of the Four Politicians. Aristides, ever the polished rhetorician, has no difficulty establishing that if the revered Four were διάκονοι then political office (ἀρχή) of its nature is a servile avocation (δουλεία, 2.152–54). He is able to reduce Plato's attack to such absurdity, however, only by declining to acknowledge the figurative uses of language in Plato. διακον- is evoked in Plato because the writer is comparing the kind of political activity espoused by the Four—the provision of popular public works— with the kind of thing a fawning servant (διάκονος) pays careful attention to in pandering to the debased appetites of a master. If authentic usage allows Aristides to exploit the connection of the words with the condition of servitude,[12] Aristides himself illustrates the undoubted breadth of usage in predicating the verb of the same politicians for purposes of depicting them as the agents by which the gods saved Greece (2.198–99): "Saviours, of course, are gods, so the gods are the ones who really saved Greece on that occasion. Nonetheless, Miltiades and Themistocles *acted in their name* [διακόνησαν], and that is man's greatest and noblest capacity—*to act for* one's betters [διακονῆσαι τοῖς κρείττοσι]."[13] Miltiades and Themistocles, in other words, carried out on the battlefield what the gods did not do in person, and in such an understanding of their historic role we see clearly that "to effect things for others" is not at all, as Aristides has earlier made out (2.153–54), the same as "to act slavishly" (δουλεύειν).

Doing Caligula in

Chaerea Cassius was the commanding officer of the imperial bodyguard in Rome in 41 CE. He was being much abused by the alarmingly unstable emperor Caligula for a streak of supposed feminity in his makeup. Caligula played on this attribute of such a prominent officer by deputing Chaerea to torture the actress Quintilia for the purpose of eliciting information that might lead to the identification of her lover. He calculated that an effeminate officer would attempt to exhibit virility by the application of an undue degree of cruelty in "carrying out his task" (ὠμότερον διακονήσεσθαι, Josephus, *AJ* 19.34). Subsequent events were to reveal another side to Chaerea's character, however, at least in the version provided by Josephus. Appalled at his own and his fellow officers' collaboration with the emperor's foul rule, Chaerea mounted a conspiracy to do away with him. The following is part of his conspiratorial address. The translation is by H. Feldman in the Thackeray edition of Josephus's *Antiquities* and is excellent in its rendering of three instances of the verb (here italicised; Caligula is given his proper name of Gaius):

> It is I, O Clemens, and Papinus here and you, more than the two of us, who are applying these tortures to Romans and to humanity at large. *We are not discharging* Gaius' orders,[1] but following our own policy if, when it is possible for us to stop him from treating his fellow citizens and subjects as outrageously as he is now doing, *we act as his agents*,[2] occupying a post as his bodyguard and public executioners instead of doing our duty as soldiers—bearing these arms not to preserve the liberty and government of the Romans, but to save the life of one who makes them slaves in mind and body. And we pollute ourselves with shedding their blood and torturing them daily, until, mark you, . . . someone *as* Gaius' *agent will do*[3] the same to us.[4]

The assassination was carried off, and Josephus reports that as word of it passed among a crowd of theatre-goers not all could credit the news; among these were mercenary soldiers, "fellow tyrants" of the emperor, he writes of them, "who *in implementing* his violent designs were being manipulated to oppress the leading citizens" (*AJ* 19.129).[5] The killing of Caligula's wife, Caesonia, rounds off the episode; Lupus was commissioned for the task and wasted no time about it, as again we read, *"carrying out the task* for those who had sent him at the first opportunity" (*AJ* 19.194).[6]

This section of Josephus's narrative thus provides several constructions expressing the idea of acting as someone's agent. Such constructions are comparatively frequent in the latter part of *Antiquities*, only occasional earlier in Josephus's writings, and occasional also in other authors. The earliest instance in fact is at Herodotus 4.154.3, where Etearchus—as discussed earlier—prevails upon Themison to get something done for him. The form of the construction in Herodotus can be set out as διακονεῖν ("get . . . done") τι ("something") τινι ("for someone"), where the dative ("for someone") is not designating a person who receives a favour or some benefit, as it does in a mother coaxing a little child over supper, "Eat up a little more just for me"; rather, its import is what a recalcitrant student would understand should he hear the forbidding Mr. Cruttwell say, "Get that disgraceful work to the headmaster for me at once," for on reporting to the headmaster the student will say, "Mr. Cruttwell sent me to you with this work." The value for translation of seeing what this dative is doing in a phrase has been illustrated in the discussion of Heliodorus 5.15; in a commentary on the translation provided there, other instances of the dative were adduced. A further instance occurs at *AJ* 18.269 in company with similar usages mainly of the kind seen at *AJ* 19.41 where the dative is not designating the person issuing instructions but is a word denoting the instructions themselves ("discharging . . . orders" and "following . . . policy" in Feldman's translation, as noted). All these phrases merit consideration.

At this point in his narrative Josephus is reporting the running controversy between Petronius and the Jews. Petronius has been appointed to succeed Vitellius as the imperial legate in Syria and had instructions from Caligula to erect a statue of the emperor in the temple at Jerusalem. He enters Palestine with a substantial force only to be confronted at Ptolemais on the northern coast by a strong deputation of Jews. In subsequent protracted negotiations both there and at Tiberias the positions of the two parties appear to be irreconcilable, the Jews arguing fiercely from their law and tradition, Petronius no less forcibly from the imperial decree. On taking up his post as Caligula's legate, Petronius "hastened to put the emperor's orders into effect" (*AJ* 18.262);[7] encountering the Jewish deputation, he insists that he has no choice but "to implement in the emperor's name programmes already predetermined by him" (*AJ* 18.265);[8] Jewish intransigence was so clear, however, that Petronius realised force of arms would be necessary if he was "to effect the erection of the statue according to Gaius' direction" (*AJ* 18.269);[9] finding bloodshed abhorrent "in the pursuit of Gaius' mad resolve" (*AJ* 18.277),[10] especially as in their stance the Jews were merely giving expression to their deepest religious sense, he decided that the only course open to him was to put their case to the emperor, even when this meant exposing himself to the inevitable anger of the emperor "for not having at once implemented his orders" (ibid.);[11] in relaying his decision to the Jews he undertook to persist "in doing everything in pursuit of their interests through his own efforts and those of his friends" (*AJ* 18.283).[12] In Rome in the meantime Agrippa has been lobbying on his people's behalf through lavish entertainment of Caligula who is moved to promise Agrippa, "Whatever might tip the scales to complete your satisfaction shall be done at your behest with my consent and support" (*AJ* 18.293).[13] Thus brought around to a favourable disposition towards the Jews, Caligula has a letter en route to Petronius to inform him that he need no longer press for the erection of the monument when Petronius's own letter reaches the court urging that very decision. Ever sensitive in all that touched on the imperial prerogative, Caligula dispatches an angry second letter to the Syrian province intimating that Petronius might best serve his own interests by putting an end to himself since he had chosen "to do everything to meet the pleasure of the Jews" (*AJ* 18.304)[14] in contravention of the emperor's own commands. Before this fatal word came to Petronius, however, Chaerea's knife had struck at Caligula.

The Piety of Petronius

Interestingly the commonest construction in this part of Josephus's narrative is the verb with the dative of a word for command or desire; the construction expresses the execution of the command or desire. As well as the six instances in the account just recorded of the dispute between Petronius and the Jews—with one of these (*AJ* 18.280) yet to be discussed in this section—five others from earlier in

Antiquities can be adduced,[1] and yet in other authors the construction can be instanced only twice, once in the orator Antipho (1.17) of the fifth century BCE, also to be discussed here, and once in the sophist Callistratus (13.4) of the fourth century CE.[2] Even allowing for the fact that the protracted context of Petronius's dispute has partly occasioned the higher frequency of this phrasing in *AJ* 18.262–304, the rarity of the construction elsewhere would seem to point to its being a mannerism of the author, although hardly one of the "peculiarities" that Thackeray saw marking the language of books 17–19 of *Antiquities* as the "unmistakable work of a 'Thucydidian' hack."[3] The mannerism does serve, however, to draw our attention to the discrepancy in frequency of the verb between *Antiquities* and *The Jewish War;* in the latter work incidentally Petronius's dispute with the Jews is related (*BJ* 2.185–87) much more briefly, it is true, but with the same drift, without one use of the verb in question—the whole work instancing only one participial use of the verb (*BJ* 4.252).

Examination of the usage resolves a long-standing problem of interpretation at *AJ* 18.280. At this stage of the confrontation with the Jews Petronius reveals a growing appreciation of the Jewish point of view. To the political considerations advanced by the Agrippa party (18.273) and to those arising from his own humane reluctance to slaughter people in the interests of what he views as a lunatic impulse of the emperor, Petronius adds his respect for Jewish piety and his fear of outraging their God (18.277). He decides therefore to approach the emperor. Although he is aware of the emperor's likely reaction, he decently resolves (18.278) "that a man who makes virtue [ἀρετή] his goal might well die on behalf of such a multitude of men."

Petronius convenes the Jews at Tiberias (18.279), accordingly, warns them of the grave dangers facing both them and himself, and discloses his decision to press their case in Rome (18.281). A roundly rhetorical period, here given in Whiston's equally nobly sustained phrasing, puts his reasoning (*AJ* 18.280):[4]

> I do not think it just to have such a regard to my own safety and honour, as to refuse to sacrifice them for your preservation, who are so many in number, and *endeavour to preserve the regard that is due to your law,* which, as it hath come down to you from your forefathers, so do you esteem it worthy of your utmost contention to preserve it: nor, with the supreme assistance and power of God, will I be so hardy as to suffer your temple to fall into contempt by the means of the imperial authority.

If Whiston's translation of διακονούμενον τῇ ἀρετῇ τοῦ νόμου (the italicised phrase in the preceding citation) had read instead, "act according to the moral values of your law," it would have been a successful version of a sentence over which commentators generally have difficulty. To obviate what he perceived to be problems, Cocceius suggested two emendations: διακονουμένων, and τοῦ θεοῦ οὗ τὸν ναόν. The first of these, which shifts the focus from Petronius to the Jews, was adopted by J. Hudson in his great eighteenth-century edition, and he then translated the phrase, "devoting yourselves entirely to such a noble law."[5] Once introduced into the sentence, this shift tends to draw the other datives (ἀξιώσει, δυνάμει) into the same relationship with the plural participle, leading D. S. Margoliouth to revise Whiston's "with the supreme assistance and power of God" to "[endeavour to preserve] also the regard due to the honour and power of God"; the final statement beginning with τὸν ναόν ("the temple") is thereby left suspended, which the addition οὗ ("whose") attempts to remedy.[6] Both emendations are accepted by H. Feldman in the Loeb *Josephus,* leading him to write in the first place, "You are carrying out the precepts of the law," which is in keeping with his reliable understanding of the phrasing elsewhere,[7] and in the second place, "and serving the sovereign of all," which is to read a second and different meaning into the same instance of the participle. The second meaning, "serving God," is especially out of place because in none of the other thirty-nine occurrences of the verb in Josephus, or indeed in any other place in profane literature, does the verb express the idea of serving God in this general sense. The English version by G. H. Maynard (ca. 1786), the French by A. d'Andilly (1700), and the Latin of Ruffinus (in the Venice edition of 1486) bypass this problem of translation by resorting to paraphrase.

The difficulties felt by translators here are resolved and the rhetoric of Josephus's period is revealed once we allow that the phrase διακονούμενον τῇ ἀρετῇ in Feldman's sense of carrying out the precepts of the law can rightly be predicated of Petronius. The context explains how this is the case. Petronius's decision to compromise with the Jews has been influenced very largely by their resolve to die rather than violate the law, and in suspending the imperial decree he is able to give expression to his sincere admiration of their attitude without appearing to them to be abdicating his responsibilities. In effect he is saying, "I can do no less than you. If you are ready to die for your law, I can at least *put some of its values into practice* by exposing myself to the danger of the emperor's displeasure." We can therefore translate this part of *AJ* 18.280: "I do not think it is right not to disregard my own security and position for the sake of ensuring that you, who are so many, are not destroyed. I would be *giving expression to the values* of the law [διακονούμενον τῇ ἀρετῇ] which you have received from of old and judge to be well worth fighting for."

The Loving Cup

To find a precedent to this usage of Josephus we need to go back nearly five hundred years to the Athenian politician and speech writer Antipho. His first oration introduces a case—real or feigned is apparently open to question—in which a certain Clytaemnestra is on trial for the murder of her husband. The case is complicated by its connections with another parallel murder. Clytaemnestra's husband had a friend who was about to discard a concubine. Getting to hear of this, Clytaemnestra cultivated the concubine for the purpose of suggesting to her that the affections of both their lovers might be rekindled if the men were to be given an aphrodisiac. She asked the concubine, accordingly, if she would be willing "to act for her" (διακονῆσαι οἱ, Antipho 1.16). The concubine agreed, and at a sacrificial banquet decided to administer the draught after rather than before the meal "to carry out the instructions" of Clytaemnestra (ταῖς ὑποθήκαις . . .

διακονοῦσαν, 1.17).[1] Being devoted to her lover, she saw to it that he received more of the draught than Clytaemnestra's husband with the result that he died on the spot while the husband lingered some weeks. In a separate trial the concubine was found guilty and executed. In the later trial of Clytaemnestra the prosecution labours to establish that while the concubine had been "the agent and doer of the deed" (ἡ . . . διακονήσασα καὶ χειρουργήσασα, 1.20), the cause of the crime lay elsewhere, and what interests us in the language here, in addition to the precedent it provides to one of Josephus's rarer mannerisms, is that for all that the story is about a plan to pour poison drinks at a dinner party, διακονοῦσα is used to designate the concubine not as a drink waiter (ἐγχέουσα is used for that) but as one who implements somebody else's plan.

Civic Duties

On this ground the verb occurs in several passages relating to politicians, bureaucrats, and public figures. The controlling principle of public life, according to Plato, should be, "Those who perform a task on behalf of the nation [τοὺς τῇ πατρίδι διακονοῦντάς τι] are to do so [διακονεῖν] without receiving gifts" (*Lg.* 955c). A sentence or two later this principle is enshrined in the law, "Do nothing for a bribe" (μηδὲν . . . διακονεῖν, *Lg.* 955d). In Apollodorus's action against Polycles the plaintiff vigorously defends his role in public office, undertaking to give a detailed account of his financial outlays and of his actions, and to show how his duties "were performed [ἐδιακονήθησαν] on schedule and in a manner beneficial to the nation" (Demosthenes 50.2). Demosthenes is cutting in his comment on the attitude of some trierarchs who had skimped their own outlay on a trireme and yet looked for returns at the expense of the man they had hired at the cheapest rates; "there is something wrong, surely," he reminds the senate, "in the trierarch's blaming the subcontractor for dropping anchor when he did and then expecting you to extend expressions of gratitude to themselves for what had been done so well for them by him [τῶν δὲ καλῶς δεδιακονημένων]" (Demosthenes 51.7). Under attack himself from Aeschines for impropriety in public life, Demosthenes puts to the jury of five hundred that he has consistently worked within principles developed within the Athenian tradition (Demosthenes 18.206): "Now what I am claiming is that such principles are your own; further, that before my time the commonwealth thought along the same lines—only that in the execution of all that has been achieved [τῆς μέντοι διακονίας, τῆς ἐφ' ἑκάστοις τῶν πεπραγμένων] I too played my part." As for what Aeschines had done to benefit Athens, he asks the jury to consider "By what foreign mission of his [πρεσβεία] or what public undertaking [διακονία] has the city's reputation been enhanced?" (Demosthenes 18.311). Underlying his own defence was the claim that he was not still technically in office when Ctesiphon sought public honours for him—the act that Aeschines' party had sought to establish constituted a contravention of the law. Aeschines granted that a distinction did exist between an office (ἀρχή) and a public commission (ἐπιμέλεια) or undertaking (διακονία, 3.13) but argued that the defendants made terms of convenience of the latter:

"What the lawmaker nominates as offices these people call public affairs and commissions [πραγματείας καὶ ἐπιμελείας]" (3.16), whereas once men have been publicly appointed to tasks their actions remain those of the office (ἄρχειν) until expiration of tenure is declared; law does not impose "the carrying out of a task" (διακονεῖν, 3.15) but "the filling of an office" (ἄρχειν).

Plato's usage reinforces the distinction alluded to by Aeschines; any action in the political field can be seen as "a task undertaken in the name of the state" (πόλει διακονίαν, R. 493d). Dio Cassius designates Licinus's procuratorship of Gaul "the commission laid upon him" (τῆς προστεταγμένης οἱ διακονίας 54.21.4); Paenius, in a phrase that is additional to the Latin of Eutropius, records that Vespasian was the general "commissioned to take the war to the Jews" (τὴν . . . διακονούμενος μάχην, 6.18.5); Aristides reports the judgement against the Greek who interpreted for the Persians: "Not even by use of the voice may one act for foreigners [διακονῆσαι . . . τοῖς βάρβαροις]" (1.122); Josephus designates the palace steward whom the patriarch Joseph put in charge of the scheme to trap Benjamin "the servant in charge of the affair" (τὸν διακονούμενον οἰκέτην, AJ 2.129).

And so our words occur across a broad range of offices and responsibilities. If we look back over the usage since the story of Caligula's death (Josephus, AJ 19.34) we find that the words have been referring to senior military officers, an imperial legate, a royal emissary, a general, another general and emperor-to-be, a procurator, a royal steward, a former ambassador of state, citizens prominent in public life, and a few others—including a slave and a concubine. The occurrences are mainly in historical narrative but are also in discursive and legal literature. In no instance is there a disparaging edge to the usage; the terms are rather designating functions performed in the main by prominent and respected figures in society of the day. The designated function is a subordinate function but not of a kind to reflect poorly on the social status of the person involved or to detract from the moral quality of the function itself; even as applying to a slave girl the term is referring to the writing of a letter to a king and, in the case of a concubine, to her attempt at renewing the dreams of love. If the words do not connote anything especially high-minded about these subordinate roles, they do not of themselves suggest anything derogatory about the kind of function they name. Demosthenes advises the city that it should seek out the kind of man who will conscientiously bring it to heel rather than "the man who, to sustain his popularity, will get every political measure through [τὸν διακονήσαντα πρὸς χάριν πάντ']" (Ep. 2.11). The phrase is damning only because it indicates the base motivation behind some forms of getting things done in the political arena.

A Little Flattery

The scope of an agent for independent action is necessarily limited; Aristides points to the difference between what an adviser or confidant might do (συμβούλου πρᾶγμα) and what is permissible to the agent (διακόνου πρᾶγμα, 2.187), the former free to review a range of options, the latter limited to a designated course of action. In an essay on how to tell a flatterer (κόλαξ) from a friend, Plutarch explores this difference (*Mor.* 48–74). The flatterer under the microscope is not the idle parasite who attaches himself to the wealthy, but a person of ability who is privy to deliberations and fills out the role of friendship with the gravity of a tragic actor (50e); he betrays himself, however, at every stage of the relationship, for he is overeager at first meeting, immediately accepts whatever is proposed, and is flamboyant in effecting it, his most damning characteristic being that he exercises no discernment as to the rights and wrongs of a project. The friend, by contrast, is undemonstrative in affection, unobtrusive in action, wise in advice, and upright in all he does.

In analysing the way these types go about their tasks, the discourse frequently uses vocabulary that expresses the notion of getting things done, the terms implying neither subservience on the part of flatterer or friend, nor favour or kindness in regard to the main party,[1] and to these belong διάκονος and διακονία. The former occurs in connection with the question of consultation. Whereas a friend never undertakes to be a collaborator or σύνεργος unless he has first had the opportunity to be an adviser or σύμβουλος (62f), the flatterer wants no part in such consultation because in his judgement immediate results are the most effective means of ingratiating himself and thus take precedence over the wisdom or propriety of a course of action; he wants only to be a doer of deeds, "a collaborator and agent" ὑπουργὸς καὶ διάκονος, 63b). διάκονος is thus part of the work-oriented vocabulary, expressing basically a person's responsibility to whatever needs to be done. In the same way, the abstract noun διακονία is linked with abstract ὑπουργία (50c; 62d; 64e) to express the idea of an undertaking in another person's name without connoting anything about good offices or beneficent action; such ideas are here expressed through other terms like χάρις (kindness), ὠφελεῖν (help), εὐεργετεῖν (doing good) (63e–f). Thus the διακον- words are terms in the analytical description of that kind of cooperation that is shorn of moral involvement and consists exclusively in acting at the behest of another; the agent διάκονος is not in a position to use his own power of decision and initiative.

At the Behest of a Deity

Aristides has said that man's noblest capacity is to act (διακονῆσαι, 2.199) for one's betters; he was writing, as we have seen, of Miltiades and Themistocles, the heroes of Marathon and Salamis, and to name them "officers and agents" of the gods (ὑπηρέτας καὶ διακόνους, ibid.) in saving Greece was his highest tribute. In the less heady sphere of personal morality Stoic teaching proposed an ideal of conformity with the will of God, and in Epictetus's expression of this teaching the good man, who trains himself to discern the will of God for human living, becomes God's διάκονος (4.7.20) by giving expression to the will of God in the way he lives.[1] A less reverent mind, that of Lucian, takes the question of good and evil back to Zeus himself and in the course of some astute cross-examination Lucian puts it to Zeus that far from being benevolent masters of a bounteous and ordered universe the gods themselves would seem to be mere "officers and agents" of Fate (ὑπηρέται καὶ διάκονοι, J Conf. 11).

"Defiled agents" (ὦ μιαροὶ διάκονοι, 4 Macc. 9.17) of a divine king is the view of the royal bodyguard of Antiochus Epiphanes in the apocryphal retelling of the martyrdom of the seven brothers of 2 Maccabees. "Most defiled of tyrants" is the term of address to Antiochus adopted by four of the brothers, and "defiled" or "unclean"—already on the lips of the youngest brother in the original account (2 Macc. 7:34)—is a leading motif in the hagiography of the mother and seven sons who suffer gruesome deaths rather than apostatise from their religion by eating unclean food (4 Macc. 4:26; cf. 8:2, 12, 23, etc.) The tyrant, who has assumed divine prerogatives, is himself unclean in this religious sense, as are those who attempt to enforce his blasphemous injunction. In addressing his torturers as μιαροὶ διάκονοι, the martyr is saying more than "hateful minions"[2] because he is reproaching these resplendent members of the royal bodyguard for defiling the dignity that properly belongs to protectors of a god king. A genuine agent of a god would emulate the young prophet Samuel who, in Josephus's version of his calling (cf. 1 Kings 3:10: "Your servant hears"), responds that "he will not omit anything of whatever charge [διακονίας] God might wish to lay upon him" (AJ 5.349).

Another worthy agent of God is the patriarch Joseph, established by God as "a father to Pharaoh, and lord of all his house" (Gen. 45:8), and described by Philo as "established by God's will to be commissioner and dispenser [ὑπηρέτην . . . καὶ διάκονον] of his gifts and riches" (de Josepho 242). In the same way, in the temple of the universe, the priests are for Philo "the angels dispensing God's powers" (τοὺς ὑποδιακόνους αὐτοῦ τῶν δυνάμεων ἀγγέλους, de spec. leg. 1.66); by the nature of his function a priest is on the borderline between the human and the divine "so that men may be reconciled to God through a mediator [διὰ μέσου τινός] and so that for his part God might extend his blessings and spread them among people through a mediating agent [ὑποδιακόνῳ τινὶ χρώμενος]" (de spec. leg. 1.116). As well as priests the creator dedicates "those souls which have not descended to embodiment on earth to be his officers and agents [ὑπηρέτισι καὶ διακόνοις] in the governance of mortals" (de Gigant. 12). The work of governance is not in fact a work that the creator sees fit to take a part in; benevolent as he always is because he is the creator of goodness and peace, his agents (ὑποδιάκονοι, de decalogo 178) are the ones to take the attack up to evil.[3]

The functions that we have seen designated by words of the διακον- group are hugely varied, yet none, so far, have been of a menial nature. The words have been designating actions of an in-between kind or people who operate in an in-between capacity, especially people (or spirits) who implement the intentions or desires of another. Before considering in the following chapter how such words also come to be used of household chores and waiting at table, we can catch a final clear ring of the words in the area of agency from a passage by Iamblichus. He is depicting the daemon of Greek cosmology as the intermediary between God and man and wishes to distinguish carefully the role of God and daemon. To do so he chooses a διακον- word. In speaking thus of what he calls "the middle race of beings" he echoes the language of Plato and other philosophers reviewed earlier in this chapter who depict the activity of intermediaries as διακονία. Plutarch attributes to Plato the statement that "those who refuse to leave us the race of the demigods make the relations of gods and men remote and alien by doing away with the power which communicates and mediates [τὴν ἑρμηνευτικὴν . . . καὶ διακονικὴν . . . φύσιν]" (*Mor.* 416f).[4] Consistently with this, the fourth-century Neoplatonist writes in *The Mysteries of the Egyptians,* in Taylor's nineteenth-century translation (*de Myst.* 1.20):[5]

> Both the visible and invisible Gods, indeed, comprehend in themselves the whole government of whatever is contained in all heaven and the world, and in the total invisible powers in the universe. But those powers that are allotted a daemoniacal prefecture, distributing certain divisible portions of the world, govern these, and have themselves a partible form of essence and power. They are, likewise, in a certain respect, connascent with, and inseparable from, the subjects of their government. . . . In short, that which is divine is of a ruling nature, and presides over the different orders of being [ἡγεμονικὸν καὶ προιστάμενον]; but that which is daemoniacal is of a ministrand nature [διακονικόν], and receives whatever the Gods may announce, promptly employing manual operation, as it were [αὐτουργίᾳ], in things which the Gods intellectually perceive, wish, and command.

7

House and Table

Finally, we come to the words as they apply to menial duties. We will be recording instances of people, usually slaves, doing things for other people around their houses and waiting upon them at their meals. We will dispose of household chores quite summarily, and in this we will note a change of method from that in preceding chapters, where frequent close studies of context have been usual for the purpose of establishing meaning; here meaning is of less import than character of usage, and in the case of household duties the review is further limited by the baldness of the statements available and by the generally scant information provided by context. This is understandable of course because it has only been in the past century or two—and then not without a whiff or two of revolution—that the stuff of life downstairs has grabbed the interest of the bourgeois writers upstairs. In the case of their eating habits the ancients have been more generous, and in their range of records, descriptions, and comment, both in literature and in the immemorial stone, we shall have a rich course of wordy fare, satisfying especially by reason of Christian interest in the ritual of tables.

Housebound

An earlier reading of *Grg.* 517–18 in the light of Plato's other usage has suggested at least two aspects of meaning when the διακον- words denote the actions of a servant in a household. One aspect is that the action involves fetching or making something available to the person who has given directions; this would correspond to the way the same words convey their meaning in relation to the doing of a message. In a passage like Aristophanes, *Av.* 1323, we get a glimpse of the kind of activity the word suggests. Peisthetaerus is waiting impatiently for a slave to fetch some baskets, and exclaims, "Bring them quickly! . . . How sluggishly you serve [διακονεῖς]. Can you move a little faster?"[1] The other main aspect is that of effecting something at someone's behest, an aspect foremost in the usage illustrated in respect of agency in the preceding chapter. In respect of menial duties the same aspect is mirrored in a passage like Plutarch, *Mor.* 677e: "We often order those who attend on us [τοῖς διακονοῦσι] to get about their task [τῆς διακονίας] more quickly," or in Aristotle 1333a7–8 where the philosopher is considering what young men in the republic may and may not do with honour in the course of their upbringing; on the principle that actions are not honourable or dishonourable in themselves but are so by reason of the purpose for which they are done, he states that "it is honourable even for free men in their youth to perform [διακονεῖν, 1333a8] many tasks which might appear to be servants' duties [διακονικῶν ἔργων, 1333a7]."[2] Plato anticipated this thought in comparing the philosopher who does not compromise his status of a free man by performing "tasks proper to a slave" (δοῦλικα . . . διακονήματα, *Th.* 175e) with the kind of man "who can perform [διακονεῖν, ibid.] these tasks smartly and promptly" but does not maintain the inner poise of the free man; here, as well, the idea that speed is appropriate to domestic duties might be apparent, as in the preceding instances at Aristophanes, *Av.* 1323, and Plutarch, *Mor.* 677e. As often as not, however, such aspects of the verb's meaning are implicit only, as when a passage speaks in a more general way of domestic chores[3] or of attending on a master.[4] Only at Aristides 2.154, 199 is the verb a synonym of δουλεύειν ("to serve as a slave") but that, as has been pointed out, is for Aristides' own rhetorical purposes in that passage. Occasionally the participle occurs as a qualification of the noun παῖς ("slave"), which indicates that the verb denotes the kind of activity more than it designates the particular status of the person performing the action.[5]

The small number of these instances of the verb—just twenty from thirteen authors among more than ninety who use the word group—is noteworthy. Some of the authors, like Plutarch and Josephus, have frequent occasion to write of indicents involving domestic service but are very sparing in their use of the verb, and the anthologist Athenaeus discusses domestic service without recourse to it except in citations that reveal special usage.[6] Thus poetry,[7] moral or philosophical discourse,[8] legal documents,[9] and reports of court life[10] account for all but three of the instances, these being one at Plutarch, *Mor.* 63e, which is perhaps a genuinely casual use but is perhaps occasioned by uses in the discussion of "flattery" we have considered (*Mor.* 48–74), and the two at Lucian, *Philops.* 34 and 35, which occur in a ludicrous parody of grand living. Such usage points to the literary character of the word; even so it enjoyed only limited currency in specialised contexts.

Staff

Even less frequent is the use of διάκονος to designate the household servant. In respect of private households the designation occurs in poetry,[1] oratory,[2] philosophical discourse,[3] and in Aristides' all-too-exquisite account of one of his illnesses when even his doctors acted as διάκονοι (1.299). In respect of royal households, excluding references to attendance at table, instances occur in Herodotus's accounts of Scythia (4.72.2) and of the regent Pausanias's retainers (9.82.2); the book of Esther uses the word indiscriminately of attendants and of the more highly situated advisers in the court of Ahasuerus.[4]

The sense of personal attendant is also infrequent, occurring only in Herodotus's account of the Scythian court (4.71.4), treated earlier, in Lucian's biography of Alexander,[5] and in Josephus's accounts of Elijah and Elisha. *AJ* 8.354 records Elisha's response to Elijah's summons, "and he was both disciple [$\mu\alpha\theta\eta\tau\acute{\eta}\varsigma$] and attendant [$\delta\iota\acute{\alpha}\kappa\text{o}\nu\text{o}\varsigma$] of Elijah"; this phrase corresponds to the Septuagint's "and he followed after [$\dot{\epsilon}\pi\text{o}\rho\epsilon\acute{\upsilon}\theta\eta\ \dot{\text{o}}\pi\acute{\iota}\sigma\omega$] Elijah and ministered [$\dot{\epsilon}\lambda\epsilon\iota\tau\text{o}\acute{\upsilon}\rho\gamma\epsilon\iota$] to him" (1 Kings 19:21), the second verb here corresponding to the Hebrew *šrt*, which, outside of cult, is used of attendance in royal and prophetic contexts.[6] This restricted usage is followed closely by the Septuagint,[7] which suggests that in $\delta\iota\acute{\alpha}\kappa\text{o}\nu\text{o}\varsigma$ Josephus is substituting for the Septuagint's formal term a formal term from his own Hellenistic tradition.[8]

Again, as was observed of the verb in the preceding section, on the basis of such infrequent and specialised uses it is difficult to see this word as having enjoyed wide currency or as having been part of the vernacular; the significance of uses in the papyri in this regard will be argued in a separate chapter. Similarly as with the verb, however, we do get glimpses of the kind of household duties the $\delta\iota\acute{\alpha}\kappa\text{o}\nu\text{o}\varsigma$ was involved in. Plato's figurative uses at *Grg.* 517b and 518c have already illustrated that the attendant is ordinarily expected to be so intent on getting a job done as to have no scope for deciding whether it ought to be done; indeed, a proverb exclusive to the Septuagint suggests that the attendant might give more satisfactory service if he is too stupid to get involved in such moral niceties (Prov. 10.4a): "An educated son will be wise, and will use a fool for a servant." In keeping with this, Aristotle remarks on the tendency of servants to rush off before understanding precisely what they are to do (*EN* 1149a27), thereby also reminding us once more that speed of movement especially characterises the $\delta\iota\acute{\alpha}\kappa\text{o}\nu\text{o}\varsigma$. In *The Birds* Trochilos—whose name, which is that also of the swift plover and the hasty wren, is derived from the word "to run"—describes himself as "pursuivant and page" of his master, $\dot{\alpha}\kappa\acute{\text{o}}\lambda\text{o}\upsilon\theta\text{o}\nu\ \delta\iota\acute{\alpha}\kappa\text{o}\nu\text{o}\nu$ (*Av.* 73), the adjectival use here specifying the kind of attendant whose typical tasks are then outlined in response to Euelpides' question, "One bird then needs another for a page ($\delta\iota\alpha\kappa\acute{\text{o}}\nu\text{o}\upsilon$)?" (*Av.* 74)

> My master does, by reason, I suppose,
> That he was formerly a man; and so,
> When he would lunch upon Phalerian whitebait,
> I run to fetch him whitebait, dish in hand.
> Soup if he craves, ladle and pot are wanted:
> I run for a ladle.[9]

This speed of movement attracted our attention elsewhere in the play (*Av.* 1323) and at Plato, *Tht.* 175e, and Plutarch, *Mor.* 677e; it was so much part of the common understanding of the words that the notion of speed was built into the ancient etymology of the words.[10] When Plato wanted to say that Pericles and his associates were "more expeditious" than later politicians in providing what Athens demanded, he wrote the unique comparative adjective διακονικώτεροι (Grg. 517b), which is reminiscent of the other unique comparative in the maxim attributed to the sixth-century BCE poet Epicharmus: "Whipped Phrygian makes a better one and one who is sharper [διακονέστερος] about his chores."[11]

In the Abstract

The abstract noun διακονία occurs occasionally as a designation of household duties in some passages of a philosophical character,[1] in historical narrative of Josephus and Posidonius,[2] and in an anecdote of Plutarch illustrating the unpretentiousness of the illustrious Philopoemen of Megalopolis, statesman and general of the late third century BCE:

> [A certain] woman, learning that the general of the Achaeans was coming to her house, in great confusion set about preparing supper; besides, her husband chanced to be away from home. Just then Philopoemen came in, wearing a simple soldier's cloak, and the woman, thinking him to be one of his servants [ὑπηρέτων] who had been sent on in advance, invited him to help her in her housework [τῆς διακονίας, Phil. 2.3] So Philopoemen at once threw off his cloak and fell to splitting wood.[3]

Other passages also merit sampling. Reflecting on the fate of the Chians, who had been delivered by their conqueror Mithridates into the hands of their slaves for transportation into exile, the historian Posidonius sees a divine retribution for their having been the first people to have used purchased slaves, although he grants that most of the Chians had fended for themselves "as regards household chores" (κατὰ τὰς διακονιάς, FHG 3.265). In his treatise on the duties of home life, Xenophon reminds the woman of the great reward in "taking someone ignorant of household management and of household duties [διακονιάς, Oec. 7.41] and making her knowledgeable, trustworthy, and efficient [διακονικήν]"—the two last adjectives referring respectively to management and duties.

An unparalleled instance of the sense "instrument" or "mechanism" occurs in a citation from Moschio at Athenaeus 208a. He is describing a ship fitted out with "woodbins, ovens, kitchens, handmills, and several other mechanisms [διακονίαι]" (Ath. 208a).

This passage has been used to support the sense "tableware" or "vessels" at 1 Macc. 11:58, where Antiochus sends Jonathan χρυσώματα καὶ διακονίαν, "gold plate and a table service" (RSV).[4] As C. J. Ball suggested last century, however, the phrase might mean "golden vessels and a suite of attendants",[5] a gift that would be appropriate at the moment when Antiochus raises Jonathan to the status of "one of the friends of the king" (verse 57). This sense is supported by the occurrence of the collective sense ("staff") at Esth. 6:3 and 5, where the phrase "those from the διακονίας" is a variant for "the king's attendants [διάκονοι]." The nouns, abstract in form, δουλεία (Th. 5.23), θεραπεία (Josephus, AJ 18.269), ὑπηρεσία (Plato, Lg. 956e), even οἰκετεία (Lucian, Merc.

Cond. 15) commonly have a collective sense, which occurs clearly for διακονία in Polybius 15.25.21, where the decadent Agathocles of Egypt appoints his courtiers ἐκ τῆς διακονίας καὶ τῆς ἄλλης ὑπηρεσίας⁶ ("from the palace establishment and the other staff"). The usage is represented also in the advice of Teles, Cynic moralist of the third century BCE, that one ought to be "satisfied with the clothing one has, with one's standard of living, and with one's establishment [διακονίᾳ, 6.10] like Laertes, who made do with an old woman for a servant to bring him his food and drink."⁷

The collective sense would seem to suit best at Josephus, *AJ* 7.378, a passage whose meaning has been obscured by the uncertain state of the text. In the account of David's instructions for the building of the temple (cf. 1 Chron. 29), the king reassures Solomon and the leaders of the people that the task will present no great difficulty because he has already gathered together money, materials and craftsmen. At this point the passage continues: καὶ νῦν δ' ἔτι τῆς ἰδίας [. . .] ἄλλα τρισχίλια τάλαντα παρέξειν ἔλεγεν ("and now from his own [. . .] he said he would provide another three thousand talents"), the general sense being that David would make additional provision from his private resources. After ἰδίας manuscripts provide various readings (following Niese):

R: ἀπαρχῆς διακονίαν
M: ἀρχῆς διακονία καὶ
S, P: ἀπαρχῆς διακόσια καὶ

Hudson, in the eighteenth century, punctuated the last reading differently (ἀπ'), and translated, "ex sui imperii reditibus" ("from his kingdom's revenue"). The Latin of Epiphanius reads "oblationis primitias faciens, tria millia talentorum" ("making as his first offering three thousand talents"), where "making" is a feasible and interesting translation of διακονίαν (or -ιᾳ).⁸ Niese has supposed that the Latin's "offering" corresponds to διακονία, and then conjectures that the text should read ἀπαρχὴν διακονίας, which Marcus translates, "as his private offering for the service," and Weil in the French, "as first fruits of his own contribution" ("comme prémices de son propre apport"); against both Niese and Weil, however, the supposition that διακονία might mean "offering" is highly improbable.⁹ If, on the other hand, the word's collective sense is borne in mind, as also Josephus's interest at this very point—unlike the Septuagint—not only in money and materials but also in the need for workmen, it is likely that David here undertakes to provide "from his royal resources *palace workmen* and a further three thousand talents." This would suppose the reading: (ἀπ') ἀρχῆς διακονίαν καὶ.

Waiters at Work

Despite the frequency with which ancient literature records feasts and dinner parties, the word διάκονος is a designation of the waiter in only some twenty of the hundred instances of the word. A survey of Athenaeus's *Deipnosophistae* or "Wise Men at Dinner" has shown that with him and among authors he cites other terms are ordinarily preferred; the same is to be observed in regard to Plutarch's *Quaestiones conviviales* or "Table Topics."[1] This already suggests that the noun διάκονος was not so generally current in the sense of waiter as to be freely used in the context of dining. Further only one of the instances in literary sources referes to the waiter actually engaged in waiting at table;[2] otherwise the term διάκονος occurs in instructions and rules of behaviour,[3] in reference to waiters engaged in duties other than attending directly on the needs of diners at table,[4] and in a series of incidents in which waiters happen to have been involved.[5]

Among the latter passages is the one where Xenophon recounts how tyrants safeguarded themselves against poisoning by having the waiters taste the food and drink (*Hier.* 4.2); when Athenaeus cites this passage, "waiters" appears not in the form of the common noun but as the participle of the verb used as a noun: "they order those attending [τοῖς διακονοῦσι] to have the first taste" (Ath. 171f). Athenaeus maintains this point of style in his own account of custom surrounding ancient symposia, "never was a slave the attendant" (δοῦλος οὐδεὶς ἦν ὁ διακονήσων, 192b). Such usage is not merely a trait of Hellenistic Greek because in the fourth century BCE Dieuchidas used the participle in expressing a similar thought, "free men are the attendants [οἱ διακονοῦντες]" (*FHG* 4.389), and the evidence of the other references to waiters at work—only eleven in number—indicates that the participle remained the preferred designation.[6] These are minor figures but they would seem to reflect an emphasis in thinking on the activity of waiting or attending rather than an awareness of the status of the waiter as servant or slave. Thus when a character is referred to as τὸν διακονούμενον ὑμῖν Αἰγύπτιον (Achilles Tatius 4.15.4), the sense is rather "the Egyptian attending on you at the time" than "your Egyptian servant."

In addition to what the use of the participle might suggest, the occasional use of a direct object with the verb reminds us that as in the areas of message and agency the activity in view is basically one of fetching. Dio Cassius records that Agrippa attempted to mollify rebels who were plundering the countryside around Rome with the following fable: "Once the limbs of the body had a serious disagreement with the belly, announcing that they laboured strenuously without food or drink to provide everything for it [ἄπαντα αὐτῇ διακονούμενα, 17.10], while it had no work to do yet was alone fully replenished." The image here is of the legs bearing the arms to a source of food and the arms then getting the food to the belly. The same usage recurs in the very different image of the lyric poet lying on a bed of lotus leaves, contemplating death, and issuing his instructions to the young god of love, "Get my wine to me" (μέθυ μοι διακονείτω, *Anacreontea* 32.6).[7]

In a few other instances the turn of phrase seems to suggest of itself the toing and froing of waiters at work. Athenaeus sketches the heroes' symposium in his idealized Homeric times (192f): "The bowls were there with the wine all mixed, and the young acolytes who did the serving [οἱ κοῦροι διακονούμενοι] would make sure that the most eminent always had a full cup." Posidonius (*FHG* 3.260) records that among the Celts "any time a number feast together . . . the attendants [οἱ διακονοῦντες] carry the drink around in clay or silver vessels . . .; the slave takes the drink from left to right and from right to left: that is the way they are served [οὕτως διακονοῦνται].'' A cook, stock figure of the comic theatre, jauntily announces, in a fragment from Posidippus (K. 3.336): "No one will see me lugging meat around any more; something really good just happened to me while I was about my business at my owner's place [διακονοῦντι παρὰ . . .].'' If the connotation these passages seem to suggest is allowed to the words in other passages, we are presented with an array of lively scenes of waiters ducking among the tables at the festival of Delos (ἐλεοῖς ὑποδύεσθαι διακονοῦντες, Athenaeus 173a); of the cook and his assistants (συνδιάκονοι) mingling with the crowds at weddings, funerals, and guild dinners;[8] of a king enthusiastically dispensing the food piled on tables (Bato, FHG 4.349); of a bevy of pretty and well-groomed slaves twice daily pampering the affluent Tyrrhenians (Diodorus Siculus 5.40.3); of Gauls, by the same writer's account, sitting on wolf skins, eating meat from the spit brought to them by their young boys and girls (5.28.4); of Heracleon's soldiers seated on the ground by the thousand with armed guards passing in silence among them with the simple food (Posidonius, FHG 3.265); of garlanded troops relaxing in a desert camp while Armenian boys attend them (Xenophon, An. 4.5.33); of the comely lads in the prime of youth smiling invitingly at the unwanted guest in Lucian's unkind satire of the type (Merc. Cond. 16); of slaves made to sit and feast for their masters to wait on them (Bato, FHG 4.350); of jubilant relatives secretly feasting the veterans of Troy (Plutarch, Mor. 301e); of the backwoods hunter reclining with his wife on a couch of brambles and skins while their daughter brings them the sweet dark wine (Dio Chrysostom 7.65); and finally, perhaps, of Odysseus dogging the steps of Cyclops until despatched to bring him his food (Euripides, Cyc. 406).

These passages cannot be pressed but together, and in the light of the previously noted preference for the participle and of the occasional use of the direct object, they draw our attention more to the kind of activity involved than to any degree of servility. Further a number of passages suggest ceremony. The possibility that a connection with formal occasions might affect the incidence of these words was raised when commenting in an earlier chapter on terms more commonly used by Athenaeus for services at table,[9] and we are now in a position to follow that up more closely.

A Sense of Occasion

The words designate waiters and their duties—among these of course is cupbearing—in the courts of rulers,[1] in the abode of the gods,[2] at dinners of an usually elaborate kind,[3] and at festivals we are about to examine. This pattern of usage of itself speaks of a special character—no doubt reflected in occurrences in poetry[4]—which does not need detailed illustration. The usage establishes that the terms are occasional designations for waiters and their functions on formal occasions. Of greater interest is evidence of the religious character of the usage in accounts of banquets and festivals.

"The God Took Joy"

Long-settled conventions controlled Greek eating, and the more formal the meal the more clearly the conventions found expression. At the same time the food, if not frugal, was plain, and only water or wine was consumed with it. Indeed, at the dinner party or banquet, the food was one thing and the drink another and more important; food and drink were consumed at separate sessions called *deipnon* (dinner) and *symposion* (drinking session), and we shall see what noble expectations could be held of both. We would distort the picture of the Greek dinner party if we made part of it the style of the more elaborate *coena* or dinner of the Romans, who drank with their food, and whose extravagances are paraded for our wonderment and amusement in Petronius's account of Trimalchio's feast in *Satyricon*: hot buttered snails, dormice dipped in honey and rolled in poppy seed, whole hog stuffed with sausages and blood puddings . . . with long demeaning carousals. Before looking at some aspects of Greek custom to serve our own limited purpose, it will be helpful to be reminded of the procedures of the meal by perusing the succinct statement by William Smith a century ago:

> As soon as the guests arrived at the house of their host, their shoes or sandals were taken off by the slaves and their feet washed. . . . After the guests had placed themselves on the couches, the slaves brought in water to wash their hands. The dinner was then served up; whence we read of τὰς τραπέζας εἰσφερειν [bringing in the tables], by which expression we are to understand not merely the dishes, but the tables themselves, which were small enough to be moved with ease. . . . When the first course was finished, the tables were taken away, and water was given to the guests for the purpose of washing their hands. Crowns made of garlands of flowers were also then given to them, as well as various kinds of perfumes. Wine was not drunk till the first course was finished; but as soon as the guests had washed their hands, unmixed wine was introduced in a large goblet, of which each drank a little, after pouring out a small quantity as a libation. This libation was said to be made to the "good spirit" (ἀγαθοῦ δαίμονος), and was usually accompanied with the singing of the paean and the playing of flutes. After this libation mixed wine was brought it, and with their first cup guests drank to Διὸς Σωτῆρος [Zeus the Preserver]. With the libations the *deipnon* closed.[1]

The next set of tables was then brought it, with dessert of fruit and confections, and the symposium began. At this point numerous other guests might arrive, and the wine was drawn from a large bowl where it had been mixed with water to a proportion determined by the master of the occasion. The time passed in conversation, games, music, and dancing.

The Greeks liked to see in such custom the continuity of their life with that of the ancient heroes in whose intercourse with the gods and struggles against adversity they felt their own identity had been established. The past bequeathed to them by Homer was what made them Greek. The present dimension of their past was exposed and celebrated in the religious meal, an exemplar of which is presented in the first book of *The Iliad*:

> When prayers were said and grains of barley strewn,
> they held the bullocks for the knife, and flayed them,
> cutting out joints and wrapping these in fat,
> two layers, folded, with raw strips of flesh,
> for the old man [priest Chryses] to burn on cloven faggots,
> wetting it all with wine.
>   Around him stood
> young men with five-tined forks in hand, and when
> the vitals had been tasted, joints consumed,
> they sliced the chines and quarters for the spits,
> roasted them evenly and drew them off.
> Their meal being now prepared and all work done,
> they feasted to their hearts' content and made
> desire for meat and drink recede again,
> then young men filled their winebowls to the brim,
> ladling drops for the god in every cup.
> Propitiatory songs rose clear and strong
> until days's ends, to praise the god, Apollo.
>   . . . and listening
> the god took joy.[2]

The elements of this ritual are recognisably what the antiquarian William Smith has described: the meat, the wine, the communion with the gods; and throughout the epic these and other significant items like the washing of hands and the mixing of the wine recur to keep us in mind of the mythical fabric of Greek life. In the third century of the Christian era, in his vast anthology of dining, Athenaeus provided further testimony of this faith (363d):

> The ancients thought of the gods in human terms and regulated festivals accordingly. . . . They set the date, and the first thing they did was offer sacrifice; this was their way of setting themselves at ease. When everyone was of the mind that the gods came for the offerings and libations, everyone took part in the gathering with the greatest respect.

Significantly it is only in the context of this religious view of eating in common that Athenaeus employs διακον-, a frequency that is the more noticeable in that in the course of the symposium which is the occasion of all his discourse he has much to say about the bringing of tables, the distribution of food, and the provision of wine; as in the following passage, he would seem to recognise an association between these words and the dignity of the ancient ritual:

> With ancients the only reason for gathering to drink wine was religion, and any garlands, hymns or songs they used were in keeping with this, and the one who was to do the waiting was never a salve [δοῦλος οὐδεὶς ἦν ὁ διακονήσων, 192b]; rather young sons of free men would pour the wine . . . just as we read in fair Sappho of Hermes pouring wine for the gods. . . .
> The bowls stood there filled with mixed wine, and the young lads doing the waiting [οἱ κοῦροι διακονούμενοι, 192f] would keep the cups of those honourable men replenished.

Accordingly, when Athenaeus later reflects morosely that among his contemporaries an appreciation of what dining together should mean had declined, and wishes to illustrate their loss of religion, he accuses them of such impiety in the manner of their address to "the pourer of the wine, the minister, and the cook" (βλασφημοῦσι . . . τὸν διάκονον, 420e) that the very god they ought to be honouring leaves their midst.[3]

Decorum

At another point Athenaeus has occasion to refer to the religious festival of Delos and by way of explaining the origin of the Delian term for those "ministering at the feasts" writes διακονοῦντες (173a), citing then the law of their amphictyons that the water for the occasion be made available by these officials as well as by "the tablesetters and such like ministers [διακόνους, 173b]." And, although he is thoroughly conversant with the vulgar boasts and daring irreverences which the comic poets attribute to the men who did the butchering and cooking on these public occasions, he is also aware that the role of these butchers-cum-cooks was basically ritualistic in character; he points to their competence in sacrifice (659d), and by means of many references to tradition and to the poets—citing Athenion to the effect that the cook's art has done more than anything else to promote piety (660e)—establishes that "the dignity of the cook's craft is of the highest order and recognized by all" (660c). For Athenaeus, therefore, the cook on duty at the festival of Aphrodite Pandemus is appropriately designated as "officiating" at that function (τὸν . . . διακονούμενον μάγειρον ἐν τῇ . . . ἑορτῇ, 659d). This is the manner of the comic poets themselves whom he cites;[1] although with them a sense of piety is not generally to the fore, their usage is nonetheless determined by poetic and probably also religious convention.

The almost rubrical quality attaching to this usage is discernible in Athenaeus's own reference to a particular custom in Sparta:

> In Lacedaemon at the evening meal which they call *aiklon* they carry in a basket of loaves and meat for all who come into the communal mess, and the minister [ὁ διάκονος, 139c] follows the person distributing the portions, announces *"Aiklon!"* and indicates the name of the donor.

In some of the historians Athenaeus draws upon, the same association between διακον- and a ritual ministry at meals can be detected. Thus when he cites a report of Demetrius of Scepsis, again concerning Sparta, that "a shrine to honour the local deities Matton and Ceraon was established by the attendants [διακόνων, 173f] who made the barley loaves and mixed the wine," these attendants are giving expression to the religious connection they understood to exist between their role and the two gods whose names reflect the tasks of "kneading" and "mixing." At the festival of Phorbas, according to Dieuchidas (*FHG* 4.389; *Ath.* 263a), it was "unholy" for slaves to be present, so that only free men were attendants (οἱ διακονοῦντες).[2] Similarly at the communal meals of the Cretans, according to Pyrgio (*FHG* 4.486; *Ath.* 143e), "the youngest stood by in attendance [ἐφεστᾶσι διακονοῦντες] and, offering libations to the gods in sacred silence, shared the food among all." Posidonius describes solemn gatherings of the Celts where a large circle is formed with the most important people in the centre, guards ranged about, and the attendants (οἱ διακονοῦντες, FHG 3.260; Ath. 152b,) carrying the drink in large vessels, serving from right to left (διακονοῦνται, 152d), and reverencing the gods.

Writers not used by Athenaeus adopt a usage entirely consistent with his. Indeed we shall see how effectively the Jewish Philo relies on the usage to illustrate the superior religiosity of his Jewish guild. Theophrastus berates both the kind of man who even at his daughter's wedding is unwilling to share the sacrificial meat with the hired attendants (τοὺς . . . διακονοῦντας, Char. 22.4) and the other who at a gathering of his clansmen for a sacrificial feast keeps a record of how many radishes are left over lest the serving slaves get hold of them (οἱ διακονοῦντες παῖδες Char. 30.16). Plutarch records that the islanders of Aegina, whose contingent had suffered heavy losses both at Troy and on the return voyage, honour Poseidon at a sacrifice which they hold in silence with no slave present in order to commemorate the feasts for the surviving veterans, which families had felt obliged to hold in private out of respect for those who mourned and at which family members themselves waited upon their fathers and relatives (διακονούμενοι, Mor. 301e); Plutarch also suggests that there is more to embodying the spirit of the Homeric feast than cooking, carving, and pouring: there must be the correct style among the attendants (τοὺς διακονοῦντας, Mor. 440b). Similarly Libanius, aware that the gods love the man whose feast is conducted with decency and taste, insists on care in the selection of attendants (Or. 53.9; cf. 28). Xenophon records how on the dreadful winter march through Armenia they came up finally to the vanguard to see the men feasting with makeshift wreaths of hay about their heads—these rude garlands introducing them to what Jack Lindsay calls "the momentary eternity of the rite"[3]—and discovered that they were making do with local boys for the serving who were wearing ordinary clothes of the region and whom they could tell what should be done only in sign language (An. 4.5.33). A timeless celebration of this same faith is the work of art described by Philostratus junior, which depicts one devotee raising his cup in song to Artemis while the other bids the attendant to bring more wine (Im. 4).

Making Friends

Of other instances of our words most occur in contexts that do not include explicit detail of religious custom.[1] The accounts are nonetheless predominantly of occasions where some formality is implied, and this for the Greek meant religious observance. Any disturbance of this ritual through dissipation or bickering meant, as Plutarch warned, that "the very reason for being together is lost, and Dionysus is mocked" (Mor. 615a). Decorum and care with the presentation, he went on, will help ensure that a benevolent spirit is maintained (616d) and that the friend-making powers of the table (612d) are given full scope to express themselves through the fundamental experiences of conversation, drinking the good health of one another, and in friendliness itself (644d). As Hagias avers there, "I don't think we invite one another for the sake of getting something to eat and drink but for eating and drinking together" (643a), and, as always, the inspiration is drawn from the past in the blending and mixing of people at the Pindaric banquets (643e); the joy and goodwill prepared people to be marked, as it were, by "the seal of friendship" (660c), and a gathering should not dissolve before each of the guests has made a friend from among his fellows (660a). An important element in protecting and fostering this harmonious atmosphere is equality of treatment for all, especially is allocating portions of food (643f).

The same practical concern is expressed in one of the laws for drinking promulgated by Lucian in the name of Cronosolon: one is to relax and eat lightly on the morning of a dinner, to bathe at noon, and, at the party, seating is to be allotted by chance, wine of the same quality is to be offered to all, portions of meat shall be of the same size, and "the attendants [διάκονοι]shall show favour to none" (Sat. 17). The Spartans, Plutarch reported, ensured the equality of portions in having their meat distributed by their leading men (Mor. 644b). The measure was not a mere strong-arm tactic to counter squabbling over food—Spartan self-discipline was its own counter to that—but expressed the deep respect for fellowship over food, which found its most striking celebration in masters feasting their own slaves. Lucian's last law of drinking applies here: "When a rich man feasts his slaves, his friends are to serve with him [διακονούντων]" (Sat. 18). In spite of Lucian's generally frivolous attitude to all things religious, he was a keen observer of what he called "the ritual of this craft" of eating and drinking (Par. 22), and in his code of the table he betrays an awareness of the ideal that Athenaeus has sought to uphold in the conduct of dinners and symposia (Ath. 192b); the figure framing Lucian's code is Cronosolon, a priest, prophet, and lawgiver for feasts, whose laws aim to promote harmony and equality at the same time as they make every allowance for pleasurable indulgence.

The ideals of fellowship enshrined in Lucian, Plutarch, and Athenaeus could not be stated so highly without the explicit acknowledgement of the presence of the gods, who were, in A. D. Nock's phrase, the "consecration of conviviality"[2] and could be thought of, as Aristides and the papyri testify in regard to Sarapis, as both guest and host.[3] Their presence made every public festival or, at will, any private dinner party the occasion of religious observance and celebration. Jack Lindsay notes that a favourite place of celebration in Thebes of Upper Egypt was the temple of Sarapis because this heightened "a sense of the beloved god's participation . . . The dinner merges with the communion."[4] The extraordinary custom enshrined in Lucian's law of masters attending on their slaves was deeply entrenched in Greek tradition and widely practised in other cultures, its most famous expression being in the Saturnalia, but the Romans themselves traced this to the Greeks,[5] whose historians record feasts at which slaves were entertained in various ways, most notably by their masters waiting on them at table.[6]

A Ritual Fit for a King

A passage that provides a helpful insight into this part of the Greek experience comes from Bato of Europe, a historian of the second century BCE, fragments of whose work survive in Athenaeus. The passage treats of the origin of the great feast in Thessaly of Zeus Peloris.[1] Bato narrates that in the course of a feast a stranger named Pelorus told king Pelasgus how new plains had wondrously risen from a lake bed. On receiving this news, the king had the festive table set before the stranger, the other guests were instructed to lay their best upon it, and the king himself attended on Pelorus and was assisted in this by leading members of the gathering. The words used here for the guests' part in laying food before the stranger are the normal words we met for such actions in an examination of Athenaeus's *Deipnosophistae* 331–410 (φέρειν, παρατιθέναι), but the words used for the role of the king in attending on the stranger and for the nobles supporting him are the following (Ath. 639f): "Pelasgus was most attentive in his ministry [προθύμως διακονεῖν] while the people of noble estate among the rest of the gathering gave him assistance [ὑπηρετεῖν]." The word designating the king's "ministry" is thus distinguished from the much less religiously coloured word, which is here used to signify the nobles' assistance to the person of higher rank rather than their attendance directly on the singular guest.[2] The king's action also marks the climax of the occasion and inspires the inauguration of a commemorative ritual, as we learn from Bato's further description of how the event was celebrated in contemporary Thessaly. In the course of sacrifice to Zeus Peloris, there was an unstinted manifestation of "kindly fellowship" with strangers being taken in, prisoners freed, and heads of households following the precedent set by Pelasgus of pious memory, "the slaves reclining and feasting with gusto while their masters attended upon them [διακονούντων αὐτοῖς τῶν δεσποτῶν]" (Ath. 640a).

The ritualistic character of the action so depicted in Bato's account is recognisable also in the imposing inscription from the first century BCE that Antiochus I of Commagene left in his remote mountain kingdom in Asia Minor (*OGI* 383). The inscription on his sanctuary at Nemroud Dagh celebrated in an authentic Hellenistic style the king's title of "God truly manifest" (Θεὸς Δίκαιος Ἐπιφανής) and then legislated for cult in honour of the gods, of his ancestors, and of himself.[3] Because his birth had been an epiphany, the decrees are themselves a divine revelation, and they establish in detail the form of cult for the anniversary of the epiphany and of his coronation. The sacred nature of the ministers and of the paraphernalia is emphasised throughout. The priest's Persian vestments originated in the king's divine ordinance: the priest crowns in gold those whom Antiochus has consecrated to the cult of the gods, then, in the midst of incense, lays the gifts on the altar and performs sacrifice in honour of the gods and of the king. The priest then lays food and wine on the sacred tables, reverently invites those present, both native and stranger, to the communal feast, and, having taken his priestly portion, dispenses liberally to the others. This part of the decree then concludes: "For as long as they are part of the gathering within the sanctuary, they are to be served [διακονείσθωσαν, *OGI* 383.159] from the vessels which I consecrated."

A Guild of Worshippers

Philo adopts this almost technical use of the verb in describing Jewish ritual for non-Jewish readers. The last half of Philo's account of the Therapeutae is devoted to a comparison of their communal banquet with the Greek dinner party. In the description of the latter (*Vit. cont.* 40–63), the emphasis falls sharply on its profanity; it is the occasion of bitter rivalries in contravention of the libations (41), it is held almost daily and is marked by an extravagance in furnishings, attendants, and in food and drink (46–54); its conversation is frivolous or debased, and the whole occasion inspires nothing but disgust (55–63). By contrast, the gathering of the Therapeutae is an expression of the worship for which they live (64; cf. 2, 10–13); it is held only once in the year, being "a major feast" (65; probably Pentecost), and is marked by joyful simplicity in its furnishings, attendants, and nourishment (66–82); it is an occasion for pious discourse, and ends in a choral vigil as it begins in prayer (83–90).

Although this summary makes it clear that the two institutions are not strictly comparable, Philo goes to some lengths to find exact parallels between the two. Thus, they are both called symposia, although the meat, wine, and ribaldry of the one are hardly the same as the bread, water, and silence of the other. Close attention is paid to the attendants. For the Greek, these are slaves, boys and youths selected for their beauty, and sumptuously adorned (50–52); the Therapeutae, on the other hand, abhor slavery as contrary to nature, and use only young men from among their number who are outstanding for virtue (70–72; cf. 81). In the description of the duties of these attendants in both accounts, customary terms for service at table regularly appear, but in both accounts also a subtle emphasis falls on carefully placed instances of our words. Thus section 50 on the Greek attendants begins with the phrase "ministerial slaves" (διακονικὰ ἀνδράποδα, 50), and this is balanced by the phrase "ministering is not done by slaves" διακονοῦνται δὲ οὐχ ὑπ' ἀνδραπόδων, 70), which opens the parallel section on attendants of the Therapeutae. At this point Philo is also adverting to the Greek ideal we have encountered that slaves render a gathering unholy (Dieuchidas, *FHG* 4.389; Athenaeus 263a) and thus impervious to divine presences, and on such a crucial matter asserts boldly in favour of the Therapeutae, "in this holy symposium there are no slaves, and service is carried out only by free men who fulfil the ministerial duties [τὰς διακονικὰς χρείας, 71] with a willing heart.

In all probability Greek readers who came across Philo's apologia for the ascetic ritual of the Therapeutae and his savage appraisal of one of their own central cultural institutions would sidestep the criticism by pointing out that the two meals were not really comparable; guests who are expected to spend the night in prayer after a meal of bread and water are just not symposiasts. They would acknowledge, nonetheless, that his account cleverly manipulated the forms and language of their own way of life, and would recognize in the attendants (διακόνους, 75) arrayed in a circle about the reclining participants the image of their own festive ministers.

Temple Liturgies

On a number of occasions Josephus also expresses Jewish ritual in these ritual terms of the Greeks. Interestingly, he seems to prefer them to the terms available to him in the source he was using, the Greek accounts of the Septuagint Bible. His usage may be seen as a minor illustration of what an eighteenth-century scholar had in mind in saying that Josephus was most solicitous when reporting religious matters of the Jews not to offend the sensitivities of his Gentile readers.[1] Recounting Josiah's reforms, he refers to the priests as "ministering to the crowds" (διακονουμένων, AJ 10.72) in the sense of distributing to them a share of the sacrificial meat, while his sources in Esdras and Chronicles used several other expressions.[2] The same meaning may be intended when a priest is said to attend to the sacrifices "and to minster" (διακονεῖν, AJ 3.155), although here the biblical account (Ex. 28) does not include a reference to the distribution of food, so that the sense may be a more general one of performing a religious rite, as in the record of the liturgical duties ascribed to the priestly families who are said "to minister for God" (διακονεῖσθαι τῷ θεῷ, AJ 7.365).[3] The idea of performing a cultic action is not expressed by the verb in any other writer than Josephus,[4] and is put to striking use in a story relating not to the liturgy of the Jewish temple but to the rape of Paulina in the temple of Isis. Paulina was a chaste woman devoted to the cult of Anubis—god of death whom Isis had preserved at a time of peril—and she received her husband's consent to pass a night in the temple as an expression of her devotion; there, however, with the connivance of the priests, Decius Mundus, who was in love with her, lay in wait. Thinking he was the god, Anubis, Paulina gave herself to him in sexual intercourse, Josephus reporting "all night she ministered to him [αὐτῷ διακονήσατο], thinking he was a god" (AJ 18.74).[5]

The abstract noun also designates ritual in the phrases "vessels for use in the *ministry* of the sacrifices to God" (AJ 8.101) and "libation cups and bowls for the *ministry*" (AJ 10.57). We are probably to understand the same when Hannah promises to consecrate Samuel ἐπὶ διακονίᾳ τοῦ θεοῦ (AJ 5.344: "to the service of God").[6] Similarly, in reporting on the annual sacrifice offered by the citizens of Plataea to those who had died for the freedom of Greece, another author, Plutarch, writes that only free young men were permitted to take part in the *ministry* of carrying the jars of wine, milk, oil, and unguents in the procession to the tombs (*Arist.* 21.4).[7] The cognate noun διακόνημα signifies "sacred vessel" in an early writer like the Cynic Diogenes (*Ep.* 37.3), "sacred task" of building a temple in the later writer Dio Cassius (65.10.2), and "sacred ritual" later still in Julian (*Or.* 2.68c);[8] at *Ath.* 274b this term designates "vessels"—earthenware or bronze as against the precious metal indulged in by some—at a meal among Romans, in a statement that aims to establish that Romans are conscious of the ancient piety they have inherited.

The Charity of Job

Like Josephus and unlike the authors of the Septuagint the creator of the Jewish pseudepigraphical *Testament of Job* does not avoid these special Greek terms but makes effective use of them in Job's account of his benefactions to the poor at the time of his affluence. With his vast herds and flocks and a landed estate extensive enough to work thirty-five hundred yoke of oxen, he never for a moment neglected to extend his liberality to the needy; for the strangers among them he permanently reserved thirty tables for use at any hour, and from his fifty bakeries he provided for the table of the poor.

When some outsiders saw how generous he was, they sought to have a role themselves in the ministering (of food; ὑπηρετῆσαι τῇ διακονίᾳ, 11.2); from the plight of the next group of helpers we are able to infer that this role included providing additional funds, and from the verb we may surmise that it consisted in overseeing the buying in of supplies and in other administrative aspects of the massive operation; other helpers, at any rate, who were without funds of their own but wished to achieve the same ministry (ταύτην ἐκτελέσαι τὴν διακονίαν, 11.3), borrowed capital from Job so that, as we have seen in an earlier section (11.4), they might undertake trade missions in the name of the poor and make the profits over to them; others again could hope to do no more than minister to the poor at Job's table (διακονῆσαι τοῖς πτωχοῖς ἐν τῇ σῇ τραπέζῃ, 12.2). From the grading of these ways of being involved in Job's programme, we see that only with the verb at 12.2 do we have helpers actually serving at tables; at 11.4 the reference is not to table at all; and at 11.2 and 11.3 the reference must be broader than service at table because in the latter instance Job's collaborators propose to take part in "the same ministry" by going abroad to raise funds. The reference to tables is more direct at 15.1 where Job avers that his own children ate only after the conducting of the ministry (τὴν ὑπηρεσίαν τῆς διακονίας, 15.1); at that time they joined their older brother, the daughters being attended by maidservants, and the sons by ministering slaves (τοῖς . . . δούλοις τοῖς διακονοῦσιν, 15.3). Job's offspring were nonetheless a worry to him so that early each morning he offered sacrifice of three hundred doves, fifty kids, and twelve sheep for transgressions they may have been guilty of, especially that of pride before the Lord in saying, "We are a rich man's sons, . . . so why should we serve [διακονοῦμεν, 15.8]?"

From Job's testament, then, it seems that our words are special significations of tasks associated with the distribution of food among needy strangers. The words are for the Jewish writer as distinctive a signal of particular values in regard to the table as the phrase [ἡ] διακονία[ ] τραπέζων ("the ministry of tables") was for the emperor Julian in respect of Christian values in their agape or love feast (305d). The Jewish scene is also strongly reminiscent of Greek festivals celebrating philanthropy, like that of Zeus Peloris (Bato, *FHG* 4.349–50), only here in the munificence of his liberality Job is far outdoing the kindly extravagances of the Greeks. Perhaps, just as we have felt that other Hellenistic Jewish writers may have pointed to a superiority of Jewish values over Greek values in matters of love (*Test. Juda* 14.2; *Joseph and Aseneth* 15.7) and of eating together (Philo, *Vit. Cont.*), this author likewise is unashamedly championing before Hellenised Jews or Greeks themselves the Torah's uncompromising call for the care of widow and orphan. Significant too for such an apologetic standpoint is the place of sacrifice in the two culturally different expressions of philanthropy. For the Greek the philanthropy expresses a fellowship with the gods that is demonstrated and celebrated in the meal shared both with them and with the strangers; for the Jew fellowship is a duty to the god who requires it and who is to be placated for its neglect by sacrifice.

More Lasting Than Bronze

In monumental inscriptions of the last three centuries BCE the common noun appears sixteen times among titles affixed to people taking part in religious gatherings of both a public and a private nature.[1] Of the material published in his time Edwin Hatch said in the Bampton lectures of 1880 that the name "seems to have been specially applied to those who at a religious festival distributed the meat of the sacrifice among the festival company."[2] Thirty years later Lietzmann similarly spoke of these people as waiters, although in the cases of *CIG* 3037 and *MB* 93 and 100 he spoke in more general terms of ministers of cult ("Kultpersonal").[3] About the same time, in an extensive examination of Greek guilds, Poland was not willing to be so precise and bracketed the word with other words from the lists ($\pi\alpha\hat{\iota}\delta\epsilon\varsigma$, "minors"; $\dot{\upsilon}\pi\eta\rho\dot{\epsilon}\tau\alpha\iota$, "assistants"), seeing them all, again in general terms, as ministerial assistants in cult ("Dienerschaft"); he also suggested that the Christian title "deacon" may have had its origin here.[4] Others also speak generally of attendants or officials in temples or religious guilds,[5] although Guerra insists that the meaning of the title is not uniform throughout the inscriptions and ranges from being a designation of devotees of a particular divinity to being the mere equivalent of the profane domestic servant.[6] The following examination of the material proposes that there is no reason to see anything more in the word than the designation of a ceremonial waiter.

The meaning of "waiter" is easily established in those lists commemorating public feasts or guild banquets in which the $\delta\iota\dot{\alpha}\kappa\text{o}\nu\text{o}\varsigma$ is mentioned in close proximity to the $\mu\dot{\alpha}\gamma\epsilon\iota\rho\text{o}\varsigma$ or butcher and cook whose role we touched on in relation to Athenaeus 659d.[7] That the function was a minor one is indicated by the position accorded to the title in the lists, where it is either last of some dozen to twenty other titles or is followed only by titles of the cupbearer, the bearer of sacred vessels, the junior affiliates, or children, or the $\iota\epsilon\rho\text{o}\theta\dot{\upsilon}\tau\alpha\varsigma$ (ritual slayer?). At the head of the lists, by contrast, are the titles of presidents of guilds, their treasurers and secretaries, and local magistrates. In one instance (*IG,* 2d ed., 451) the names of the $\delta\iota\dot{\alpha}\kappa\text{o}\nu\text{o}\varsigma$ and cupbearer are inscribed in a hand different from the hand that inscribed the titles and other names on the stele, and also, as happens occasionally elsewhere, without the patronymic which is the style for the civic dignitaries and other figures higher in the list, both signs perhaps that these two were of lesser consequence than their associates, even to the extent, might we say, that their names had been forgotten and could be added only at a later date to assuage who knows what hurt feelings. The order of the titles varies among the inscriptions, and from that alone it is most unlikely that the titles refer to any kind of hierarchy.[8] The majority of the lists record no διάκονοι at all—none for example occurs in the numerous lists from Olympia, where the cook and cupbearer appear frequently—nor can the number be made up by taking ὑπηρέτης as a synonym,[9] for that is used in a technical sense approximating to "clerk" attached to the γραμματεύς or secretary.[10] The word does not occur in such documents as rules for guild banquets,[11] or in those dealing with the Hellenistic religious associations of Egypt or the popular Kline of Sarapis.[12] Given the commemorative nature of the inscriptions, it seems that this word, which is comparatively rarely attested in literary sources but is there almost always a formal or religious term, was judged to be suitably honorific; in this it may be compared with some of the other titles which are elaborate even to the point of pretentiousness.[13] The role itself may have been to an extent ceremonial, and the hard work of getting around among the affiliates may well have been at the expense of the juniors. As an "office," it was of course ephemeral.

Some inscriptions require particular comment. The first of these is *CIG* 1800, which is dedicated to Sarapis and four other Egyptian gods by τὸ κοινὸν τῶν διακόνων, and the list of names begins with that of a priest (ἱερεύς), continuing with those of eight men without any further indication of title. Guerra takes διακόνων as a synonym of θεραπευτῶν in that word's sense of devotees of a god and, noting particularly the presence of a priest, speaks of the "group" in terms of a religious community.[14] Several difficulties attend this interpretation. Although there is ample evidence of a κοινὸν τῶν θεραπευτῶν dedicated to the cult of Sarapis,[15] there is none to indicate that διάκονοι were ever so called because they were devotees of a god and engaged in cult; in literary sources, the διάκονος of a god is the god's messenger or agent. Nor does the word κοινόν necessarily signify a tightly knit organization, for in this connection the term means anything from casual gatherings for a particular occasion to corporate bodies of a religious or political nature;[16] in this instance Poland is very doubtful about a technical meaning ("society"), and allows for the possibility that the διάκονοι might be no more than an informal grouping of cult personnel like the συμπορευόμενοι or fellow marchers of a procession who also inscribe themselves as a κοινόν.[17] Finally, the presence of a priest is very likely to be irrelevant to the status and function of the διάκονοι because a priestly figure, designated by any number of terms, often identical with municipal titles, is recorded in many inscriptions simply as an occasional president.[18]

Some kind of parallel is probably to be seen in the act of piety reported by Demetrius of Scepsis (Athenaeus 173f) when the διάκονοι of the Spartan messes raised a religious shrine to heroes connected in tradition with their functions. Similarly, the dedication in *I Mag.* 217 records that, during the term of a certain priest-magistrate, the *komaktores,* the heralds—as we have noted in connection with *IG* IV.824a, their title can cover a reference to the cook—and the διάκονοι have set up a hermes or pillar capped with the head of the god; the first named of this small group (the Latin "coactores"?) may have raised the subscriptions for the statue, while the two last would have officiated in the sacrificial banquet at its dedication; the priest may merely have honoured the occasion by his presidency.

A more public function seems to have been the occasion of a thank offering to the mother goddess (*Michel* 1226). The inscription names the hipparch or cavalry commander in office at the time, then two leading municipal or borough officers (the διοικητής, being the chief administrator, and the secretary), then five men listed under the common title διάκονοι, and finally one οἰνοφύλαξ or keeper of the wine. The latter may have been a permanent municipal officer but his presence suggests that he was also responsible for the wine at the religious ceremony; this in turn creates a suitable context for the five men before him on the list, who would thus be neither assistants to the borough secretary[19] nor members of any kind of hierarchy, but individuals fortunate enough to have their role in the festive banquet commemorated.

Finally, in *CIG* 3037 we have two different but related lists consisting of, in each case, the priest of "twelve gods," who is the same person in each inscription; one priestess, who in the second instance is a person identified as the priest's mother; one female διάκονος, who in the first instance is the same priest's mother; and two male διάκονοι, who are a different pair in each inscription but in the second are identified as the priest's brothers. The pattern of family relationships led Boeckh in *CIG* to speak of hereditary priests and ministers. Comparable with these two inscriptions are *MB* 93 and 100, the first in particular recording a priest, a priestess who is his mother, one female διάκων,[20] and five male διάκονοι. Each of these lists might seem to suggest that the διάκονοι were attendants at a shrine with general cultic duties or even that they were ministers within a hierarchy; if so, the usage would be unique. A more likely interpretation is one based on the function of ministering to participants at sacrificial banquets. The fact that a family connection can be traced between some of these ministers or that one or other like the priest's mother may have graduated to a priestly role need not imply that they all belonged to a special ministerial caste; the case could be that the heads of family may simply have been able to confide an occasional and privileged role to their own dependants at the presumably religious feasts being commemorated.[21]

8

A Question of Diplomacy

An entry in the thesaurus of the Greek language we have been referring to as Stephanus—and the entry goes back beyond the nineteenth-century editors to Henri Estienne himself in the seventeenth century—raises the question whether our words are ever technical terms for officebearers and their functions in diplomacy between states. Our earlier survey of the words within the area of message might appear to have left a door open for such a further specialized application of the words, which could then, as Stephanus himself points out, be seen as being reflected in Paul's phrase about "the ministers of God" (2 Cor. 6:4, *AV*) in the role of bearing the message of the gospel. Stephanus instances only Thucydides 1.133 and Pollux 8.137, 138 as diplomatic language and provides readings like "embassy," "envoy," "to act as an ambassador." His interpretation arises from the idea that a legate carries out the commands of a prince or civic authority with the result that this service *(ministerium)* on behalf of such a public authority is in reality a public office *(officium; munus);* thus a phrase in Thucydides—which we shall be examining in context but for the moment can cite only in the Greek—ἐν ταῖς πρὸς βασιλέα διακονίαις (1.133) is taken to refer to "the public office of an embassy" ("de Legatione, quae et ipsa munus publicum est," 2.1184a). The idea underlying the use of the common noun in the sense of envoy is presented in a similar way (2.1186d): "just as the head of a household sends his servants to different places with his orders, so a head of state or the state itself despatches an envoy." In other words, on Stephanus's understanding this so-called diplomatic use is not related to the kind of material we have examined in the area of message but it is a figurative application (he says "metaph.") of uses in the area of domestic service; the thesaurus does not in fact register any instance of "messenger," "going on a message," and so forth.

"Authorised Representative"

Stephanus's suggestion that his diplomatic sense might be significant for a deeper understanding of early Christian phrases about "ministers of the word of God" was taken up in a work by Dieter Georgi, which has already been referred to more than once as a novel treatment of the words for these decades.[1] Noting what Stephanus had overlooked—that in Aeschylus, Sophocles, and Aristophanes διάκονος means "messenger" ("Bote")—Georgi translates "envoy" ("Gesandter") in the instances adduced in Stephanus; turning then to Epictetus, he finds in the description of the wandering Cynic preacher as "minister of God" (διάκονος θεοῦ, to which is to be compared the Christian expression "ministers of Christ," 2 Cor. 11:23, AV) the idea of a religious or godly envoy: "the envoy of God in the sense of being his authorized representative."[2] Thus the word διάκονος is presented by Georgi as a technical term shared by Cynic preachers and early Christian missionaries, an area where other scholars had previously been able to advance only parallels of a general nature.[3]

Such a point of linguistic contact between what Georgi describes as rival religious systems would have interesting implications for the history of the early Christian mission, and Georgi has amply illustrated their scope in respect of Paul's experience as registered in passages like 2 Cor. 11–13; what he has made of them may not have proved universally acceptable to other scholars but the linguistic point itself has been widely conceded.[4] The main question to be addressed at this stage, accordingly, is whether we would be correct to follow Georgi in seeing "the διάκονος of God" in Epictetus as God's "envoy"; preliminary to this, however, is the question whether Stephanus had grounds for seeing in Thucydides and Pollux the idea of envoy in a profane diplomatic sense.

Thucydides 1.133 we have already examined in sufficient detail to establish that the Spartan regent Pausanias has been despatching "messengers" on "missions" to the Persian king. The terms here are those used by the last of the regent's couriers in the very act of betraying Pausanias to the Spartan ephors. If we look more closely for indications in the passage of the status of this courier, who was playing for his life, we find that he was a man without a name, probably very young and certainly a one-time "favourite" (παιδικά) who had also once been a much trusted member of the regent's entourage; he is the last of a number of couriers, the rest of whom had been eliminated at the completion of their missions, and his only role is to carry a letter of whose contents he is supposed to remain ignorant. These are not the marks of an ambassador. As proposed earlier, the two sole instances of the words in Thucydides are likely to have been derived from a Spartan source and to reflect its legal or conservative style; they are not descriptive of diplomatic status or activity.

The other passage adduced by Stephanus is from *Pollucis Onomasticon,* a philological work whose tortured textual history makes it difficult for us to be confident what the mind of this second-century grammarian may have been in regard to specific points of language.[5] On diplomatic terminology his listing at 8.137 begins with the cluster Πρέσβεις, πρεσβευταί, πρεσβεία. These are the conventional terms for emissaries representing states in dealings with their foreign counterparts, the first two terms meaning equivalently "ambassadors"—the Greeks practised multiple representation—and the third, "embassy." A century ago Franz Poland showed that subsequent to Herodotus, who in the diplomatic arena shared an archaic style with Homer and the tragic poets, no ambassadors or diplomatic missions of Greek states were designated by other than these terms.[6] The list in the *Onomasticon* continues through a series of cognates to conclude with the statement: ὁ δὲ πρεσβευτὴς εἴη ἂν καὶ ἄγγελος διάκονος, which I translate, "The term πρεσβευτής will also appear in the sense ἄγγελος and διάκονος." To translate in this way is to understand the statement as saying that the two new terms introduced here by Pollux are not to be taken as synonymns of "ambassador" but indicate in what ways one of his terms for ambassador can also be a designation of persons acting in precisely a nondiplomatic capacity.[7] To understand the statement, on the contrary, as extending the diplomatic terminology by two items would lead us to anticipate encountering the terms in the language of diplomacy, whereas we do not. In the instance of ἄγγελος, in the first place, Poland further illustrated that the word did not have a diplomatic value for Greeks, who tended to reserve it as a designation of emissaries of foreigners. Accordingly, if Pollux has aligned the word with διάκονος, it must be for a purpose other than illustrating diplomatic language. In fact the two words are equivalent only in the area of message (as in Sophocles, *Ph.* 500; cf. Aeschylus, *Pr.* 942–43), and there, especially, message for a god, so that it is likely to be this area that Pollux's entry is meant to cover, an area where, as Bornkamm has observed in respect of πρεσβευτής, there is no question of plenipotentiary powers of a diplomatic kind.[8] In respect of διάκονος and its cognates, secondly, our literature provides no trace of a technical use.[9] What we do notice, however, when πρεσβεία and διακονία occur in the same context is that the latter term either has an entirely nontechnical sense[10] or, more tellingly, is set in contrast with the diplomatic signification of the former.[11] διάκονος and its cognates are not part of the terminology of diplomacy.

Stoic Attitudes

If we were to find Epictetus championing the Cynic missionary in these terms as God's envoy, we would thus have to say that his usage was unprecedented. Georgi, we notice, recognised that in one instance (4.7.20) the reference is to the characteristically Stoic idea of doing God's will,[1] and we are going to see that in the language of this Stoic teacher this is the exclusive reference of the common noun in this passage as at 3.24.65, 3.26.28, and 3.22.63—that is, without any connotation of the designated person travelling on a mission from God. Far from meaning "envoy," the term does not designate a messenger of any kind. By contrast the abstract noun, which occurs only once in a discourse on Cynicism (3.22.69), does designate a commission in the area of message, this being the Cynic's call to be a travelling preacher. In the light of this the Stoic context of all Epictetus's thought is to be noted. He did not himself adopt Cynicism as a vocation (4.8.43) or advise others to adopt it (3.22.107), nor would he countenance any pretensions to Cynicism that were not founded on a sound Stoicism (e.g., 3.22.93–95). His picture of the Cynic in 3.22 is idealistic; he was aware of the Cynic mobs of the time and found their poses and uncritical protest repugnant (3.22.9, 50, 80).[2] Only when he says so, accordingly, is Epictetus speaking of the Cynic rather than about his own way of life, which was Stoic.

Two Stoic ideas put his use of the term διάκονος in perspective. The first of these concerns knowing one's place in life (τάξις and χώρα: station, position).[3] The world constitutes a commonwealth of god and man (1.9.4), and man is of the stock of the gods (ibid., 22); men, however, are like children before the great truths (3.19) and have great difficulty comprehending their place in the great system (3.24.92). The first responsibility of the Stoic, therefore, as "the good and excellent man," is to come to an understanding of his place (ibid., 95): "the good and excellent man, bearing in mind who he is, and whence he has come, and by whom he has been created, centres his attention on this and this only, how he may fill his place in an orderly fashion, and with due obedience to God." To conduct his life in accord with this knowledge, the Stoic must call into play his highest faculty (2.23.27) and subordinate his will to him who administers the universe (1.12.7), cultivating his moral purpose "so as to make it finally harmonious with nature" (1.4.18). The process will involve him in a severe schedule of moral education, during which he will understand that in the hardships he endures God is training him (3.12).

When the Stoic has achieved his conformation to the natural order of things (4.5.5), a second idea operates, and it is the idea of giving witness to one's situation in life. Whereas in other Stoics, however, the idea is of witnessing to Stoic doctrine after the manner of the legal advocate, in Epictetus it is the idea of witnessing to God himself, and the witness is given not in words but in the way that Stoics live, so that they are "examples to the uninstructed to prove that He both is, and governs the universe well" (3.26.28).[4] The hymns of praise the good man sings "to one or to many" (ibid., 30) are not proselytising sermons but are a praise of providence in the hymn of life itself which all men should sing "publicly and privately . . . as we dig and plough and eat" (1.16.15–21). A public commitment to preach, we discover, is no part of the Stoic's vocation. Indeed, mindful of the maxim "Lions at home, but at Ephesus foxes" (4.5.37), the Stoic is dissuaded from preaching for it smacks of the boast (1.29.64):[5] "Must I proclaim this to all men? No, but I must treat with consideration those who are not philosophers." Instead of committing himself to preaching, therefore, he will realise that the good life is itself an exalted office (ἀρχή, 3.24.117), and that he can authenticate it among men by the quality of his own living (ibid. 118; cf. 3.13.12): "Only make no display of your office, and do not boast about it; but prove it by your conduct; and if no one perceives that you have it, be content to live in health and happiness yourself."

Thus the witness of deeds, rather than of words, is integral to the Stoic life. The tract "On Steadfastness" (1.29) is insistent on the point; God says, "Go you and bear witness for me" (ibid., 47) but the witness is not to be through words—

the writings of Stoics are full of those; what is needed is "the man to bear witness to the arguments by his acts" (ibid., 56).

Civil Servants and Agents in God's Commonwealth

It is against the background of these two ideas about taking up one's true place in life and of giving witness through living in that station that we are to understand Epictetus's use of the two words διάκονος and ὑπηρέτης. These words and their cognates speak of creatures being the instruments of the immanent divine purpose for each; animals are born inherently equipped for their roles (1.16.4; cf. 2.8.6) just as the sense faculties of humans have a naturally instrumental role in respect of the higher human faculty (2.23.7–11).[1] This higher capacity sets humans apart so that they are not in the service of any other creature (2.10.3) but are established in service under God (1.9.16), which is an absolute human condition and more than subservience,[2] being in fact an active participation in the commonwealth of God and man (ibid., 4–5). A further brief discussion will show how within this concept the meaning of one of the two common nouns in question is to be differentiated from that of the other.[3]

As already mentioned, man's service under God (ὑπηρεσία) is conceptualised as an office (ἀρχή). This is possible because, as well as enjoying divine friendship through kinship with the gods, man is understood to share in the divine work of the world's governance. Man is thus by nature God's officer in the cosmos ὑπηρέτης, 3.22.95).[4] The ethical implication—necessarily to the fore with the Stoic teacher—is that each individual is to be the executive in his or her life of God's design for human conduct. This teaching is expanded in the dissertation on the central Stoic doctrine, "That we ought not to yearn" (3.24). Because man is able to comprehend the harmony of the universe and to compose within himself the discords of life's griefs and loves (ibid., 1–22; 38–66), he can if he chooses conform himself to this pattern, taking his moral stance in his God-given station in life and ready to die a thousand deaths before abandoning it (ibid., 99); living or dying, he carries out the duties proper to him as a free man and as God's officer (ὑπηρέτης. ibid., 98). His consolation in the midst of the labours that this commitment brings him is that he is exercising his office to the full (ibid., 113–14; 117–18):

> He brings me here, and again He sends me there; to mankind exhibits me in poverty, without office, in sickness; sends me away to Gyara, brings me into prison. Not because He hates me—perish the thought! And who hates the best of his [officers: ὑπηρέτων]?[5] Nor because He neglects me, for He does not neglect any of even the least of His creatures; but because He is training me, and making use of me as a witness to the rest of men. When I have been appointed to such a [commission: ὑπηρεσίαν],[6] am I any longer to take thought as to where I am, or with whom, or what men say about me?
>
> . . . [So what of you] who have received so important an office [ἀρχήν] from Zeus? Only make no display of it and do not boast of it; but prove it by your

conduct; and if no one perceives that you have it, be content to live in health and happiness yourself.

Occasionally in discussion of this kind the word διάκονος also occurs to much the same effect as ὑπηρέτης, although the two words are not entirely synonymous. Where the latter is here associated with the idea of office and draws on the notion of civil service, the former is associated with the idea of domestic attendance and draws on the notion of going forth to do the master's will; the word thus implies no less a responsibility than that of the officer of government to get things done or to carry out a programme. So we read in the discourse "On Freedom from Fear" (4.7.20): "I wish rather the thing which takes place. For I regard God's will as better than my will. I shall attach myself to Him as a servant and follower [διάκονος καὶ ἀκόλουθος], my choice is one with His, my desire one with His, in a word, my will is one with His will." This is not merely, or even primarily, an expression of a private spiritual disposition before God but is a clear statement of resolve to act constantly in the manner indicated by God; by embracing events—"the thing which takes place"—the Stoic will identify himself with the divine will which is the cause of events. In similar fashion, addressing those who experience anxieties in their worldly affairs (3.26), Epictetus observes that they are not called to give a witness to success in life but to an acquiescence in the way the world is conducted by God (ibid., 18), who does not neglect "his prime creatures, the agents [διακόνων] and witnesses" (ibid., 28), namely, those creatures of his—human beings—who carry out the immanent purpose of the universe and in so doing demonstrate "that He exists and governs the universe well" (ibid.).

Of the two other instances of this noun in Epictetus, one occurs in the Stoic discourse "That we ought not to yearn" (3.24.65) and the other in the discourse "On Cynicism" (3.22.63), both in reference to the Cynic as the exemplar of the Stoic principle that should govern the relationship of love. Because affection more than anything else in human experience is likely to take such a hold on man that he depends on "something other" than himself (cf. 3.24.58), Epictetus illustrates by way of Socrates that, provided man remains first and foremost a friend of the gods, love need not restrict the freedom necessary for the exercise of true morality (ibid., 60). Choosing then the more pertinent example of Diogenes the Cynic, he asks (ibid., 64): "Come, was there anybody that Diogenes did not love, a man who was so gentle and kind-hearted that he gladly took upon himself all those troubles and physical hardships for the sake of the common weal?" Underlying the answer to this grand query was the knowledge that God had trained Diogenes to a pitch where neither circumstances nor people could impair his freedom to act, think, and feel like God's man (ibid., 56–59), so that even in loving with true concern he loved within limits set by God (ibid., 65): "As became a servant of Zeus (τοῦ Διὸς διάκονον), caring for men indeed, but at the same time subject to God."

In the discourse "On Cynicism" (3.22) Epictetus brings this thought to the attention of a young man who wonders whether it would be correct for a Cynic who had fallen ill to accept an offer of hospitality from a friend (ibid., 62); instead of answering directly, Epictetus points out that the Cynic has no friends except those who share his principles, which owners of houses patently do not (ibid., 63): "But where will you find me a Cynic's friend? For such a person must be another Cynic, in order to be worthy of being counted his friend. He must share with him the staff and the kingdom and be a worthy servant [διάκονον] if he is going to be deemed worthy of friendship."[7] This reply takes up the thread of the discourse prior to the young man's intervention. Then, Epictetus was disabusing young men of their illusions about the Cynic life by giving them an uncompromising account of its priniciples, aims, and severity. A common misconception on the part of one with pretensions to the life is that one has only to take up the cloak, wallet, and staff—symbols of Cynic life—to find oneself provided with all one's needs (ibid., 50), whereas the true Cynic is fashioned within by God's hard training, which makes of him a man who fears no ruler of this world (ibid., 56) and whom others recognise as "king and master" (ibid., 49), one who is "worthy to bear the staff of Diogenes" (ibid., 57). To be a friend of such a man one must wield the same spiritual staff of total detachment, come under the authority of the same inner kingdom, and embody that kingdom's moral imperative in being its faithful agent.

As applied to the Cynic, therefore, the term διάκονος (3.24.65; 3.22.63) says no more than in Stoic doctrine (4.7.20; 3.26.28), and in either of these contexts it is but a variation of the term ὑπηρέτης, both terms designating a person who acts in total accord with Stoic moral teaching. In this the usage mirrors usage that defines the role of sense faculties in respect of higher human faculties (2.23.7, 8; 3.7.28). In no instance does the term carry the sense of messenger, not even in the context of witness (3.26.28); the witness under discussion there is not given by word of mouth but by the conduct of the agent denoted by this term.[8]

The Cynic's Mission

The tract on Cynicism also includes Epictetus's only use of the abstract noun. The young man to whom the remark on friendship was addressed in the preceding discussion next asks what part marriage can play in a Cynic's life. In an ideal society, Epictetus replies, marriage would play its normal part because all parties involved would be Cynics; because no such society exists, however, but only a society that is a mix of conflicting needs and personalities, the married Cynic would almost inevitably be caught up in domestic responsibilities that would cut across his responsibilities as a Cynic. He says, in the translation we are using (3.22.69):

> But in such an order of things as the present, which is like that of a battlefield, it is a question, perhaps, if the Cynic ought not to free from distraction, wholly devoted to the service of God [τῇ διακονίᾳ τοῦ θεοῦ], free to go about among men, not tied down by the private duties of men, nor involved in relationships which he cannot violate and still maintain his role as a good and excellent man, whereas, on the other hand, if he observes them, he will destroy the messenger, the scout, the herald of the gods, that he is.

These last three designations of the Cynic's role, none of them common words in Epictetus,[1] refer to the Cynic as the wandering preacher; two of the terms are taken up from earlier in the discourse (3.22.23–26) where the Cynic is the god-sent messenger (ἄγγελος) who shows men how to distinguish between good and evil as well as the scout who, like Diogenes, has gone ahead through life's battlefield and is able to bring back to the combatants a reliable account of what is in store for them. These are not the only images under which the Cynic's role is presented; he is also the community teacher and private tutor (ibid., 17), the brother and father who must speak openly[2] to his kinsmen (ibid., 96), and the general who must inspect his troops (ibid., 97). In the main these highlight the Cynic's role as instructor, even though Epictetus does not lose sight for a moment of the first responsibility of this God-sent man to give witness by example: "God has sent you the man," he announces to those who are looking for the secret of a contented life, "who will show in practice that it is possible" (ibid., 46). As herald, however, he is the one who must be able "if it so chance, to lift up his voice, and, mounting the tragic stage, to speak like Socrates: Alas! men, where are you rushing?" (ibid., 26), and the one of whom men will then ask, "tell us, Sir messenger and scout" (ibid., 38).[3]

As both teacher and "good and excellent man," then, the Cynic is called by God to go out among men, and such a vocation is truly a διακονία or sacred commission as we have come to recognise it elsewhere.[4] As a Cynic preacher he is singular among men; as "a good and excellent man," embodying in his life the Stoic wisdom, he lives out a divine commission that differs only in degree from the commission (ὑπηρεσία) or office (ἀρχή) that lies on every man and commits him to a life wholly devoted to God (3.24.114). The difference between these two terms for the divine commission of the Cynic as preacher and the Stoic as creature is that διακονία in its context connotes more strongly than ὑπηρεσία the idea of going forth, and that where the latter connotes bureaucracy the former has deep roots in the language of the gods.

9

In the Language of the Papyri

Papyri of the century or two either side of the first century of the Christian era turn out to be a disappointment if we approach them thinking they will provide down-to-earth illustrations of how our words fitted into the everyday language of Greeks or come up with valuable sidelights on literary usage. The half dozen or so instances of the verb in these often fragmentary documents and the solitary instance of the abstract noun only marginally extend our material and offer no surprises. In earlier discussion we have gleaned what we can from these instances[1] and, while it has been interesting and helpful for our understanding of the special place our words held in the language of the day to read of the lover's desperate spell in *P. Warren* 21, we have had very little other indication of any place the words may have held in everyday language. Instances of the common noun, on the other hand, are more frequent, numbering about forty and deriving in the main from the second and third centuries of the Christian era, but far from illuminating ancient usage they present a set of problems of their own that form the matter of this chapter; solutions here and there are tentative, and are likely to remain so.

This is a conservative approach and could come as a surprise to those who still have in mind Schweizer's phrase about "the few hints of a new development of the Greek usage in the papyri,"[2] but Schweizer's view would seem to owe something to Moulton and Milligan[3] whose own judgement on precedents for Christian usage rests not on papyri but almost entirely on the inscriptions that were examined above in chapter 7. Of papyri Moulton and Milligan cite only the business letter *P. Flor.* 121, which is the only document cited also in both Liddell, Scott, and Jones *(LSJ)* and Guerra's study.[4] In other works Lemaire refers to but offers no comment on *P. Mich.* 473, *O. Mich.* 1046 and one of the instances in the *Charta Borgiana*,[5] while Hans Lietzmann cites only *BGU* 597.[6] There is thus good reason to present a fuller collection of material containing the common noun, which in these sources is more commonly of the form διάκων, and to attempt to extend our understanding of the word there.

BGU 597 is one of only two private letters attesting the common noun, and for its meaning there Lietzman suggested "cowherd," by force as we shall see of the context, although he was aware that such a meaning was out of character with uses of the word in literature and epigraphy. More in line with his view of the word as belonging to higher ranges of language is the instance cited in chapter 4 where the word designates an agent of the law (*SB* 7696) because there the word possesses the formal character reflected consistently in literary and epigraphical sources and paralleled in instances in the papyri of both the verb and the abstract noun and in the one Egyptian inscription of the common noun (*SB* 7871). Exhibiting the same characteristic is the occurrence in *P. Mil. Vogliano* III, 188, a papyrus from early in the second century.[7] This document records expenses incurred by priestly ministers (παστοφόροι), probably of the temple of Soknebtynis,[8] in relation to the food, drink, regalia, and facilities for washing during special celebrations for the month of Pachon. The account includes payments of four obols for a "bath attendant" (βαλανείου διάκονος, i.11) and thirty-two drachmae for "hire of waiter" (μισθοῦ διάκονι, i.15), the latter being doubtless the ceremonial waiter of the kind attested in the inscriptions. Of interest is the great difference between the two payments (the obol being but a fraction of the drachma), and it is hardly just the attendant's lowly station in the bathhouse or the longer hours of duty of the waiter that account for this; the difference must owe a lot surely to the superior social standing of the kind of person the priestly celebrants would want to have attending on them at a religious festival. The religious character of the occasion thus colours the occurrence of the common noun at i.15, while the fact that the bathhouse is in the temple precinct possibly explains why the word also occurs at i.11; the word does not occur among terms for temple clergy in Egypt.[9]

We can allow then that the instances in *P. Mil. Vogliano* III, 188, have recognisable connections with Greek literary usage. The same has already been established for an instance of the common noun in a transcript on papyrus of legal proceedings (*SB* 7696), and further discussion will perhaps show that the connection holds for instances in a business letter (*P. Flor.* 121) and in one at least of two personal letters (*BGU* 597; *P. Mich.* VIII, 473).[10] In other papyri, however, which are documents like receipts, tax lists, and population and property surveys, the word is attached to a personal name, and in these cases, which form the bulk of the material, interpretation is problematical. Certainly editors have not taken a consistent line. Some speak in terms of an avocation and others of a personal name; in the cases of *O. Mich.* 1046 and *P. Mich.* 596 the editors have fallen back on the Latin term "diaconus." This would seem to echo uncertainties about the meaning of the word expressed in 1788 by Nicolas Schow in his edition of the first papyrus ever published, the *Charta Borgiana*. He wrote: "I have not tried to translate the word διάκων because I do not know what it means: I just hope that no normal person will start thinking about Christian deacons."[11] Trying to sort out what the word does mean in these instances will be the main part of the discussion that follows. If the word points to an avocation, we will want to know what kind of avocation and how the usage ties in with what we know of literary usage; if, as is mainly argued here, the word is commonly a personal name, an examination of how this could have come about should reveal something more about how ancients viewed the word. The enquiry will also clear the ground for any consideration of the sense "deacon" in the papyri.

A Personal Name

There is no mistaking the common noun is being used as a personal name when it appears as part of the conventional formula for double names.[1] A clear example is *BGU* 1046.ii.24 (second century CE):

ὁ δεῖνα τοῦ δεινὸς ἐπικαλούμενος διάκων.
N, son of N, called Diakon.

This entry occurs in a list of names submitted to higher authorities for duties in connection with tax collecting. In another list providing names of those who enjoyed the privilege of paying tax in lieu of shifting an allotted quantity of earth, an alternative formula is used that also clearly establishes the presence of a personal name (*P. Bour.* 42 v.ix.598; second century CE):

ὁ δεῖνα τοῦ δεινὸς ὁ καὶ Διακων Δημητριου.
N, son of N, also known as Diakon, son of Demetrius.

A double name has also been registered by the juxtaposition of two names without a linking formula.[2] Rare earlier, this style became increasingly frequent but can be difficult to recognise; if the second name is abbreviated or damaged so that a case ending does not show, one might not know immediately whether the name is a patronym, as it would normally be, or the second element of a double name. An indisputable instance, however, is *SB* 7621.52 (fourth century CE):

Αυρηλιος Διεκων.[3]
Aurelius Diakon.

The instance is clear because the bank receipt in the prescript of which these two names appear later refers to the client as "the aforesaid Diakon" (ὁ αὐτὸς Διέκων).

When lists of names regularly include patronyms and the shortened form διακ appears among the latter, we have an indication that it too is to be read in the genitive case. There would be five instances of the name being registered in this manner, all from the middle of the second century CE, one of them being *P. Bour.* 42r.xiii.304:

Πτολεμαῖο(ς) Διάκων(ος)
Ptolemy, son of Diakon.[4]

In one other list of names of the same century and in three receipts of a century or two later the shortened form of the word stands alone after a name and one can probably never know whether it is a patronym or the second element in a double name.[5]

In all these thirteen instances, notwithstanding inconsistencies in the published material in respect of lower or upper case for the initial letter, the common noun looks like one name among others. The fact that it is not ordinarily acknowledged as such is probably due to the fact that the word has no history in the mainstream of Greek nomenclature.[6] The nomenclature of Ptolemaic and Roman Egypt, however, has characteristics very much its own by reason of the province's ethnic diversity. Scribes had to register in Greek not only traditional Greek and Macedonian names but Latin, native Egyptian or Coptic, and foreign semitic names. An individual could bear names that mixed these types, as happened commonly when a native adopted a Latin name on entry into military service. Again, a native who was registered in official documents under a traditional Greek name might be known in his village only by an Egyptian name; he might be registered elsewhere under either type of name or in a combination of both. The Egyptian name itself might be presented in a transliteration of the native Coptic or in what was considered to be a Greek equivalent;[7] if the latter, the substituted name might be recognisable both to Greek and native as a name—for example, if the name of a Greek god was substituted for the name of an equivalent deity from the native culture—or it might be a meaningless word to the native and unrecognisable as a name to the Greek.[8]

H. C. Youtie has analysed several interesting names in the extensive tax rolls from Karanis,[9] and some of what he has had to say would seem to have a particular relevance from our point of interest to one entry on the rolls, *P. Mich.* 224.1546 (172–73 CE; the brackets in the following are not a modern editor's but were made on the papyrus by the scribe):

Κάρανος Ἡρακλείδου (Ἥρων διάκων).
Karanos, son of Herakleides (Heron diakon).

Karanos is the landowner, and the following line in the roll shows the land tax he paid. Whereas those sections of the rolls recording poll tax show precise personal details of the individuals being registered, the sections recording tax on land merely identify the landowner; his tax is often then recorded as having been paid through a lessee. Thus in *P. Mich.* 223.1693 and 359C.29 Karanos pays "by the hand of Tapais" (διὰ Ταπαῖτος). This arrangement can also be recorded by the use of brackets around the name of the lessee, although the brackets may serve other purposes like including information about occupation, relationships, and addresses or giving the native name of the landowner in addition to the name already registered.

Youtie points out that while the names of proprietors of land are always given conventional Greek endings and those of lessees are repeatedly left in their Egyptian form, there are among the names of lessees a few instances of Greek translations of Egyptian names that bear no relation to standard Greek nomenclature.[10] A nice example is Κύλλος Τάλαντον, "Kullos, son Talent," where "Kullos" is really "Clubfoot," but our interest is in the name of the father. His father is recorded elsewhere in the rolls not in the Greek translation of "talent" but in the transliteration of the Coptic word meaning talent, namely, *kinkiol*. Youtie attributes the unusual instances he discusses to the learned eccentricity of the clerk, but the clerk, of course, did not invent this system of recording names so that we can expect to find traces of it elsewhere. In the case of *P. Mich.* 224.1546, already cited, we can probably safely surmise that the clerk was following custom in representing a particular Egyptian name in the Greek form of "Diakon." Whether "Diakon" was part of an alternative name for Karanos or the second element in the name of a lessee or the name of Heron's father is immaterial to our purposes.

In the index of *P. Mich.* IV Youtie himself lists this instance under professions, trades, and occupations with such words as "priest, doctor, herdsman," but in fact such designations were particularly favoured by Copts as personal names. G. Heuser wrote that "in the formation of Coptic names a great role was played, above all, by avocation, business, social position, and title,"[11] and he pointed out that the later Coptic style was certainly inherited from an earlier native nomenclature,[12] which included as well names signifying relationship with a god; names that express the idea of servant in respect either of man or of god also occur,[13] as do hybrid Egyptian-Greek formations.[14] There are ample grounds, therefore, on which to explain "Heron Diakon" as a personal name in the Egyptian style.[15]

A Coptic Nickname

Three instances of the word in the *Charta Borgiana* would seem at first sight to conform to the same style; we read in quick succession from *SB* 5124.207–9 (192 CE):

Σεσιβελᾶς διάκων
Πᾶσις διάκων
Κανᾶς διάκων.

Juxtaposed names of such a kind, however, cannot be said to occur elsewhere in the document, whereas a common pattern is for a name, with or without a patronym, to be followed by words in the nominative case meaning ordinarily such things as "weaver, thresher, greengrocer, baker." Three other instances of the common noun fit this pattern precisely (73, 107, 283):

Ἡρώδης Εὐδαίμονος διάκ(ων)
Πρωτᾶς Τούρβωνος διάκ(ων)
Εὐδαίμων Τούρβωνος διάκων.

These would not seem to be alternative proper names because for that purpose this document uses the formula "called";[1] in a village society, nonetheless, the designation of avocations would have been a helpful and easy way to tag individuals, much as we might identify a person by saying "Wilson the gardener." In addition the information thus provided may have been useful to the clerk or clerks[2] compiling the document, which was not a formal census but a simple record of those who had completed their work on the canals.[3] The difficulty with our word, however, is that it is not otherwise known as a term for an avocation after the manner of words like "weaver" or "donkey driver." But it is certainly not a euphemism for "slave" because fifteen of these (δοῦλοι) are carefully documented, in each case the name of the owner being added in the genitive case— Wilcken in fact noting that the distinction between bond and free is a leading characteristic of the document.[4] And then, in the simple society this list of agricultural labourers and tradespeople bears witness to, it is difficult to think of people subsisting as "attendants"—unless this were in the grand houses of large estates, when they would in any case no doubt have been registered as slaves. Wilcken suggests "manager"[5] and although literary sources provide no parallel for such a professional designation it may just be that the word—which does after all have multiple uses in the sense of "agent"—was judged a suitable translation of a native word that did signify some such role. The fact that the word was also current as a personal name means that it would easily settle as a kind of nickname, in contrast with a word like $ὑπηρέτης$ which might readily apply to a middling administrator or bookkeeper but which was already saddled with clear significations and strong connotations from the civil service.[6]

Odd Jobs

BGU 597 is a personal letter which, just like so many of our own, is hard for an outsider to come to close grips with; unless we are familiar with the people, events, and customs we easily miss the connections and implications. In this case spellings and damage to the papyrus create other trouble spots. The translation that follows gives us a workable context; the two main omissions are of only four words, mainly names, and of a garbled sentence of some eight words.

> Chairemon to my brother Apollonios,
> many greetings and good health.
> Would you please arrange with Petheus $τῷ$ [line 4] $διάκωνι$ to deliver[1] the calf before the sheep. [. . .] and if he says he has brought it, get Hatres to take it into the country. Also get Hatres to pay over the bag of wheat, which you can then take into Hephaistiadas where Pasion will exchange it with you for good seed. Don't make other arrangements. But do make an agreement with Panesneus the . . . about the cattle, just as I said, until I come and give him the money. And since you are still a soldier, watch yourself with the corn collector in the matter of travel expenses. [. . .] Don't make other arrangements. But if you wish, arrange with Kaisaras son of Papontos about the calf, since he told me he too would take it into the country.
> Tell everyone in the house to look after themselves—and to keep an eye on the door when it is open. Tell Iranuphis . . . and Orsenuphis . . . that I'll be up soon. Keep well.
> 4 December 75 [CE]

All this seems a bit of a rush, with Chairemon doubling up on his arrangements and clearly preoccupied with the calf—as well he might be if, as the indications seem to be,[2] cattle were rare in comparison with the smaller beasts valued for food and produce (sheep, goats, and pigs) and with the larger beasts domesticated for work (the camels and asses). But what is the role of Petheus here? He has been seen simply as "the servant,"[3] but would Chairemon really need to identify Petheus in this way to a brother who had been privy to arrangements in his household for some time? That the word indicates some role is certain from the presence of the definite article. Lietzmann, as we have seen, proposed "cowherd,"[4] but that makes no sense in terms of the history of the word and not a great deal of sense in this letter where Petheus ends up with little to do in moving the animal. If Petheus did not in fact belong to the household but was somebody else's servant or drover, one would really expect identification of him by mention of his master's name, but the difficulty with the choice of term would remain. One of Avogadro's observations perhaps leaves room for a different approach. It concerns peculiar arrangements that obtained at times in the management of livestock. "Sometimes," she wrote, "between the two groups of owners and herdsmen there is a kind of middleman, one who, in the interests of the owner, oversees the flock which in the physical sense is in the care of the herdsman."[5] Without wishing to particularise from such a general observation, perhaps the designation given to Petheus here does set him apart in some way from Chairemon and his household. One might think of something like a rural agent. As such the term would not signify a profession but a function within a profession. The following document would seem to point in the same direction.

P. Flor. 121 is again a letter, the fourth of seventeen in the surviving business correspondence of the rural agent (φροντιστής) Heroninos. His task was to attend to the interests of landowners, both civic and private, in the area around Theadelphia, supervising and coordinating work in regard to crops, herds, transport, and so on.[6] The position was not an exalted one. In this letter from the landowner Alypios, written in 254 CE, Heroninos is informed of a decision on the part of, it seems, higher authorities to provide him with the assistance of a person designated τὸν διάκονον Ει[. . . (line 3). The letter is damaged here so that the name cannot be read; it seems, however, that the person is to be available for a year, and the letter notifies Heroninos of payments in cash and kind that will have to be made. What is not clear is the nature of the man's duties, but with the mention of both ὑπηρεσία and οἰκονομία—both standard terms in the language of management—one can probably assume that the duties lay on the clerical side. We notice that in the previous letter from Alypios, in July of the same year, advice was forwarded of an assistant designated Δ . . . τὸν βοηθόν—also a technical term[7]—who was to take some of the administrative load during harvest. In the διάκονος of *P. Flor.* 121.3 we seem to have the same kind of appointment, a clerical assistant, nominated in general rather than in bureaucratic terms.[8]

A precise designation in vocational terms, attendant in a household, is in the following notice issued from one of the highest sources in Ptolemaic Egypt (*P. Lond.* 2052):[9]

> Memorandum to Zenon from Sosikrates of the slaves formerly belonging to Apollonios the ex-Dioiketes but now the property of Paideas. If anyone comes across them, arrest them and write to me. Pindaros, a Lycian, servant [διάκονον, line 7], aged about 29, of medium height, honey-coloured, with meeting eyebrows, hook-nosed, with a scar under the left knee.

Descriptions of three more slaves follow. Apollonios had been the Egyptian treasurer for many years towards the middle of the third century BCE under King Ptolemy II Philadelphus and was now retired, dead, or, as Rostovtzeff argued, disgraced;[10] Zenon had been the manager of his estate. We must understand that Pindaros had been an attendant within the household of this powerful man. The style of the description matches that in the other notices: for each of the four slaves name, age, and brief personal description are issued; for one among them an Egyptian alias is given, signifying he is a native, and for the three foreign slaves an ethnic designation followed by a word indicating the slave's function: "coachman," "shampooer," and for Pindaros "'attendant."

Some information from two other papyri in the Zenon archive on what was probably the earlier career of Pindaros could give a sharper picture of his role on Apollonios's estate. At least six years earlier a Pindaros appears as a young slave ($\pi\alpha\iota\delta\acute{\alpha}\rho\iota o\nu$) in a list of Apollonios's slaves and sailors (*P. Cairo Zen.* 59677), and in 249 BCE a docket covering a consignment of priced articles belonging to various people is addressed to a Pindaros (*P. Cairo Zen.* 59319). If, as is commonly accepted, this is the same Pindaros who later made his bid for freedom,[11] he would have received an apt designation in the wanted notice because the word would then signify the kind of personal attendant who was entrusted with the conduct of personal affairs.[12]

Mistaken Identity

A document that raises a different sort of question is *P. Berl. Leihg.* 4 *verso* (165 CE). This is a grain collector's register of his daily intake, and 4v.ix.16 begins: $\Sigma\omega\tau\alpha\varsigma\ \Delta\iota o\sigma\kappa o\rho o\upsilon\ \delta\iota\alpha\kappa($), which, in his corrections,[1] Kalen resolves to $\delta\iota\acute{\alpha}\kappa(\omega\nu)$, probably for the meaning "Sotas, son of Dioskoros, attendant." Of the fifteen other persons who furnished grain on this day, all of whom with one dubious exception are identified only by name and patronym—that is, without mention of avocation, three did so through a deputy or agent, thus (12): "by the hand of ($\delta\iota\alpha$), Ischurion." This suggests that Sotas, instead of being given the additional tag of $\delta\iota\alpha\kappa$, may really have been recorded as having done his business "by the hand of ($\delta\iota$') Ak . . . ('Aκ[ ])." In fact, *P. Col.* V,2, a similar document from the same location and century, has the following entries:

193. $\Pi\tau o\lambda\epsilon\mu\alpha\iota o\varsigma\ \Delta\iota o\sigma\kappa o\rho o\upsilon\ \delta\iota\ A\kappa[\epsilon\iota(o\upsilon\varsigma)]$
     Ptolemy son of Dioskoros by the hand of Akes
208. $\Sigma\omega\tau\alpha\varsigma\ \Delta\iota o\sigma\kappa(o\rho o\upsilon)\ \delta\iota\ A\kappa(\epsilon\iota o\upsilon\varsigma)$
     Sotas son of Dioskoros by the hand of Akes
218. $\Sigma\alpha\rho\alpha\pi\iota\omega\nu\ \Delta\iota o\sigma\kappa(o\rho o\upsilon)\ \delta\iota\ A\kappa\epsilon\iota(o\upsilon\varsigma)$
     Sarapion son of Dioskoros by the hand of Akes

The editors of these tax returns from Theadelphia, J. Day and C. W. Keyes, consider that this Sotas of *P. Col.* V,2.208 is probably to be identified with the Sotas of *P. Berl. Leihg.* 4v.9.16 and suggest that the three sons of a Dioskoros in *P. Col.* might be brothers acting through the same agent, Akes.[2] This curious set of congruencies suggests that the same Akes may have done further business for the same Sotas with the grain collector who recorded his transactions in *P. Berl. Leihg.* 4 *verso*.

A similar doubt about the reading adopted by editors arises in regard to *P. Mich.* VIII, 473.12. This papyrus is a letter of the early second century CE from Tabetheus to her brother Tiberianus laying claim to money which she wants to use as a ransom for a son convicted of murder. The papyrus is damaged, the spacing of the letters is irregular, the Greek is common, ellipses occur at awkward points, and the editors confess to difficulty in reconstruction. The passage of interest to us has no apparent connection with the grave issue Tabetheus is mainly to press with Tiberianus and reads in the published translation: "I was able to send you the robe this year; I did not send it last year (?), but I sent and sold them to Cabin the attendant." For "last year (?)" the editors print διὰ πέ[ρ]υσι, which balances with "this year" but, as the editors note, the initial preposition is otherwise unattested (in fact "last year" occurs without the preposition at line 31); they note accordingly that "the context also permits διὰ Πε[κ]ῦσι." The word they translate "the attendant" they print as δ[ι]άκων[ι], which includes a degree of conjecture.[3] Another conjecture might be the following. If their suggestion of "through Pekusis" is taken up, and the same preposition is also read at the beginning of the word they translate as "attendant," we would then have Tabetheus saying: "I was able to send you the robe this year; I did not send it through Pekusis, but I sent and sold them to Cabin through . . ." The missing name would require further conjecture,[4] but the mention of a person in the phrase beginning with "but" balances naturally with the preceding phrase.

Panopolis

The last document we need to examine was published in part in 1962 by V. Martin as a "topographical survey of some metropolitan real estate (*P. Gen. inv.* 108)."[1] The city was Panopolis, a major centre within the Thebaid of Upper Egypt, and the date proposed by Martin for the drawing up of the survey was about 250 CE. Other pieces of the survey were held in Berlin (*P. Berl. inv.* 16365), and in 1975 both sets of remains of the survey were published by Z. Borkowski.[2] He reconstructed a roll of eighteen columns, with large gaps throughout, and with parts of another five columns which he could not locate within the eighteen, and estimated that the undamaged document had consisted of twenty-eight columns. Because each column contains about thirty-five lines and a property is recorded in nearly every line, the survey presumably registered something approximating to one thousand properties. These are listed in the order of their position along the route taken by the surveying party; they include homes, vacant blocks, workshops, depots, and other unoccupied buildings, and shrines or temples. Each privately owned property is further identified by the name of the owner; this name might stand alone or be followed by a patronym—extending even to a grandfather—or by an indication of a trade or profession. In very many cases damage to the papyrus has meant that such information has been lost, but we are nonetheless provided with the names of some 350 owners—many other unnamed owners being included under a hundred instances of expressions like "sons [=heirs] of," "brothers of," "associates of"; of this number something like 40 percent are listed with addition of an avocation.³ The question for us is what to make of six occurrences of the common noun διάκονος attached to names of owners.

The six instances read as follows:

i.19 ἄλ(λη) Βουβάλου διακόνου
Another [house] belonging to Boubalos *diakonos*
ix.29 ἄλ(λη) Νεμεσᾶ διακόνου
Another [house] belonging to Nemesas *diakonos*
xi.7 ἄλ(λη) Χώλου διακόνου
Another [house] belonging to Cholos *diakonos*
xi.32 ἄλ(λη) φορτουνάτου διακόνου
Another [house] belonging to Fortounatos *diakonos*
xiv.30 οἰκ(ία) σὺν ψιλῷ Ἀθᾶ διακόνου
House with adjacent block belonging to Athas *diakonos*
A.iv.10 ἄλ(λη) Ἀπόλλωνος διακόνου
Another [house] belonging to Apollon *diakonos*

Borkowski is straightforward in regard to this term. A section of his index is for Christian religious terms, and it is made up entirely of the six instances above and of one instance of the word ἐκκλησία (iii.27), the arrangement indicating that these words are to be taken as meaning "deacon" and "church." In addition, when considering the date of the document, he writes of the two terms as "indisputably Christian," and uses them to illustrate that Martin's date of 250 CE is too early for the inclusion of terms evidencing so settled a Christian community and hierarchy in remote Panopolis.⁴

While there is room for some surprise at the firmness of this opinion, especially as it is put forward without reference to views about the spread of Christianity in Egypt at this era, Borkowski is fortunate that the argument for a later date does not really rest on the Christian signification of these two terms but on information provided by T. C. Skeat, which establishes that the survey could not have been made before 298 CE.⁵ This later date, however, does not necessarily strengthen a case for the Christian provenance of the words. Certainly it brings the document much closer to the period when Christianity burgeoned in Egypt, but it brings this report of deacons closer still to the period when the new religion came under the bitter proscription first of Diocletian from 303 CE and then of Galerius and Maximian until the peace of Constantine. Information about the impact of these measures upon Christianity in the Thebaid is hard to come by, as Borkowski notes, the most telling item to have survived in the papyri being perhaps the submission of the illiterate (from the Greek point of view) lector of the former church in Chusis that no gold, vestments, animals, or land were in the possession of the church and that its bronze gate had already been delivered to Alexandria, but this is from Middle Egypt (*P. Oxy.* 2673).⁶ The imperial proscription remains, nonetheless, a major historical experience of the religious movement and can be reliably supposed to have created a climate that discouraged claims to hierarchical status; in Milman's old world phrase, "bishops, presbyters, and deacons were crowded into the prisons intended for the basest of malefactors."⁷

An awareness of these conditions has inclined Borkowski to set the document between 315 and 320 CE, but he is reluctant to advance the date much further into the true era of Egyptian Christian institutions for the same reasons of palaeography that led Martin to date the document nearly a century earlier and H. C. Youtie, who tightened the case from Skeat for a date after 298 CE, to suspect an earlier rather than a later date.[8] If the survey was in fact made in a period when Christians of Panopolis had already established their hierarchical structures, however, the odd thing would be to find the structures represented only by deacons. About deacons, after all, one characteristic is clear historically: they did not operate alone; the early canons, distilling the essence of a two-century-old relationship between members of the church order, are at least witness to the total integration within established Christian communities of bishops, presbyters, and deacons.[9] Even with the gaps in the papyrus of the Panopolis survey, we might have expected that if it could come up with one instance of a Christian "church," it could have provided other evidence of Christian hierarchy than deacons, in particular, some sign of presbyters. More curious, however, than the claim that deacons lived in Panopolis—of course they may have done—is the supposition that they would have gone on public record either in a period of suppression or in the first decade or so of peace merely as "deacon." Such a bald designation points to a professional clericalism more suited to the Byzantine era; we shall see, in the light of other evidence from the papyri of Christian origins in Egypt, how singular such testimony would be at the beginning of the fourth century.

Meanwhile, on the other side of the street in Panopolis, so to speak, by all the indications the real life of this pagan city pulsed on. Named the city of Pan, it was really the city of the Egyptian Min, immeasurably more ancient than his latter-day Greek counterpart, and a god of fertility joyously kept within the city's consciousness in the Panopolite games, which even Herodotus had written about. Something of their vitality comes to us in the following flourish of a signature to a claim before the Oxyrhynchus senate at a date just preceding that under discussion (26 July 289 CE):

> I the Astounding Aurelios Herakleios called Nikantinoos, citizen of Antinoopolis and Panopolis and Hermopolis and Lykopolis and Oxyrhynchos, Victor in the Olympic, Capitoline and Pythian Games, Victor of Many Games, First Officer of the Sacred Artistic Grand Society, under the Patronage of Diocletian and Maximian, have signed and sealed for Aurelios Hatres, son of Peteesios, son of Nechthenibis, an Oxyrhynchite, on his enrolment as Highpriest in my Presence in the Most Noble City and Most August City of the Panopolites, in the 7th Pythiad, during the Presentation of the Sacred Eiselastic International Dramatic and Scenic Games of Perseus of the Sky at the Great Festival of Pan.[10]

Pan or Min was one of the local triad of gods; the other two, Triphis and Kolanthos, like many gods are so frequently represented in the names of inhabitants registered in the survey that Martin claimed the place could be identified as Panopolis by that alone;[11] the ten shrines or temples and the other evidences of religious cult recorded there perhaps do not say as much about the spirit of the place as do these names. This fits well with the reputation of the city as retaining well past this era a hearty dedication to its traditional religion whose voice was still vibrant a century or two later in the exuberant testimony of the *Dionysiaka* of its son Nonnus.[12]

If, in line with this, Jack Lindsay can assume "a strong literary and philosophic development in Panopolis throughout the 5th century . . . in which pagan systems of symbolism were carried vitally on,"[13] there cannot be much to be said of any Christians of Panopolis in regard to what E. M. Forster called "the great spirit of aggression the new religion everywhere displayed as soon as Constantine labelled it as official"[14]—especially if the powerful hierarchy was represented in the city shortly after 300 CE by almost as strong a body of deacons as was considered adequate at that time for a Christian centre like Neocaesarea.[15] J. van Haelst has presented the evidence from the papyri of the rapid repression of pagan institutions in "the explosive epoque" of Christianity following 330.[16] Of twenty-two references to pagan priests and temples after 300 (he was writing in 1970), only two date from after 342; of twelve temples registered in Oxyrhynchus in 295 there are only traces after 350: one is a prison, one a public monument, then a church, and another has been replaced by houses; some land registers of 330–40 show half a dozen pagan priests holding less than 10 arures each, while the bishop of Hermopolis holds more than 470; the last pagan priest appears in papyri of 371 CE.

Observations of this general kind are possibly what Borkowski had in mind in stating that speculation about these "deacons" and "church" would be premature,[17] but if we do not have enough information to be sure about who and what they were, a sensible approach would seem to be to put what information we do have into some sort of historical perspective. And if from those considerations of language, history, and culture a record of Christian "deacons" in an early fourth-century survey of Panopolis would already be surprising, the sparse evidence we have from the papyri of the development of the institutional life of Egyptian Christianity would make any such record remarkable.

In the first place, the fragmentary phrase in the Panopolis survey, οἰκία ἤτοι ἐκκλισίας σ[(iii.27: "house or of *ekklesia* s[") brings to mind the same word in a comparable and almost contemporary document from Oxyrhynchus where two streets take their name from "the north" and "the south" *ekklesia* (*P. Oxy*. 43v.i.10; iii,19; 295 CE). Because most scholars recognise in the Oxyrhynchite document designations of Christian churches, that is, church buildings,[18] and because the word certainly occurs in this sense in the lector's report of "the former church" of Chusis already referred to (*P. Oxy*. 2673.9; 304 CE), one might expect that the Panopolis document is also registering "a house used as a church." J. D. Thomas has followed this line of thought,[19] but the first editor, Martin, could not, under the influence, no doubt, of his early dating of the papyrus.[20] Borkowski, while certain of its Christian attribution, is unable to say whether the word designates a building or a church group; this is largely because the poor state of preservation of the line does not allow him to decide between the readings "the *ekklesia* of" and "of the *ekklesia*."[21] This being the case, one might wonder what determines its Christian signification; the word or words missing from the line are crucial to the meaning of the term. Préaux—but not Martin—was content to see it not as a religious term at all but as designating a building for civic gatherings,[22] which is the meaning one scholar at least sees in "the north" and "the south" *ekklesiai* of Oxyrhynchus.[23] All other occurrences of the word in a Christian sense are from years of secure establishment following 330.[24] And of course it is not possible to infer a Christian meaning for ἐκκλησία from the presence in the same document of the word διάκονος when the Christian character of that term is itself under review.

Thus, although the word ἐκκλησία in this document could designate a Christian church building or church group, we cannot be sure. Comparable material is very scarce, and even information about the developing institutional life of Christian communities in Egypt at this time is not abundant. That Christian communities existed and were developing is richly evidenced in biblical and other recognisably Christian literary remains—C. H. Roberts, following E. G. Turner, puts the possibility of a scriptorium for Christian books in Oxyrhynchus as early as the second and third centuries[25]—but for bishops, presbyters, and deacons evidence is more meagre. The first relating to church order concerns dealings between a Fayum community and one of its members on business in Rome, which were facilitated in some way by Maximus, bishop of Alexandria (264–82 CE; *P. Amh.* 3a).[26] The earliest mention of deacons in the papyri would be *P. Oxy.* 1162, which is a letter of introduction for one member of a community to another; dated by its editors in the fourth century, it is placed by van Haelst, with others of the type, on grounds of palaeography and style, in the last decades of the third:[27]

> Leon, presbyter, to the presbyters and deacons who share the local service,[28] beloved brothers in the Lord God, fullness of joy. Our brother Ammonius, who is coming to you, receive in peace; through whom we and those with us greet you and those who are with you kindly in the Lord. I pray for your health in the Lord God. Emmanuel is my witness. Amen.

Perhaps the next earliest occurs in the prescript of a letter from some time after 309 (*P. Giss* 103),[29] with two others in letters of about 335 (*P. Jews* [=*P. Lond.*] 1913, 1914), the second of which records the sufferings of a bishop, priest, and deacon at the hands of Athanasius. These, of course, are all documents that circulated within the Christian community and are cast in the terms of religious respect and endearment that were reserved for fellow Christians and were intelligible largely only to themselves. Other early references to deacons between 330 and 350 occur in documents addressed as appeals or declarations to civic authorities and using a formula more or less all their own: "N, deacon of the catholic church from the village N."[30] The style would suggest that the deacons are concerned to present themselves in the most favourable light to authorities who might otherwise have known of them only through their profane craft or trade.[31] In a register of householders, like the document from Panopolis, by contrast, it must remain unlikely, especially in the era of persecution or in the years succeeding it, that Christian deacons would declare their role within their sacred rituals or be publicly—much less officially—identified by their title.

We would thus say that the six men listed as "N *diakonos*" in the Panopolis survey are not Christian deacons. Some other significations of the term can also be excluded but we ought to note first that the term would not seem to be distinguishing the six men from other men of the same names because in the large sections of the document to have survived three of the six bear names not otherwise recorded. Second, the term will not be a straightforward designation of an avocation among the secretaries, goldsmiths, bakers, trainers, and army captains of the city. With over half the individuals on the list bearing no such indication of trade or profession, we do not have to assume that this word designates a vocation, and of course the word does not do so elsewhere; Pindaros was called διακονος (*P. Lond.* 2052.7) because his function as a slave on the estate of a wealthy man was being differentiated from that of other runaway slaves, but unless the word occurs in some such context its meaning in vocational terms cannot be known. Third, the term will not be a double name after the style of "Aurelius Diekon" (*SB* 7621.52), because with the exception of the wealthy Roman Ulpius Claudianus no one in this list is known by a double name. Fourth, it will not be a patronym as in "Ptolemy, son of Diakon" (*P. Bour.* 42r.xiii.304) if only because no other patronym occurs more than twice—even with a name as popular as Besas only once in its thirteen occurrences is it a patronym; as well, if we were to attempt to read six instances of "N, son of Diakonos," we might expect to encounter Diakonos also as the name of some of the offspring.

The material most helpful for comparison is probably the instances of the same term in the *Charta Borgiana* (*SB* 5124), in considering which we concluded that the term was likely to be a Greek equivalent for a native Egyptian word that designated an avocation in the line of management (cf. also *P. Flor.* 121.3) but that acted at the same time as a personal designation. In his commentary on the term in the Panopolis document Borkowski seems to intend to exclude this possibility in noting that none of the names of the six men in question is Egyptian in character.[32] The comment is hardly adequate, however, in the light of the much closer work on the nomenclature of the document by Martin, who emphasised the heavy preponderance there of indigenous elements and drew attention in particular to Hellenising tendencies evident both in the adoption of Greek names and in their use to represent native names; he even called for the collaboration of Egyptologists to help determine the instances in which this process may have been occurring.[33] Given that and the kind of usage we have recognised in the *Charta Borgiana*, we can suppose that διάκονος[34] in this document is a scribal convention to represent in a Greek word, which after all does exist in its own right in Egypt as a personal name, a native name that derived from a vocational role perhaps peculiar to native sociological conditions. Martin himself estimated—for reasons no doubt different from those advanced here—that the word might be designating administrators of a junior level.[35] One other possibility could also be put, and it is suggested by the name of one of the six, Apollo. This name was the standard equivalent of the Egyptian divine name, Horus (e.g., *P. Oxy.* 1299.4), and allows us to envisage that "Apollo *diakonos*" might represent two components of a single native name meaning "Servant of Horus," even in the sense that Hermes was the διάκονος of Zeus.

If the approach which has been taken here to the occurrences of the common noun in the papyri is sound, it illustrates that in this part of the Greek world also the word is confined to certain formal modes of expression: insofar as Greek nomenclature is concerned—unlike native Egyptian or even Anglo-Saxon—no ordinary words were elevated to share rank with names deriving from the names of gods, heroes, and the graces of life. If this impression is correct, it confirms the impression already gained from uses in literary and epigraphical sources that the word and its cognates could hardly ever have been part of the vernacular.

III

FIRST CHRISTIAN WRITINGS

We have now to see how the words sat in the language of the first Christian writers. At the beginning of our study we saw that in recent decades many things have been said about the special way Christians used the words. To remind us of these views we return to some pages by Eduard Schweizer in *Church Order in the New Testament* that discuss the nature of office in the minds of early Christians and in the process provide interesting and now widely held opinions on the character of the words and on why Christians first began using them.

> As a general term for what we call "office," namely the service of individuals within the Church, there is, with a few exceptions, only one word: διακονία. Thus the New Testament throughout and uniformly chooses a word that is entirely unbiblical and non-religious and never includes association with a particular dignity or position. . . . In the development of Greek the basic meaning, "to serve at table," was extended to include the more comprehensive idea of "serving." It nearly always denotes something of inferior value. . . . The very choice of the word, which still clearly involves the idea of humble activity, proves that the Church wishes to denote the attitude of one who is at the service of God and his fellow-men, not a position carrying with it rights and powers. This new understanding is the continuing testimony to God's action in Jesus of Nazareth. The fact that it was in lowliness that God revealed himself as God implies for the Church that through being itself prepared to be lowly it must become separated from the world, to which indeed all kinds of ceremonial associations with imposing dignitaries belong. . . . Special ministry takes place in the Church only in special subordination. Acts, as well as Paul, likes to describe all special activities, particularly the preaching of an apostle or some other church member, as that of "ministry" or "service," and the person who performs it as a "servant" or "slave," with God, Christ, or men appearing as those to whom the service is rendered.
>
> It is nowhere forgotten that such renunciation of titles, honours, and offices testifies to the Church's newness in contrast to the old religious or secular order.[1]

These passages thus remind us of the striking fact that words that we found only rarely enough in non-Christian sources over many centuries became central to the early expression of what makes Christian community. Beyond that, however, nearly everything in the extract about what the words meant in non-Christian sources is at odds with our reading of the sources. Thus, the words show no signs of having developed in meaning over the course of changing literary eras, the sense "to serve at table" cannot be called "the basic meaning"—in fact that sense has to be perceived as a particular application of a word capable of signifying doing messages and being another person's agent—and "the more comprehensive idea of 'serving' " is vague and inadequate. If the words denote actions or positions of "inferior value," there is at the same time often the connotation of something special, even dignified, about the circumstance, and though they may truly be said to be unbiblical, they have wide religious connotations. Because the root idea expressed by the words is that of the go-between, the words do not necessarily involve the idea of "humble activity" at all, and never express the idea of being "at the service of" one's fellow man with what that phrase implies of benevolence; in commonly signifying that an action is done for someone, the words do not speak of benefit either to the person authorising the action or to the recipient of the action but of an action done in the name of another. This, which applies also to actions done in the service of God, means that the words do not speak directly of "attitude" like "lowliness" but express concepts about undertakings for another, be that God or man, master or friend. In accepting such undertakings or in having them imposed on him, the agent has a mandate as well as a personal obligation, and, if he is thus deprived of the exercise of his own authority and initiative, he has whatever "rights and powers" the mandate extends to; in the case of a Roman procurator, to take an extreme illustration, the mandate reaches over life and death. Finally, from what we have seen, where one of the words has been an age-old tag of possibly the most loved god, it is difficult to know what the words were likely to tell early Christians about "renunciation of titles."

In coming to Christian writers, then, after a close inspection of the words in non-Christian sources, one question that will be of interest concerns what Schweizer has called here "the Church's newness in contrast to the old religious or secular order." We will be assessing the extent of the newness in the use of these words. But our overriding purpose is much more direct; it will be to bring to bear on Christian usage insights gathered from our earlier reading.

10

Spokesmen and Emissaries of Heaven

Paul provides the earliest and most ample evidence of our words among the first Christian writers, and the statements in which the words occur are made largely in the course of his controversy with Corinthian Christians about who could claim apostolic rights among them. Here then is a good place to begin. The terms are examined from two points of view. A first section called "The Apostle as Spokesman" assesses the importance that Paul attaches to the word διάκονος in three passages of the two letters to the Corinthians, coming to a preliminary understanding of the word as "spokesman" (1 Cor. 3:5; 2 Cor. 3:6 with 6:4; 2 Cor. 11:23). A second section, "The Spokesman as Medium," goes over much of the same material in the light of the use made of διακονεῖν and διακονία in 2 Cor. 3–6, and adds to the idea of spokesman an aspect of mediation that is seen to be used there to authenticate claims to apostleship; this then contributes to elucidating some questions about Paul's opponents. Finally in this section related usage in Colossians and Ephesians is examined. Other briefer sections follow on Paul's designation of "The Apostolic Commission," which leads into Luke's language about "Paul and the Twelve in Acts." The chapter ends with looking at "Other Messengers and Emissaries of Heaven" designated in these terms by Paul and other Christian writers.

The Apostle as Spokesman

In examining the way Paul uses the word διάκονος in 1 Cor. 3:5, 2 Cor. 3:6 and 6:4, and 2 Cor. 11:23, we shall recognise affinities between his usage and that of non-Christian writers that require us to understand him to be speaking not in an imprecise way about "servants" of God or of Christ but about messengers on assignment from God or Christ.

1 Cor. 3:5

In the first four chapters of 1 Corinthians Paul is concerned at how a congregation of unsophisticated and spiritually immature people (1:26; 3:1) has come to foster parties claiming allegiance to the doctrines and persons of various evangelists (1:10–31), and is at pains therefore to establish the unity and integrity of the gospel and to bring the Corinthians to an understanding of how the gospel has been imparted. Because the gospel is not like human wisdom, it cannot be understood or debated at that level (1:17–24), and it had not in fact been delivered to them initially as a matter of wisdom but as "Jesus Christ and him crucified" (2:1–5). At the same time, the gospel may be said to have a wisdom of its own, but this is reserved for the mature, "a secret and hidden wisdom of God" (2:6–7); Paul's own understanding of this has come only by God's direct revelation through the Spirit (2:10–13), and while he may claim on this account to have "the mind of Christ" (2:16), he has never spoken in these terms to the Corinthians, who even now continue to display their immaturity by raising factions around the persons and teachings of, for example, Paul and Apollos (3:1–4; cf. 4:6).

In such circumstances the first necessity for Paul is to define as sharply as possible the role of the preacher of the gospel. This he does at 3:5 in terms of διάκονος: "What then is Apollos? What is Paul? Servants [διάκονοι] through whom you believed . . ." To the Corinthians this word conveys at once that Apollos and Paul belong to a god, that they have been entrusted with the god's message, that they have the duty to pass it on and the right to be heard and believed, and that their rights and duties are equal; in what they say there is no more and no less than what they have been commissioned to say, and their conformity to "the mind of Christ" is therefore assured. The additional phrase, "through whom you believed," brings the thought closer to the underlying idea that it is God who saves those who believe (1:21) by virtue of the fact that in faith there is a "demonstration" (ἀπόδειξις) of spirit and of the divine power itself (2:4–5); this means that the preaching of the evangelist is merely the vehicle of God's own effects—lest flesh should boast in the presence of God (1:29)—and suggests that the word at 3:5 (which *RSV* gives us as "servants" but *NEB* as "agents") denotes that kind of messenger who mediates between heaven and earth.

The phrase at 3:5 would seem then to summarise the teaching of the preceding two chapters of the letter on the processes of evangelising and of being saved, establishing in fact the connection between the two. It also exposes the futility of any attempt to raise parties around Paul or Apollos, for true believers recognise in the message they have received from whatever διάκονος the in-pouring of God's power. Paul realises, however, that aspects of the evangelising activity witnessed by the Corinthians remain to be explained, principally the fact that he had done a different work from that done by Apollos (or other preachers). He notes, therefore, that to each of these messengers the Lord may give differently (3:5), and that the difference is determined by the stage at which each preacher is called, one in the season of the planter, the other at the time of the watering, neither, however, being anything in himself, but only God who gives the growth (3:6–8). Amplifying this by a change to the metaphor of building, he points a number of lessons behind which aspects of the historical situation are no doubt to be discerned, but the apostolic identity of Paul and Apollos has been established at 3:5.¹

If we ask why Paul states his identity by means of this word, the answer must lie in the affinities between what he is saying and what non-Christian writers said respecting a messenger of the gods, message in general, and the notion of mediation itself. No other background to the word throws light on his emphatic use of the word here. His usage would thus be Hellenistic rather than peculiar to himself or something derived from a newly emerging Christian idiom. If the non-Christian background helps elucidate his statement and add vigor to his argument, we have no reason to make assumptions about Christians coining a new terminology of apostleship or ministry. Two other aspects affecting interpretation may be noted. First, the context is profoundly religious, which also accords with the predominantly religious connotation of the word in non-Christian sources, and, second, although Paul's primary interest in the word seems to be that it says something basic about the process by which the message is delivered, he may also be hinting at the intimacy existing between the divinity and its διάκονος. As remarked, the statement at 3:5 relates to the message about "Jesus Christ and him crucified" which, in Paul's words, is but the milk of Christian doctrine; the solid food he is holding back until the Corinthians are mature belongs to the realm of mystery (2:6–7), but this too is in his gift because God has communed with him (2:9–12).

2 Cor. 3:6; 6:4

The problem under discussion here is somewhat different, for now it is not only that the Corinthians harbour rivalries with respect to preachers of the gospel, but that preachers have set themselves in rivalry against Paul. Here only so much of the situation will be described as seems necessary for an appreciation of the place the word διάκονος holds in the two passages that are reminiscent of 1 Cor. 3:5, namely 2. Cor. 3:6 and 6:4.

Paul's position is difficult for, among other things, he does not have the letter of credence from human authorities that has led the Corinthians to hold his rivals in such high esteem. A large part of his apologia, accordingly, consists of a vigorous assertion of the spiritual dimension of a new covenant that overrides the authority of letters, just as it exceeds the authenticity of God's own lettered covenant (3:5–18). Authority now resides in God's word spoken "as from God in the sight of God" (2:17; *RSV* writes "as commissioned by God"). Paul, consequently, can only hope that this kind of authority is evident to the Corinthians in their experience of the faith he has preached (3:2–3; 4:2; 5:11), and in the rigours he is able to endure to see that it is preached (6:4–10).

Although this line of argument is, as Paul observes, commendation of himself and would appear to have no objective value, it is more than mere persuasion. For as well as engaging the sympathies of the Corinthians by professing affection, utter dedication (δούλους ὑμῶν, 4:5: "your slaves"), and complete disregard for his own person and fate, in regard to his position under God he does not confine himself to general sentiment but produces a specific entitlement. That this is as διάκονος is strongly suggested by 3:5–6 with its repeated emphasis on "sufficiency" (ἱκανότης) issuing in the statement that God "has qualified [ἱκάνωσεν] us to be ministers [διακόνους]," and by the importance that Paul attaches to the cognate διακονία at this point through the four interrelated phrases at 3:7–9. When he then concludes the section by cataloguing his credentials to be counted among "God's ministers" (θεοῦ διάκονοι, 6:4), there can be no further doubt that this is the word by which in a controversy about his apostolic authority he chooses to designate himself and to be designated. The way he expresses himself at 6:4 bears a further comment. He is not saying, as the Jerusalem Bible (*JB*) put it, "we prove we are servants of God by . . ." or even, as the New Jerusalem Bible (*NJB*) has revised this, "we prove ourselves authentic servants of God; by . . ."—although the latter is at least an attempt to recognise the special place of the term in Paul's rhetoric at this stage—but that "in the way of ministers of God" he is commending himself on grounds of the following catalogue. On this understanding the phrase "ministers of God" sets a norm or pattern to which Paul claims to conform; to put it another way, both he and the Corinthians knew that θεοῦ διάκονοι were men of the kind he goes on to describe. In addition, we notice that Paul does not use any other noun, such as apostle or ambassador (cf. 5:20), to designate his role. His position is stated in διάκονος. What the term meant to him will be clearer from a closer examination of these passages, but what is already clear is the intimate and singular connection between this word and the announcement of God's revelations; as at 1 Cor. 3:5, it indicates Paul's role as the authoritative mouthpiece of God, and for his readers has precise connotations of a person entrusted with God's full message, charged with the duty to deliver it, and endowed with the right to be heard and believed. The phrase at 6:4 is the technical expression of the claim made earlier that Paul speaks "as from God in the sight of God" (2:17).[2]

2 Cor. 11:23

Here the title receives even greater emphasis by reason of the greater intensity of Paul's controversy with the rival apostles (2 Cor. 10–13), the issue being so grave that Paul fears the gospel itself is at stake (11:4). The matter is also much more complex and must receive more extended treatment. The precise issues are obscure, but the many specialised studies that these chapters have attracted are a sign of the importance scholars attach to them.[3] The following discussion necessarily concentrates on the meaning of the word in question and attempts to do this in a way that throws light on the literary, historical, and theological contexts as a whole.[4] This may appear to be an oversimple and limited objective,[5] but there would seem to be some virtue in the procedure in as much as 1 Cor. 3:5 and 2 Cor. 3–6 have already indicated in general that Paul can use the word διάκονος to say something central and specific in regard to apostleship, whereas 2 Cor. 10–13 is generally discussed without adequate account being taken of what this fact might imply not only for the literary but also for the historical and theological questions the chapters raise.

In the handling of a dispute about apostleship it is to be expected that Paul would invoke the idea of authority, and at 10:8 this appears in the phrase "our authority [ἐξουσία], which the Lord gave for building you up." The phrase is a significant one, and is repeated in full at the conclusion (13:10), which suggests that authority has been one of the main themes throughout. It is possible, however, for this to be obscured by several other issues, and particularly by that of his weakness. Paul had personally resolved the conflict between the exercise of authority and the practice of Christian humility. The latter had become for him an essential component of his theology of apostleship (cf. 1 Cor. 2:1–5), and it receives considerable emphasis in his argument in chapters 10–13. Thus at 12:5–10 he expounds the kenotic principle that Christ's power is made perfect in his weakness, and at 12:10 views his own hardships as evidence of this weakness. Many would therefore agree with Roloff in seeing in this notion of weakness the nub of Paul's argument, and in διάκονος, taken in the sense of servant, a precise expression of it.[6] Such an interpretation, however, runs into several difficulties. The first arises from the supposition that the word means "humble servant," which not only runs counter to the general history of the word but is also not the notion expressed by Paul himself at 1 Cor. 3:5 and at 2 Cor. 3:6 and 6:4. The second is the supposition that Paul would hope to sway the Corinthians in such an urgent situation by an argument based on the paradox of weakness. A third difficulty follows from this, which is that in the course of his argument Paul also evinces more than one attitude to weakness and these have to be distinguished.

In his appeal in 10:1 the idea of weakness appears to be used with a certain amount of irony, and this would seem to indicate that the theme is not entirely of his own choosing and that, in view of the jibe reproduced in detail at 10:10, it was in fact occasioned by reports of what people are saying about him: "his bodily presence is weak and his speech of no account." Paul immediately rejects this with contempt as an irrelevance (10:11–12), though he once more becomes ironical on the subject at 11:19–21a. It would appear that he has had a great deal more to say on a different basis before he seeks to turn weakness to his advantage (11:29a; 11:30–33) by expounding the kenotic principle (12:5–10) and letting this colour also his firm but conciliatory remarks (12:21; 13:4,9) before he returns to and concludes with the original thought of "authority" (13:10). If it is the case that Paul begins by finding the Corinthian charge of weakness a threat to his authority, his ultimate championing of weakness is likely to be convincing only when that authority has been firmly established on other grounds. What grounds were these?

It may be noted that ἐξουσία (10:8; 13:10: "authority") is not a usual term for Paul when discussing his apostleship. He uses the word of the moral rights of the person of the apostle (1 Cor. 9), but elsewhere and with particular emphasis of the absolute and divinely founded authority of the state (Rom. 13). Deputed authority of this kind is also at issue in 2 Cor. 10–13, and Paul has prepared for the thought with a forceful statement about the all-conquering power of the divine knowledge that he bears (10:4–6). But because his right to it is challenged on the grounds of his personal deficiencies by those who feel secure enough to use this kind of argument in support of their own authority (10:10–12), Paul has first to select a basis other than an appeal to weakness on which to establish his own authority. This appears at 10:18: "it is not the man who commends himself that is δόκιμος [accepted], but the man whom the Lord commends." This statement intends more than that the man whom the Lord commends "is to be accepted" (*NEB*), for that would beg the question. What would give a polemical edge here would be for δόκιμος to have a sense that takes the question into the human forum, that is, "judged by men to be acceptable";[7] Paul would then be saying that the appointed man is recognisable, and would be inviting the judgement of the Corinthians as to how clearly he is this man.

Certainly, throughout 2 Cor. 10–13, Paul is pursuing such a line,[8] and for him a large part of the folly of the situation stems from a realisation that he is laying facts before the Corinthians that are actually staring them in the face, and which should have led them to a decision about him: "I ought to have been commended by you" (12:11). This principle goes deep, for if a man "is confident he is Christ's" (10:7) he should be able to judge whether Christ speaks in another (cf. 13:7). He appeals to this principle in 2 Cor. 3:2–3 in a statement whose figurative language merely emphasises the reality and importance of "conscience" (συνείδησις) in the field of Christian experience (cf. 2 Cor. 4:2; 5:11), and 1 Cor. 1–3 is itself to a considerable extent an effort to educate the Corinthians in this precise discernment of the gospel in themselves as in their preachers. That this effort failed is clear from the situation confronting Paul at 2 Cor. 10–13; the wayward enthusiasm of the Corinthians has led them instead to accept a gospel (11:3–6) other than the one preached with an evident singleness of purpose (11:7–11) by him alone (10:13–14; 11:2).[9] Because their judgement is so distorted that they do not discern the man appointed by God, he proceeds to lay before them what are his qualifications.

His Jewish background, for whatever reason it may have been important, is not going to be challenged and is not elaborated (11:22); his status as one of the διάκονοι Χριστοῦ (11:23, *RV:* "ministers of Christ") receives close attention. The phrase itself is given the same kind of prominence as the similar one at 2 Cor. 6:4 and, as there, is the only expression that has the character of a title or self-designation. Clearly, therefore, the phrase bears directly on the question of who has the ἐξουσία or "authority" of an evangelist, and it is thus important to determine in what way the catalogue of his labours and hardships that follows is meant to characterise this διάκονος and to support this ἐξουσία.

The catalogue begins with κόποι (11:23: "labours"), moves into imprisonments and assaults on his person, and then to perils from the elements and from people in the course of his journeys. To these he adds his daily concern for the churches, a thought leading to the sentiment, "Who is weak and I am not weak?" (11:29). In this question there is no suggestion of the earlier ironical attitude to weakness (11:21); indeed, after a similarly sincere expression of sympathy and solidarity with those who are made to fall (11:29b), the thought of the weak reminds him of his own weakness, in the treatment of which there is no hint of the hostility displayed at 10:11. It is as if, after a consideration of the facts recorded in 11:23–29, the subject of his weakness can now safely be mentioned. That his own weakness is an aspect of his life and work quite distinct from that which is illustrated by the labours and hardships of 11:23–29 and not a continuation of it is indicated also by the way the subject is solemnly introduced (11:30–

31). By his oath to God, he would affirm that, pitifully different as this next event is from those he has just recorded, it is nonetheless true. The event is his flight from Damascus, evidence, he maintains, of human weakness. The implication is that the other events are evidence of the contrary, that is, of persistent endeavour, even of strength.

We notice Paul pointing the same paradox in the account of his revelations that follows (12:1–9): the man who has seen Paradise is contrasted with the man of weakness. Here again, he refers summarily to his hardships (12:10), relating them to the kenotic principle just discussed there (12:5–10). If they are to be seen in the light of that, he cannot here mention items like journeys and labours—for that could suggest endeavour and strength—and must confine himself to those that illustrate the limitations of human effort. That he should present two such similar catalogues in such opposing lights need cause no surprise. Paul consistently uses catalogues to different ends, even, as here, in the course of a single argument. Thus, while the kenotic notion given expression at 12:10 ("when I am weak, then I am strong") inspires the mention of the Damascus incident at 11:32–33, and is elaborated in the same way at 2 Cor. 4:7–12 ("the transcendent power belongs to God and not to us"), elsewhere he uses his hardships to point other lessons. At Rom. 8:35 they are the measure of his attachment to Christ, whereas his account of them at 1 Cor. 4:9–13 is highly ironical, the labouring apostles being presented as almost ludicrous figures, truly "a sight to be seen," while those whom they have evangelised rest on their laurels. Finally, at 1 Cor. 15:30–31, in protesting that he is in peril every hour and dies every day, he indicates the extremities of his endeavours on behalf of the gospel. This is also the motif of 2 Cor. 6:4–10, where there is no suggestion that the humiliations are evidence of the weakness through which Jesus works, as in 2 Cor. 4; rather they are the sign of endurance, and are the things which the διάκονος is prepared to do. Indeed, they are bracketed with his virtuous actions, and together these things illustrate the direct path which as διάκονος he takes to the minds and hearts of the Corinthians, inviting from them as direct a response (6:11–13). The catalogue of labours here aims to show that Paul is one of the θεοῦ διάκονοι (6:4, *RV:* "ministers of God"), and as such it is noteworthy in that it occurs in an argument that uses another catalogue (2 Cor. 4:7–12) to illustrate weakness.

Similarly, 2 Cor. 11:23–29 is best understood as a factual, if impassioned, record of what Paul has done in the interests of the gospel and as evidence that what he says is true. This corresponds with the idea of the διάκονος as a faithful emissary and spokesman, and gives an edge to the boasts at 11:21b–23, which otherwise might appear to be merely rhetorical or even feigned. It also explains why at 11:30 he is truly abashed. On the other hand to suppose that 11:23–29, 30–33 constitutes a single whole and is all concerned with the subject of weakness is to overlook the evident transition at 11:30 and to expect readers to take from the passage (11:23–29) a notion so unlikely as weakness in an account of strenuous labors, arduous journeys, and intrepidity in the midst of dangers.[10] Likewise, the suggestion that the catalogue is all about suffering—which "greater labours" (11:23), the leading idea, does not obviously illustrate—makes of suffering a major idea here of which there is no trace in the rest of the argumentation; the "thorn" in the flesh (12:7), like the crucifixion of Jesus himself (13:4), contributes to the theology of weakness developed at that later stage, not to a theology of suffering.[11] Further, on neither supposition of a motif of suffering or of weakness would the catalogue serve the declared purpose of establishing Paul's higher credentials to be a διάκονος; that after all is the title in dispute—not "apostle," claims to which in a Christian sense may or may not be suitably laid on grounds of suffering or of weakness. Nothing in the picture of the διάκονος of the gods, however, or of anybody else, points to suffering or weakness as making up part of the credentials; rather, this word brings with it from that world the strength of its own rhetoric, which requires that what follows a claim to the title will be a telling illustration of the nature of such a singular figure. Labours and journeys against high odds are just that. If we look to catalogues in non-Christian literature, it will not be to use them for determining Paul's intention here or in any particular catalogue because Paul uses catalogues for varying purposes. To none of these does Epictetus's catalogue cited earlier (3.24.113) correspond; for this Stoic, trials and journeys are a school in which he learns to become a living example of indifference to worldly circumstance. On the other hand, a parallel of sorts to 2 Cor. 6 and 11 is to be found in Lucian's account of Hermes, also cited earlier (*DDeor.* 24):

> "And now here am I only just back from Sidon, where he [Zeus] sent me to see after Europa, and before I am in breath again—off I must go to Argos, in quest of Danae, 'and you can take in Boeotia on your way,' says father, 'and see Antiope.' I am half dead with it all."

The humour of this piece arises from the self-pity ("I am the most miserable god in Heaven") of Hermes, who is honoured by men for his exploits and expeditiousness as διάκονος of Zeus. Paul is certainly not cavalier about his own exploits, but, like Hermes, he can use them to show how busy a god's emissary and spokesman must be.

The idea of the emissary and spokesman, which this discussion has sought to establish at 2 Cor. 11:23, corresponds with the idea already examined at 1 Cor. 3:5 and 2 Cor. 3:6 and 6:4. In each of these instances reference is to the emissary of Christ or of God. At 2 Cor. 11:15, by contrast, we read of Satan's διάκονοι, and in coming to an understanding of the term here we are helped in the first instance by a parallel between the phrases "apostles of Christ" and "his [Satan's] διάκονοι," which requires us to read the term as at least a designation of emissaries, and then by a second parallel between the phrases "an angel of light," namely Satan, and "διάκονοι of righteousness," where the parallel arises from the notion of delivering a message from an unworldly realm and requires us to read the latter term as a designation of spokesmen.[12] What is of interest in this is not the association of the term with Satan, as if that were to debase its currency,[13] but rather its suggestion of the other world, of the transcendent and the unknown, and of mediation between that world and this. We have already seen the suggestion that something of this connotation is discernible at 1 Cor. 3:6 in the reference to "what the eye has not seen" (2:6–16), and we will now go on to examine how

The Spokesman as Medium
2 Cor. 2:14–6:13

The earlier brief consideration of this passage aimed to establish that at 3:6 and 6:4 the common noun is an important term for the designation of an authoritative mouthpiece of God. We can now see that these instances are two in a series that stretches from 1 Cor. 3, through 2 Cor. 3–6, to 2 Cor. 10–13; the series can be seen as such even if the middle passage is not recognised as part of a letter that antedated 2 Cor. 10–13 by a year or so because the former is a separate discussion in any event. The frequency, however, of the abstract noun and verb in 2 Cor. 3–6, and the intimate connection there between these words and the notion of revelation itself suggest that "mouthpiece" or "spokesman" might not exhaust the meaning of the common noun in this section of Paul's Corinthian correspondence.

The problem addressed by Paul at 2 Cor. 2:14 is that of rival claims to spokesmanship based on legal technicalities which he is unable to dispute. Paul accordingly argues on a level where he is able to bring into play principles that are strictly theological and, at the same time, experiences that both he and the Corinthians have had of these principles at work in themselves. This experiential element is a strong factor in the argument. Nor is it restricted to the underlying appeal to "conscience," by which the Corinthians would know the truth of what they have heard (cf. 4:2; 5:11) and by force of which, accordingly, they should attribute to Paul the position he claims, for the appeal to experience occasions also the kind of language that an analysis of revelation necessarily elicits; that is to say, any account of knowledge from another world must speak by analogy with what is known in our world, so that language is used that alludes to sensitive areas of life as we know it. Thus from the outset Paul speaks of "the fragrance of the knowledge" of revelation, and of evangelists as "the aroma of Christ to God among those who are being saved" (2:14–16). Apart from any liturgical colouring of this image, Paul is setting up a comparison of some kind between the process by which an evangelist communicates the gospel and the process by which we know things by the sense of smell. As the path of odours is direct and unimpeded, so, he goes on to say, the gospel must be spoken "as from God in the sight of God." The analogy at once puts disputes about technicalities in a poor light (cf. 2:17) at the same time as it underlines the immediacy of God's word. Numerous other analogies to the process of evangelisation occur in terms of light (brightness, glory, enlightenment, image, manifestation) which convey even more effectively the idea of God making contact with the believer, and of the preacher's word, which necessarily stands between the two, as the medium by which they meet. Of interest in such a context is how διάκονος and its cognates also serve this end.

At first sight the words appear to be merely something of a literary motif. Thus the participle occurs in an early phrase describing the Corinthians as "a letter from Christ delivered [διακονηθεῖσα] by us" (3:3, RSV), which is an intricate figure, almost a conceit, and other cognates then recur nine times in the rest of the section (3:6–9; 4:1; 5:18; 6:3, 4); in one respect worth noting, however, the phrase cited from 3:3 is less boldly figurative than is sometimes supposed for in regard to the participle the context alone has suggested to some translators the meaning "delivered" (so RSV above), and the correctness of this view is confirmed not only by other transitive uses of the verb in contexts of message,[1] but by the close parallel at Chariton 8.8.5. At the same time the image of the letter and its delivery is meant to be a vivid illustration of how revelation reaches directly to the believer despite the intervention of an envangeliser.

This image had arisen from the commendatory letters that Paul's rivals held and the use of which, in Paul's mind, impeded the passage of God's word. Commendation, he argues, is rather to be found in the possession by the believer of the word itself, which, in his figure, is written in spirit on his heart and forwarded by him; accepted in faith, it speaks its own testimony about Paul to the Corinthians.[2] The self-effacement implicit in Paul's position does not arise from any menial connotation of the word he uses for "delivered"; his participle expresses only the notion of a connection ("go-between"), not technically (as through a postal service), or even from a perspective that might be called horizontal (Paul to Corinth), but in the perspective of the vertical (God to man).

The idea of a letter gives rise in turn to that of a written covenant. Like the letter, the covenant is written in spirit and, as with the letter, Paul is the bearer (3:6).[3] In the comparison that then follows of the covenant of spirit and the written covenant, Paul uses phrases about the διακονία of spirit and death and of condemnation and righteousness (3:7–9). The translations that speak of "dispensation" come closest to the meaning here but must not be taken to mean—what the *Living Bible*'s paraphrase has taken it to mean at 3:8, "the Holy Spirit is giving life"—that it is God who dispenses.[4] What the abstract noun is expressing is the passing on or mediation of death and spirit.[5] The word implies the existence of a mediator, which in the one instance is the law itself and in the other is Paul, but designates a function abstractly; it cannot designate a function of the deity.

Paul's conviction that in his role in this God's glory passes by way of him is the ground of his own "boldness" (3:12: παρρησία), and the point of 3:18, even without the mirror metaphor, is the inevitability as much as the ineffableness of such glory residing in those who believe. If he can convince the Corinthians that by virtue of his preaching the glory has reached them unimpeded and resides in them unimpaired, his status as God's authorised preacher is established. His claim is explicit in the phrase "having this ministry (διακονίαν)" (4:1), and its validity can be tested in the awareness (4:2: συνείδησις) of every man, for that experience will recognise a manifestation (φανέρωσις) of the word of God. The claim is nonetheless a large one, especially as διακονία at 4:1 also expresses the notion—lacking at 3:7–9—of commission or charge; to counterbalance it Paul invokes evidence of his personal incapacity in the afflictions he endures (4:7–15; cf. 4:16–5:10).

Paul returns to his claim at 5:11 with another appeal to the conscious experience of believers as the ground on which the claim is to be substantiated (5:12), and with a restatement of the gospel of the heavenly Christ (5:14–17; cf. 4:5). In 5:18–6:13 is found a concluding but substantial reflection on his role in this gospel about the coming together of God and man in Christ. Precisely because the gospel is about a coming together, nothing must intervene between God and believer, so that although the union is achieved by reason of a word which can only be delivered by representatives (5:20: "ambassadors"), the word remains God's word (5:20: "God making his appeal through us"), and Paul's part is merely a διακονία or transmission of God's reconciling activity (5:18). Again, therefore, the type of function Paul has in mind is paramount but at the same time, as at 4:1, the idea of a commission is also connoted, and Paul is also clearly thinking of this commission as one involving the delivery of a message.

These thoughts in turn lead him away from the consideration of revelation in the abstract to the concrete situation of himself as a messenger of the divine, labouring strenuously to get his message through, "working together" with God (6:1) and, as "ministers of God" (θεοῦ διάκονοι, 6:4) should, presenting the credentials of strength in adversity and of integrity (6:4–10). Even so he seeks no status for himself in this but only proclaims the efficiency by which the grace of God is channelled through him. This thought is expounded at 6:3: nothing that he does can be construed as an obstacle to the process of the διακονία or task with which he is charged to transmit God's word. He concludes with an appeal to the Corinthians to "widen" their hearts (6:11–13), which, in the light of his conception of revelation as the passing of grace and power from God to believer, is a fitting expression. His mouth is open, his heart is wide, and God's word passes untrammelled to those who will receive it.

The preceding review of 2 Cor. 2:14–6:13 has bypassed several matters of theological and historical interest as well as certain problems of interpretation in order to concentrate on the theme of revelation and on indications from the context that διακονία and its cognates express notions relevant to this. The fact that the words serve similar purposes in non-Christian discursive writing points to Paul's familiarity with such usage but has not been used to determine Paul's own meaning. As in non-Christian usage, to which specific reference will be made in the following section, the expression of the notion of mediation (3:3, 7–9) does not preclude the simultaneous expression of other ideas, so that we also encounter notions of the mouthpiece and courier of the deity (3:6; 6:4), delivery of message (5:18) and transmission of some other effect of the deity (ibid.), commission or charge (4:1; 5:18; 6:3). It seems reasonable to suppose that the words occur in this section of 2 Corinthians precisely because they signify such a set of interrelated ideas affecting both the process of revelation and the role of men under God within that process. The importance Paul attaches to the words both here and in the other passages already examined is to be measured also by the emphasis that falls in particular on the word διάκονος (1 Cor. 3:5; 2 Cor. 3:6; 6:4; 11:23), and from the fact that these words have been introduced into a crucial argument under the pressure of controversy; Paul is either defending his role within revelation against rivals (2 Cor. 2–6; 10–13) or he is protecting revelation against the distortions of human wisdom (1 Cor. 2–3).

The Opponents at Corinth

From what we have seen of Paul's usage so far we conclude that our words are saying something about the communication of the gospel rather than about service to the Lord or to the brethren. This was first emphasised in modern scholarship by Georgi in investigation into the nature of the opposition to Paul at Corinth. His book has been the occasion of our chapter 8. Reviewing his position briefly here, we recall that he correctly drew attention to the occurrence of the sense "messenger" for the common noun in the classical tragedians but then presented what he saw as a sharply defined development of that sense whereby in Thucydides and Pollux the term is a technical designation of the diplomatic envoy; turning then to Epictetus's description of the wandering Cynic preachers, he found the term signifying envoys in the style of a god's definitive manifestation, and concluded that in the syncretistic but highly competitive religious climate of the first century certain Jewish Christian missionaries adopted this nomenclature in order to present themselves to the Corinthians as the authentic envoys of Christ in a direct challenge to the apostolate exercised by Paul. In this they would be giving expression to the conviction that they were, after the manner of Moses and Christ, "divine men"; as such, they saw themselves as partners of the deity enjoying their share of the divine power and equipped with knowledge of the cosmic processes and with a capacity to override nature in miracle.

Our earlier critique of Georgi's presentation insofar as that relies on a particular understanding of the word διάκονος has shown, first, that the word was never a technical designation of diplomats and, second, that while the term applied in the writings of Epictetus to the Stoic moral order it never applied there to religious knowledge communicated by the deity or to its propagation. Outside of Epictetus the term had no currency in Stoic or Cynic circles other than as designating household servants and waiters. In other words there is no evidence from the non-Christian world that the term designated a vocational type within the field of religious mission, and it would be surprising if it did for, although the word signifies vocational types like waiters and valets within the domestic sphere, it has no vocational overtone such as words like "courier" and "herald" can have in the sphere of message; a military officer, for example, might have a pool of couriers whereas a διάκονος is nominated ad hoc. Thus no ancient Greek would have said in regard to religious propagandists, "I saw five διάκονοι enter the city today," although he could have said, "I saw five men today and recognised them [from what they said or did] as διάκονοι of God." There is no way, in sum, in which the common noun can be made part of an argument aiming to show that Paul's adversaries considered themselves to be "divine men."

If Epictetus, however, is not talking about couriers of the gods, in some way Paul undoubtedly is, and that is with a strong emphasis on the notion of being a medium for the passage of heavenly knowledge to earth at the expense of the notion of taking a message from heaven to earth or from place to place. In his use of the common noun and its cognates within the context of his preaching mission Paul gives no direct expression to ideas of movement from one place to another,[6] which in other literature to do with message is so characteristic of the usage. Accordingly if parallels are to be sought for Paul's usage, they are not to be found in anthropomorphic accounts of gods voyaging from heaven on errands for Zeus but in accounts of the deity communicating directly with certain mortals for the purpose of transmitting divine knowledge to others,[7] in accounts of the deity achieving effects through the medium of daemonic beings,[8] and in statements of any natural effect being achieved by means of any medium.[9]

The last type of material referred to there reminds us that parallels between Paul's usage and other Greek usage are linguistic in character and not mythological or even theological; thus a passage speaking of the "interpretative and ministerial" (διακονικὸν φύσιν) role of daemons (Plutarch, *Mor*. 416f) helps us to appreciate what Paul was saying about his role as a medium in the communication of heavenly mysteries but is not telling us what kind of medium or διάκονος he was—as if someone were to infer that he saw himself as a daemon. He saw himself of course neither as daemon nor as a divine man embodying heavenly knowledge and power but as a man passing on heaven's word, and he chose the term διάκονος to depict his role of go-between in the same way as the philosopher chose the cognate verb in attempting to speak of the voice carrying thoughts from one soul to another (διακονούσης αὐταῖς τῆς φωνῆς, Ammonius, *in Cat*. 15.9), or of the moon transmitting the power of the sun without itself possessing the power or being consumed by it (Alexander, *Mixt*. 5.16), or of how another body must stand between the sense faculty and its object if there is to be sight, hearing, or smell (Themistius, *in de An*. 125.9). Paul himself says as he prepares to discourse on the process of revelation—on the διακονία of death and the spirit and on his role as minister of a new covenant—"God spreads through us the fragrance of the knowledge of him" (2 Cor. 2:14), and it is difficult not to see in his self-designation as διάκονος of revelation a mirror image of the philosopher's "ministering body" (σῶμα διακονοῦν), which stands between us and the things we come to know through our senses. One would conclude that here lies the root meaning of Paul's weakness. With no power (ἀρετή) of his own to boast of, he stands at the opposite pole to "divine men."

In applying these words to the situation of the Christian preacher Paul expresses a profound and possibly original Christian thought but he cannot be said to have given the words a specifically Christian meaning for he makes use of meanings already current in non-Christian literature. He uses the words to such advantage, however, exploiting their connotations even when he is not disputing with his adversaries (as at 1 Cor. 3:5), and defends the title διάκονος with such passion that he is unlikely to have been inspired in its use only by what his adversaries had made of it. As we have seen, there is no known group of propagandists who used the word in connection with religious knowledge or with proselytising activity; all we know of it in this connection is what Paul says, so that his adversaries' conception of it can be measured only by that. Although it would be possible to suppose that the adversaries appropriated the title in the first instance from Paul, it is more likely, given the currency of the term in many types of religious situations, including that of divine message, that Paul and his adversaries made independent use of it. The difference between the two uses may well have been that whereas the adversaries were content to parade the word as a title in the sense of messenger of Christ, Paul probed more deeply into its meaning in order to show that any genuine message from the heavenly Christ is conveyed directly in spirit, and that those messengers who boast of rights, skills, and works interfere with this process and cannot be the kind of medium that a διάκονος by definition is.

In whatever way Paul and his adversaries differed in their understanding of the title, the fact that they shared it must have its own significance for the history of the Corinthian opposition. The overriding fact is that though the opponents are Jews their title is indisputably Hellenistic. If they also claimed the title "apostle," which may be assumed to reflect the Jewish rather than the Hellenistic side of their interests, Paul manifests little interest in that claim (2 Cor. 11:13), quickly transposing the debate onto the level of διάκονος (11:15), where he concentrates both his thought and his rhetoric (11:23–12:10). In this, the need to put his argument in a way that the Corinthians would appreciate it is certainly not to be overlooked, but had the adversaries been Judaisers Paul is not likely to have omitted a treatment of the objections traditionally raised by Judaism against his Gentile mission, or in such circumstances is he likely to have made so much of a purely Hellenistic title. For Paul at least the import of the title lies in what it says about access to revelation so that this would seem to be the contested issue; in that case the issue is likely to have been raised by Hellenising rather than by Judaising Jews because for the latter revelation may be said to have been a closed book. The Hellenist, on the other hand, had every reason to accommodate what he had received to the spirit of the age; as a consequence distortions of the gospel as Paul conceived it could well occur.

There are only obscure indications of what such distortions may have been, but if the contest about who has rights to the title διάκονος can rightly be said to give a unity to 2 Cor. 10–13, this is itself a sound reason for identifying the apostles of 11:5 with those of 11:13, and for finding in the satanic falsity of the latter the measure of the heterodoxy of the "different" gospel at 11:4.[10] Should we leave ourselves open to underestimating how gravely Paul would have weighed his phrases about another Jesus, spirit and gospel? Unethical behaviour of the intruders would not of itself put these great matters at risk, and Paul would manifest such urgent concern about them only if they were at risk. There would seem to be a real divergence of doctrine somewhere, and something of its nature is perhaps concealed in Paul's reference to another "Jesus." One possibility here may be considered.

To argue that the use of the name Jesus necessarily refers to the earthly Jesus is in the end unconvincing, but the fact that the name lacks any title may have a significance in another direction. The Jesus preached by Paul was "Lord," a person with a specific past in death and resurrection but also with a powerful present in the spirit, and of these elements the latter is what makes the gospel real for the believer. Unless the knowledge and power of the "Lord" is communicated, the gospel is as vain as if Christ was not raised. The Jesus preached by the intruders, on the other hand, may not have been "Lord" but a *heavenly* figure of another kind, whose nature, functions, and powers can now only be guessed at. Surely he is not likely to have been anything less than the Lord of Paul, and that he was something more may be indicated by the curious epithet ὑπερλίαν (11:5).

It is customary—and correct—to draw attention to the ironic emphasis this word adds to Paul's mention of the apostles—"the derisive import of this extraordinary adverb," as one writer puts it[11]—but at the same time we are perhaps not meant to overlook another aspect of the word. λίαν itself indicates not only degree, as represented in translations here: "very chiefest" (*AV, RV*), "preeminent" (*RV* margin), "superlative" (*RSV, NEB*), but superfluity or excess (*LSJ*), which is a condition compounded by the prefixing of ὑπέρ, a word Paul called on to say that Onesimus was "more than" a slave (Philem. 16); Paul's composite adverb might then be implying that these apostles were "more apostle" than they had warrant to be, an aspect reflected perhaps in the translations "archapostles" (*JB*), "super-apostles" (*NIV, NJB*), and effectively expressed in a phrase from a note to the *RV:* "only too apostolic." That these men had great (excessive?) facility in speech is well attested, but in the same breath Paul comments also on their knowledge (γνῶσις, 11:6), and the implication might be that in their preaching they exaggerated the position and functions of Jesus, placing him in a category different from Lordship, exaggerated too his manifestations in believers and in the preachers themselves. They preached more than was believable, or were too clever by half. In saying that he does not fall short of such preachers, Paul is being ironical indeed, for he has known greater mysteries than his opponents (12:1–4)—a claim eliciting his second reference to the super-apostles (12:11)—and has taken such complete command of all knowledge (10:5) that there is not a legitimate thought to spare with which they might embellish or pervert the gospel.

Epaphras and Paul

Paul calls Epaphras "a faithful διάκονος on our behalf" (Col. 1:7). The phrasing and what facts are known about Epaphras makes this a significant description, a significance that is enhanced by the application of the term to Paul himself at 1:23 and 25 for effects reminiscent of uses in 1–2 Corinthians. Epaphras is the founder of the church in Colossae, that is, he has been the preacher of the gospel to a group who had not yet seen Paul (2:1). For whatever reason Epaphras came to be emprisoned with Paul (Philem. 23), there are reasons for thinking that he had initially come to Paul for advice on problems that beset his community; he may even have been discarded by his community, over which Archippus now seems to preside (4:17; Philem. 2). Paul nonetheless continues to commend Epaphras in the highest terms, some of which are inspired by what Epaphras had done in establishing the Colossians in the gospel (4:12–13). This suggests that by naming him διάκονος at 1:7 Paul is paying him a weighty tribute. That Paul does not use the term lightly would appear from the crowning emphasis which he gives to the concluding phrase at 1:23, "I, Paul, διάκονος." These aspects of the context invite a closer inspection of the place of this term there.

The encomium of the gospel to which this self-designation by Paul forms a conclusion (1:9–23) is preparatory to Paul's discussion of the relative merits of cosmic philosophies and Christian revelation, and aims to enrich the Colossians' understanding of the latter so that they will give worthy expression to it in their conduct. The emphases accordingly fall on the revealed knowledge whose lack had occasioned their earlier evil ways and separated them from God, and on the power by which God maintains them in Christ's victory over cosmic forces. Knowledge and power are thus two aspects of God's achievement and presence in the gathering of all under Christ. They are, however, the effects of "the invisible God," and "under heaven" are within reach of experience only by means of a spoken gospel; this the Colossians have already heard from Epaphras. Paul's statement about himself would seem to be saying that in a preeminent way, or at least prior to others, he himself was constituted a διάκονος of the gospel, that is, a bearer of the message from an invisible world who stands in a singular relationship to the God who originates it. As far as Epaphras is concerned, he has been a substitute ("on behalf of us" is a better reading at 1:7 than "on behalf of you") for Paul-διάκονος.

The strong revelatory context in which the last phrase occurs of itself suggests that the bearer of the word is also the medium by which knowledge and power came, and from 1:24 we have other indications that this is indeed the case. Just as Paul's sufferings for the sake of the Colossians are a complement of Christ's sufferings and are the continuity in his flesh of Christ's work, so continuity is maintained also between the mystery hidden eternally in God and the faith of the Gentiles by means of the word that Paul carries. Although he says he became a διάκονος "of the church," (1:25), that is, of "the body" (1:24), this connection contributes little to the meaning of the term in comparison with the resonant phrases that are attached to it in the Greek: "a minister according to the divine office which was given to me for you, to make the word of God fully known, the mystery hidden for ages." What characterises this minister is not church, which is merely the locale of his activity, but the word-filled role that encompasses Gentiles in the glorious mystery by manifesting its knowledge and power. Effectiveness resides in the word, not in the minister, who teaches strenuously but by vicarious energy (1:28–29), and the process by which the mystery emerges is "according to Christ," who overrides or dispenses with any process that is "according to human tradition" (2:8). Thus the passage is not merely about a message from the other world but also about the difference between the process involving an authentic διάκονος and that adopted by "one who preys on" the human mind (ὁ συλαγαγῶν, 2:8). The method of the minister of the gospel is manifestation, not demonstration, and leaves the way open for God's direct working through faith.

Only one other point is taken up from this rich bed of terminology of mission and is the "divine office" given to Paul (1:25, *RSV*). This is the Greek's οἰκονομία of God, which is the working out in time of a plan eternally conceived. It forms a link with a similar line of thought in Ephesians where "the eternal purpose" (Eph. 3:11) is actualised in an οἰκονομία or temporal operation (3:9), and the process of revelation is again presented but in even more thoroughly dynamic terms. Knowledge of the mystery is an outflow of God's own comprehension of how the times are brought to their fulness in Christ (1:8–10), but the outflow is this controlled dispensation or οἰκονομία by God through Paul (3:2), who is accordingly διάκονος of the gospel (3:7), again with the emphasis on the vicarious energy by which he operates (ibid.).

Paul's Sacred Mission to the Gentiles

In the letter to the Romans such processes of revelation are not Paul's theme; his purpose is rather to present grounds on which extension of the gospel to Gentiles is to be understood. Although he celebrates his status of apostle (Rom. 1:1–6), enveloping it in a splendid aura of worship with his ministry and the Gentile response being now a vast liturgy to the praise of God (15:15–16), he does not seek out occasion to justify or explain the precise role of the preacher. At one poignant juncture he attempts to meet the paradox of the failure of blessed Israel to come to faith by undertaking to display to Israel the rich store of blessing among the Gentiles, professing that the harder he works among them the richer the store of their blessings will be and the more likely some of Israel will be attracted to claim their own ancient inheritance (11:11–16). In presenting himself as going out from Israel in this cause he writes, "Inasmuch then as I am an apostle to the Gentiles, I magnify my διακονίαν" (11:13), and means by this abstract noun the sacred mission with which he has been charged to venture with the word of the gospel to lands that are not Israel.[1]

The word recurs with a lot of this weight at 2 Cor. 11:8 but with a narrower reference to the particular mission to the Corinthians. In the course of insisting that his work among the Corinthians had not been a financial burden to them, he notes that he had taken expenses from other churches (11:8) and that once he was among them he had received further help from the Macedonians (11:9). To report his position in this way is to propose that Paul is accounting for two sets of expenses, one lot to cover his journey to Corinth (11:8) and the other to subsidize his residence there (11:9). When we turn to the Greek, Paul's claim does not seem to divide so neatly as this. We read that he accepted support πρὸς τὴν ὑμῶν διακονίαν (11:8: "for the ministry of you"); this is a useless phrase in English and an unusual one in Greek, but with the Greek we can appreciate at least that the abstract noun can designate a mission to somewhere and that the unusual element is only the genitive "of you."[2] Perhaps Paul has been lazy here but perhaps also we can discern what has happened in the process of his writing. His phrases here read: "taking support for the ministry of you and being among you and in need"; this contains two phrases beginning with πρός ("for" and "among") in close proximity, and to express himself conventionally about his "ministry to you" he would have needed a third in between them (to replace "of"). He went for the genitive instead. His meaning is still recoverable: "I received support for the mission to you."

If these uses of the abstract noun are strong and effective within the field of apostolic mission, we might ask why in Romans and other letters Paul has not drawn also on what was his preferred self-designation in 1–2 Corinthians, the common noun διάκονος. The answer is likely to lie in the nature of the questions addressed. We have noted the shift of focus in the view of apostleship in Romans as compared with 1–2 Corinthians. In Galatians we confront law again and, although Paul indicates that he personally draws on a revelation (Gal. 1:12), his argument is built on the accord with the Pillars of Jerusalem, on a theology of faith expounded from the scriptures, and on what Christ's death must mean for a religion founded on law. These are not contexts eliciting words about messages from a heavenly source; when Paul says he was received as "an angel of God" (Gal. 4:14), he is not attempting to characterise his role but to provide a measure of the esteem in which he had been held: it was a term from Judaism (see 3:19) and was entirely appropriate in a debate about Paul's authority vis-à-vis the law of Judaism. The discussion in Philippians, by contrast, does enter into the mysteries of revelation but celebrates these, and it does not constitute an apologia in which Paul might need to defend heavenly knowledge against worldly wisdom or his own position as a purveyor of revelation against rivals. He drew on the words when he was pressing one or other of these points about his role; his other uses of the words are incidental, as at Rom. 11:13, except for the sustained uses in the passages dealing with the collection for Jerusalem. These will be examined separately in the next chapter. Before that we will look at what the author of Acts means in speaking of the διακονία of Paul and the Twelve.

Paul and the Twelve in Acts

In his farewell to the elders of Ephesus Paul expresses a strong desire to complete his course and the διακονίαν which he received from the Lord to testify to the gospel (Acts 20:24). There are several indications here that we are dealing with a sacred commission to preach as in Rom. 11:13: the task involves testifying to the gospel, it is imposed by the Lord, and both the circumstances of the address and the association with the course he is running connote an itinerant mission. The meaning is different when the term next emerges in Paul's address on his arrival in Jerusalem; he relates to James and the elders "one after the other the things God had done among the Gentiles through his διακονίας" (Acts 21:19). The concluding phrase reads most naturally as "through his agency," and a consideration of the context will support this.

The passage records the return to Jerusalem of the apostle of the Gentiles, an incident that is as momentous for the church as it is for Paul personally, and Luke's careful drafting of the events leading up to it leaves no doubt as to his sense of the occasion. The statement reports only in general terms ("one by one") on Paul's activities since his previous meeting with James and the elders, details that the author Luke has no need to repeat; whatever else Paul may have included in his address were not to Luke's purpose either, for his bald statement concentrates on what "*God* has done." The same formula occurs in a similar situation at 15:12 and is the expression of Luke's conviction that the word of God is a vital power in itself (6:7) and that it is God who opens the door of faith (14:27); the statement concludes with "through them," the clear instrumental sense of which elucidates 21:19, where διακονία would not be an express designation of mission but part of a phrase signifying God's mode of operation. In a passage that intends to expose God's action directly to view and stresses that human capacities are inadequate for the effects achieved—the elders respond by praising *God*—the phrase is to be seen as speaking of the mediation of the divine power and authority. That

Luke has used this phrase in reference to Paul at 21:19, but only the colourless prepositional phrase "through them" in reference to Paul and Barnabas at 15:12, is an expression of the special relationship that Luke establishes between Paul and the διακονία of the Twelve. This takes us to the beginning of Acts.

The opening theme of Acts is the commission entrusted to "chosen apostles" to be witnesses of the resurrection to the ends of the earth. For this the gift of the Spirit is necessary but, given Luke's special interest in the Twelve, a prior requirement is to fill the place of Judas who, as Peter says, "was numbered among us, and was allotted his share in this διακονίας" (1.17). The word "this" is interesting here for while the "ministry" which it qualifies must embrace the commission to go forth as witnesses (1:8), "this" in 1:17 cannot be connected syntactically with a commission in 1:8; in the immediate context "this" can only refer to apostleship (1:17: "numbered among us"; "allotted his share") so that διακονία is part of the role reserved for those who are chosen by the Lord (1:24) and enrolled with the apostles (1:26). Apostleship and διακονία come together in fact at 1:25: "the place in this διακονίας and apostleship." This phrase is not hendiadys; the latter term, used only here by Luke, is a specific designation of the office of apostolic witness and takes its colour from the Lukan history of the Twelve, whereas the former, in accord with general usage and similarly to the instance at Rom. 11:13,[1] designates a commission to go forth under a divine mandate.

Acts 6:4 is the third and last instance of the abstract noun in connection with the Twelve, who here undertake to dedicate themselves "to prayer and the διακονίᾳ of the word." Luke's meaning here appears from the phrase in which he combines the abstract noun with "the word" of the gospel, the combination establishing that Luke is writing of the apostolic commission of 1:17, 25 to which the Twelve themselves allude (6:2), but the objective genitive "of the word" alters the connotation from one of preaching journeys to one of the transmission of the word, in the manner of Paul's connotation of the mediation of covenant and reconciliation (2 Cor. 3:6–9; 6:18), and in line with Luke's own usage at Acts 21:19. In this we are not to overlook how Luke expresses himself in regard to "the word." He does not, for example, speak of giving up "preaching the word" (6:2, *RSV*) but of "giving up the word," just as he subsequently writes that "the word of God increased" (6:7), "grew and multiplied" (12:24), "spread" (13:49), and "grew" (19:20); "the word" has its own power, makes its own progress through the regions and into believers, "through my mouth," as Peter says (15:7), and brings to mind a view of transmitting the word of the gospel that is not far from Philo's view of the transmission of the word of the law (*de vita Moysis* 84). The Twelve are under commission to "the word" and are the channel by which it communicates itself.

Such usage aligns Luke so closely with Paul and with non-Christian discourse in this area that we need to explain why Luke does not also use the common noun διάκονος to designate those commissioned within the διακονία. One part of the explanation lies in Luke's dislike for a term that was current as a title for Hermes and other messenger gods. That Luke was familiar with Hermes-διάκονος is clear from the incident at Lystra where the enthusiastic crowds, proclaiming that the gods had come down, identified Paul in particular with Hermes because he was "the principal speaker" (Acts 14:12); this phrase can be no more than a gloss on Hermes' title, and the kind of "divine man," which such a comparison with Hermes could make of Paul, is the antithesis of what Luke has in mind for an apostle. More can be said of this aspect of Luke's attitude. If the apostle is not an emissary from heaven, neither is he the bearer of a purely heavenly word but of a word that "dealt with all that Jesus began to do and teach . . . during all the time that the Lord Jesus went in and out among us" (Acts 1:1, 21); the historical experience of these things in conjunction with a personal commission from the Lord—the case of Matthias being made to meet the latter requirement (1:24–25)— is what constitutes the apostle. The word of Paul, to whom alone among all other evangelisers Luke attributes a διακονία from the Lord after the manner of the Twelve, is essentially the same as theirs; like them he is "witness" (1:8, 22; 22:15; 26:16) of the resurrection, and although he has known the Lord only in visions (16:9–10; 18:9; 22:17–21; 23:11; 27:23) Luke understands that these experiences are an extension in time of the kind that authenticated the original apostleship. Here lies the main part of the reason why Luke calls neither the Twelve nor Paul διάκονος; the διάκονος of a god does not bear a message that has emerged from within history but one that is from the realm of spirit, and his word is not that unique amalgam of history and faith that "the word of God" is in Luke's mind.

How carefully Luke enforces this distinction between the old religion and the new is seen from the fact that he designates Paul—to go back to the translation of the Authorized Version *(AV)* and the Revised Version *(RV)*—"a minister and a witness" (26:16), the latter term expressing an element fundamental to the makeup of an apostle on Luke's understanding and included in each of the other accounts of Paul's original vision (9:15; 22:15), whereas the former term would seem to coincide with Paul's fundamental conception of himself as the διάκονος, except that Luke's Greek term is ὑπηρέτης; now while more recent translations give the colourless word "servant" for this, it does in fact commonly designate specific officers in the middle and lower levels of legal, military, and governmental bureaucracies, as we have more than once had to take note of in the interplay between this word and διάκονος.[2] Luke's uses of the word, after the manner of John, the other synoptic gospel writers, and Paul, are exclusively technical. What this makes of the "minister" of 26:16 is an officer charged with a message that is as truly recorded as the legal documents authenticated and delivered by clerks of court and similar functionaries. What this officer is to witness to are, in the Lord's words there, "the things in which you have seen me and . . . those in which I will appear to you," but these visions are for Luke of a kind that takes Paul back to an original apostolic word delivered to Luke by "eyewitnesses" and other ὑπηρέται or "officers of the word" (Luke 1:2). Whether as "officer," as "witness," or in his διακονία, Paul is the equivalent of any member of the Twelve; that he is not also in Luke's mind the διάκονος that Paul thought himself to be is a minor result of their different approaches to the question of revelation. Paul writes out of the personal experience of vision and under the pressure of controversy, whereas Luke with no less an interest in revelation writes with an eye to the church and to the necessity of maintaining the link between the church and the Lord who broke into history.

The Later Christian Mission

The author of 1 Timothy writes that Christ appointed Paul εἰς διακονίαν (1 Tim. 1:12). This is not an appointment "to his service" (*RSV* and commonly) but an appointment "to mission," that is, to the sacred mission of going forth with the gospel. Indications of this sense are clear: Paul is entrusted with the glorious gospel (1:11), he is endowed by God with strength for the task (1:12), and he is faithful (ibid.), which is the sine qua non of a minister charged with a message; these are the grounds on which Paul argued the authenticity of his own claim to the mission in Corinth (2 Cor. 11:23), and the reference here is to the same apostolic mission as at Rom. 11:13 and Acts 21:19. Timothy himself is told he will be a worthy διάκονος of Christ Jesus (1 Tim. 4:6) if he maintains himself in good doctrine and if he passes on to the community instructions he has received; again the hallmark is fidelity, and the difference between this minister and Paul is that in these "later times" (4:1) when the community is exposed to "deceitful spirits and doctrines of demons" (4:1) and "godless and silly myths" (4:7), the concern has shifted from channelling faith into living hearts (2 Cor. 6:11–12) to transmitting the faith across the generations. Two second-century apologists produce phrases relating to the process. Aristides relates that after the first disciples "went out into the provinces of the empire" they were succeeded by those who were "still purveying righteousness" (διακονοῦντες δικαιοσύνῃ, 15:2),[1] and Tatian upbraids nonbelieving Greeks for rejecting "the διάκονον of the God who suffered," namely, the one charged with his message.

Prophets and Angels

Hermas's phrase "prophets of God and his διάκονοι" (Sim. 9.15.4) refers to pre-Christian prophets; we notice that the ministers are not the prophets' ministers but God's, and we are to understand them as being authentic bearers of God's message who were not honoured with the traditional designation of "prophet." The prophets and ministers stand in the same relation to one another as the apostles and teachers (ibid.); the canonical title comes first. The same prophets are said in 1 Peter to have passed on predictions of the sufferings and glory of Christ: διηκόνουν αὐτά (1 Pet. 1:12). This passing on is not as from prophets to future generations but as from the unseen world to their own; the author is specific about this for the prophets had to "search and enquire" about the secret knowledge to which "the Spirit of Christ within them" then testified, so that in prophesying subsequent to this revelation they were communicating knowledge from the other world. Translators have had little success with the concept, Moffatt even discarding the verb in favour of a Rendel Harris conjecture to provide the sense, "they got this intelligence not for themselves but for you," which ironically is what the author is talking about; *The New Jerusalem Bible (NJB)* would be adequate if the words in the square brackets in the following were omitted: "it was for your sake and not their own that they were [acting as servants] delivering the message which has now been announced to you." The usage derives not from talk about servants but from discourse about seers and divination where Plato has set the norm (*Plt.* 290c); his phraseology is reflected in Tatian's rebuke of pagan expectations: "What is the art of divination?" he asks. "It is but the vehicle [διάκονος] of your worldly desires" (Or. 19.2).

The author of Hebrews also writes within the idiom to depict angelic beings as bearers of message from heaven to earth. He is arguing that the Son, who for a little time was made lower than the angels (Heb. 2:9), was superior to them (1:4), and one facet of the argument is a comparison of the messages they bear. The Son has a word of power to uphold the universe (1:3) and to announce salvation (2:3), while the angels delivered a law that did not save (2:2) but was delivered only for the sake of a salvation yet to come (1:14). The angels are only creatures of message—winds and flames of fire (1:7)—who are "ministering" spirits (λειτουργικά, 1:14) because they are in the service of God and are εἰς διακονίαν (ibid.), not because they are "sent forth to serve" (*RSV* and commonly) but because they are "sent on mission" or are in transit between heaven and earth; the phrase "for the sake of those who are to obtain salvation" depends on "sent," whereas translations often give the impression that the angels were "to serve for the sake of those." We met the same prepositional phrase when Paul was "appointed to mission" (1 Tim. 1:12).

11

Emissaries in the Church

The preceding chapter has illustrated that a substantial part of Christian usage relates to the passing on of revelation and in so doing conforms to the general usage where the words appear in connection with message from the gods and with the notion of mediation. The present chapter examines usage relating to travelling in the name of a community or in the name of an evangelist. In this Christian usage continues to parallel in all respects non-Christian usage.

Collection for Jerusalem

The abstract noun and the verb are prominent among terms used by Paul in discussing the collection for the poor in Jerusalem.[1] The modern fashion for their interpretation was set by Kittel's *Theological Dictionary* over fifty years ago in Beyer's statement that the words refer to "the general service of love which Christians evince to one another as saints."[2] This view is reflected in most contemporary translations, although a confusing array of renderings indicates that the words have still often proved problematical. Thus the Authorised Version's "the ministering to" for the abstract noun at 2 Cor. 9:1 has in more recent times become "aid" *(NEB)* and "the offering for" *(RSV);* similarly the Authorised Version's rendering of the participial phrase at Rom. 15:25 "to minister unto the saints" has become "with aid for the saints" *(RSV),* "in the service of the holy people" *(NJB),*and "on an errand to God's people" *(NEB).* These are not all compatible, and the problems of translation are not removed by referring to the *Testament of Job* to illustrate that "Greek-speaking Judaism had come to use these words in a technical sense for supplying the needs of the poor";[3] the *Testament of Job* exhibits its own subtleties, as we have seen, but not a "technical sense" of aiding the poor. The interpretation of Paul's usage presented in the following pages, by contrast, derives from the understanding we have built up of general Greek usage

217

in the area of errand, and sees the words expressing ideas about going on a mission to deliver something for a delegating authority. Interestingly one element of this is apparent in *The New English Bible*'s phrase cited previously "on an errand to" (identical to Moffatt 1935), and Georgi, as we saw in discussing early Christian attitudes to works of mercy in chapter 2, came close to another part of it in attempting to maintain in these passages the concepts of envoys and legations we examined in chapter 8, although he confessed to difficulties in applying such notions consistently.[4] We will first attempt to show how the idea of a formal mission is appropriate to the historical circumstances described by Paul.

Aspects of the historical situation are to be found in Rom. 15 and 2 Cor. 8 and 9. Basic to Paul's thinking is his conception of the collection as a corporate gesture on the part of the Gentile churches. Given his position as the founder and leading apostle of these communities, however, it was not always easy for him, especially in the face of any kind of opposition to the project, to avoid giving the impression that the collection was something he was imposing on the communities. While therefore he insistently appeals to the Corinthians to get their own contribution to it organised, he is equally insistent that he is not derogating from their authority in the matter, making the point that it is not he who will appear as the major benefactor even should they agree to his joining the delegation.

Thus the instructions to the Corinthians about the λογεία (which is the only term meaning specifically "collection of money," 1 Cor. 16:1) include the information that Paul will send those whom they accredit to carry the gift to Jerusalem. The Corinthians, however, are slow to put their gift together and, because the Macedonians meanwhile have gathered money and are keen to take part (2 Cor. 8:4), are urged by Paul to complete their arrangements. He sends Titus to assist in this and along with him the brother famous for his preaching of the gospel. At this juncture (8:19–20) he emphasises that the responsibility for the collection is not his own but that of the churches and points out, lest criticism about extravagance of the gift should be directed against him personally, that this brother has been chosen for the journey by the churches themselves. He endorses the point by implying that a project of this kind must be duly constituted if it is to appear honourable to men (8:21) and by describing the brethren not as his own agents but as "apostles of the churches" (8:23). In this way Paul hopes to disarm Corinthian opposition to the arrangements.

The first statement from Paul with a word of interest to us is at 2 Cor. 8:4 where Paul is explaining to the Corinthians how the Macedonians "had kept imploring us most insistently for the privilege of a share in the fellowship of τῆς διακονίας [τῆς εἰς] to God's holy people" *(NJB);* in the light of the historical and personal circumstances just outlined and in view of the multiple instances of this abstract noun and its cognates expressing action in the name of another,[5] we are to understand Paul as speaking of "a share in the fellowship of the mission to God's holy people." Such a reading of the abstract noun provides no difficulty either contextually or syntactically *(NJB* has had to force "service to" out of the Greek preposition "into") but enhances the whole rhetoric. And so through the rest of the passages Paul finds the terms suitable in discussing a gift that was to be managed and delivered by him or other appointed persons. The extent of his own involvement in the operation is pointedly conveyed by two phrases at 8:19 and 20. To take their meaning we need to draw on our familiarity with the syntax of "the letter delivered by" Paul (2 Cor. 3:3) and with the illustrative Greek usage adduced there in the preceding chapter, and as well we need to place the phrases in the context of Paul and others being already engaged in passing from church to church gathering the gift, and we arrive just a nuance beyond the perception of the seventeenth-century translators of the Authorised Version *(AV)* who wrote of the gift and of the abundance "administered by us" (διακονουμένη in each instance, 2 Cor. 8:19, 20). Numbers of translators have in fact retained "administer," and the word lacks only the connotation of movement, which the Greek in such a context conveniently supplies.

Paul takes up the topic of the collection again in 2 Cor. 9 in what is probably a section of a later letter, and he is again attempting not to step outside the role of adviser in a matter that lies within the competence of the Corinthians themselves. The matter is urgent, nonetheless, because the Macedonian section of the delegation is practically on the road and it would be an embarrassment for all concerned if the Corinthians had nothing to hand when they arrived. Accordingly he urges on them the virtues and rewards of liberality, not least among the rewards being the gratitude in the Jerusalem community, which will redound to the praise of God and, indeed, to the esteem of the Corinthians themselves. Their generosity, he writes, will produce "through us" thanksgiving to God (9:11), and that this does not mean their reward will be in seeing Paul make thanksgiving is apparent as he goes on to write, "because the rendering of this service not only supplies the wants of the saints but also overflows in many thanksgivings" (9:12), that is, thanksgivings in Jerusalem; "through us," then, relates to "the rendering of the service"—that is, through Paul and and his associates—that underlines the role he has assumed as an agent of the various communities in the management and delivery of the collection and allows him to sidestep any charge of direct intervention in the monetary affairs of Corinth. The translation "rendering" *(RSV)* of διακονία represents clearly the idea of the execution of a task—compare the inappropriate translation "help" *(NJB)*—but cannot represent the important connotation that this task is an errand undertaken in the name of certain parties, namely the churches, and is thus a mission or delegation.[6] The objective genitive "of the service" (λειτουργίας) is the translation of a word meaning more precisely "public undertaking."

The concept of the gift to the needy church of Jerusalem as a delegation from the churches of Asia is vital to Paul because in Jerusalem he hopes to receive the acknowledgement of an act of fellowship that has been demonstrably public on the part of the churches founded by him. If the act is seen to be such it will demonstrate that the Gentile communities are authentic churches, subject to the gospel; by participating in the mission, therefore, the Corinthians will establish their credentials as a church. The idea of the community proving and testing itself by contributing to the collection generously has been prominent throughout (8:2; 8:22) and crowns his appeal here: "because when you have proved your quality by this διακονίας"—mission—"they will give glory to God for the obedience you show in professing the gospel" (9.13, *NJB*).

A third passage to consider is Rom. 15. In explaining that his visit to Rome will be delayed Paul states: πορεύομαι εἰς Ἱερουσαλημ διακονῶν τοῖς ἁγίοις (Rom. 15:25, *NEB:* "I am on my way to Jerusalem, on an errand to God's people there"). The conjunction here of the present participle of the verb with the main verb of motion signifies that the participle is referring to being on an errand of some kind, as the translation correctly represents.[7] What kind of errand this is and in particular what part in it "God's people" play will be important for us to clarify. The following verse is helpful: "For Macedonia and Achaia have resolved to raise a common fund for the benefit of the poor among God's people at Jerusalem." This statement explains how it has come about that Paul is on an errand, and if he is on an errand expressed precisely as διακονῶν we can only infer that he is acting at somebody's behest, namely, that he is acting as a courier for the churches of Macedonia and Achaia; it is clear that the churches have made their own decisions about this as their resolution is reported in successive verses (15:26, 27), and it is from here that Paul derives his sense of obligation to them. Because he is on a mission in their name he will not be free to undertake his journey to Rome and Spain until he has signed over the gift in Jerusalem (15:28), an act that will mark the end of his "mission to Jerusalem" (διακονία ἡ εἰς, 15:31 Moffat).[8] He expresses concern that it will be favourably received.

Thus Paul has used the words in this passage to indicate that he is acting in an official capacity as the appointed courier of the churches. His responsibilities are primarily technical and it is these that he urges in order to explain his delay in visiting Rome, and the instances of the verb and abstract noun at 15:25 and 31 speak directly and exclusively of his commitment to this charge from the churches in Macedonia and Achaia. This has an important bearing on the identity and role of "God's people" in the earlier translation: "an errand to God's people," that is the people in Jerusalem. These are rather to be understood as the people of the churches who have commissioned Paul so that we should translate: "on a mission from God's people." That this is appropriate to the context goes without saying but it is also required for stylistic and syntactical reasons. The construction of a present participle with a verb of motion in the present tense is the natural way of saying that Paul is performing the action of "ministering" to or for the people while he is on the move, and the only people these can be are of course those who sent him on the journey; he is carrying out the journey "for" them. The first unequivocal reference to the people of Jerusalem comes in the following verse: "a common fund for the benefit of the poor among God's people at Jerusalem" (15:26); these are indubitably the people who are to benefit from the mission and we notice that Paul describes them fittingly: "the poor among God's people." If he had had these people in mind in the preceding verse we could have expected him to identify them just as clearly, whereas he has written only "God's people" ("there" of the *New English Bible [NEB]* is not expressed in the Greek). But these are stylistic considerations. The determining factor is syntax. A dative of person after a verb in a context of errand indicates the person in whose name the errand is undertaken.[9] The only possible understanding of the phrase διακονῶν τοῖς ἁγίοις in this context is that Paul is on a journey at the behest of people called saints who, we learn, are communities in Macedonia and Achaia attempting to provide material support for a deprived community of fellow saints in Jerusalem; that the journey is a mission or delegation we learn from the broader history of the event but can also infer from the formalities surrounding it, especially its conclusion (15:28, *NEB;* "under my seal"; 30–31: prayers for a suitable reception), and from the meaning and character of these special terms, which are eminently suited as well to express relationships between religious communities.

The Delegation from Antioch to Jerusalem

The abstract noun occurs twice in Luke's account of the mission from Antioch to Jerusalem (Acts 11:27–30; 12:25). He reports that with the onset of the famine foretold by Agabus "the disciples determined, every one according to his ability, to send relief to the brethren [εἰς διακονίαν πέμψαι τοῖς . . . ἀδελφοῖς] who lived in Judea; and they did so, sending it to the elders by the hand of Barnabas and Saul" (11:29–30, *RSV*). At 12:25 Luke adds: 'Barnabas and Saul returned from Jerusalem when they fulfilled their mission [τὴν διακονίαν]" *(RSV)*.[1] These statements are clearly telling us that one community rendered practical assistance to another in a time of hardship. The translation "to send relief" (11:29) is very old in English, going beyond the Authorised Version *(AV)* (1611) at least to Rheims (1582) and perhaps through that is indebted to the Vulgate's "in ministerium mittere"; the Revised Version *(RV)* (1881) retained it but was probably not happy with it, noting in the margin the not very helpful "for ministry." We notice above that when the abstract noun returns at 12:25, in what is essentially the same context, the Revised Standard Version *(RSV)* switches to "mission," and this is the concept that Luke would in fact seem to be emphasizing throughout. The idea of assistance he disposes of with the phrase 'every one according to his ability" (11:29); after that there is no explicit reference to the need of the brethren in Judea (the "it" in *RSV*'s 'sending it to the elders" is not in the Greek). By contrast the picture of a delegation is carefully built up. The verb "determined" is the solemn ὥρισαν and denotes decision at the highest level (hardly a groundswell); it is a remarkable word in the circumstances and is used by Luke elsewhere only of divine ordinances in the scheme of salvation.[2] The words that follow, "the brethren who lived in Judea," are suitably formal and certainly do nothing to elicit an image of a community languishing in famine. Further, neither the infinitive "to send" nor the participle "sending" has a direct object in the Greek, indicating that the verb is being used in its normal sense of sending persons and that Luke is writing of mission rather than of the despatch of goods. "Sending" is followed by "to the elders," pointing to the high level at which the interchange is to take place, and in Barnabas and Saul two delegates are clearly recognisable. Such a context requires that εἰς διακονίαν (11:29) be read as we have read it at 1 Tim. 1:12 and Heb. 1:14 (the latter instance with the same verb, "sent") and meaning "to send [representatives] on mission." The sense is so natural to the context— and indeed the only possible one—that, as noted previously, *RSV* translates it at 12:25.

Emissaries of the Apostle

In the letter to Philemon from his prison, Paul writes warm words of commendation of Onesimus, the Christian slave he is sending back to Philemon, the rightful owner. We read that Paul would really prefer to have kept Onesimus, in order (to use *RV,* which is closer to the structure of the Greek here) "that in thy behalf he might minister unto me in the bonds of the gospel." The question is what kind of ministering (μοι διακονῇ, Philem. 13) is intended. Commonly it is taken as "to look after me" *(NEB),* "to help me while I am in the chains that the gospel has brought me" *(NJB),* and this would suit the personal circumstances Paul was in and would seem to meet the Greek. Aspects of the Greek sentence, however, make us realise that Paul is thinking further afield. We notice firstly that the phrase "in the bonds of the gospel" is rendered in more recent English in a way to apply unequivocally to Paul, as in "while I am in the chains" or *(RSV)* "during my imprisonment for the gospel,'.' whereas the Greek phrase follows immediately after the word for "minister" so that the ministry rather than Paul is more naturally understood as being in the bonds of the gospel; connected with this is the idea behind Paul's word for "keeping" Onesimus, for that is a strong word meaning "hold back" or "detain" and could almost amount to "imprison," so that Paul can be seen as playing with the idea of detaining a slave for work in bonds of another kind.

Such a sentence, then, would be speaking of more than a butler for a gaoled apostle, and we are led to read "minister for me" in the sense of going out on errands[1]—here, in the interests of the gospel—with the dative "for me" having the same grammatical function as the dative "for the saints" at Rom. 15:25, namely, to designate the person organising and authorising the activity. The plan, conceived by Paul but recommended now to Philemon for its implementation, is to use Onesimus as a liaison officer in affairs of the gospel. Philemon is in a position to appreciate what Paul is urging because he is one of the significant figures in Paul's circle called "co-worker" and like other co-workers, Prisca and Aquila, who had a role in the extension of the mission,[2] had a Christian congregation based on his household (Philem. 1–2); he would be familiar with the needs and methods of the mission and with the constant interchange between the churches, and could be expected to cooperate with Paul's request despite the disadvantage he would suffer with the loss of his slave. The letter is addressed also to Archippus, who is presumably head of the community at Colossae and as such is informed of the situation. Onesimus is to be set aside as "more than a slave"; he is to be a "brother" (16) in the community and take on a role that Philemon the co-worker had already filled—as we are to infer from the phrase "on your behalf" (13)—on numerous occasions.

We have a glimpse of the role in the activity of Tychicus, who is sent to Colossae, Onesimus accompanying him, to inform the community there of Paul's affairs and to encourage their hearts (Col. 4:7–9). He is called "faithful minister [διάκονος] and fellow slave in the Lord" (4:7). As "minister" he is without doubt and in the first place the bearer of the letter,[3] but he is clearly also an emissary charged to speak of personal and Christian matters. From neither of these points of view, however, is the usage specifically Christian,[4] although it reflects a respect that is typically so. Tychicus is Paul's minister, a role reflected in the qualification "in the Lord," which is substantially different from the phrase attached to the title of Epaphras earlier in this letter, "of Christ on my behalf" (1:7); the lesser qualification attaches directly to the designation of Tychicus's role in Eph. 6:21 where an almost identical commendation is made. He is mentioned at Acts 20:4 among those travelling with Paul.

A strong tradition in the Greek manuscripts and ancient translations of 1 Thessalonians indicates that Paul designated Timothy also "God's minister in the gospel of Christ" (διάκονον τοῦ θεοῦ, 1 Thess. 3:2, RV) and this is the reading preferred by the earlier modern editors Tischendorf, Westcott and Hort, Von Soden, Vogels, and Merk. Although a weaker manuscript tradition replaces the phrase "God's minister" with one that designates Timothy "God's co-worker" (συνεργόν)—various combinations of both readings also occur—most recent opinion has been that "co-worker" is what Paul wrote. Changes in the textual tradition are explained along the following lines: in the ears of some members of early Christian communities the expression "co-worker of God" would have been too high-sounding a commendation of the young convert Timothy, and someone ended up tampering with the text to replace "God's co-worker" with the simpler "our co-worker" or to substitute the innocuous word διάκονος to provide the sense "servant" (RSV) or "slave" (NJB note).[5] From what we have seen of the latter term, however, as the designation both of the privileged bearer of the divine revelation and of the delegate of the apostle, this kind of argument against its originality does not hold water. With such more elevated usage in mind we can in fact see how well the word sits among the phrases "we sent," "in the gospel of Christ," and "to keep you firm and to encourage you about your faith" as the designation of another of Paul's delegates. We ought to note as well that the expression "co-worker of God" would be unusual for Paul, who, as observed in relation to Philem. 13, normally infers that co-workers are co-working with himself; the expression "minister of God," on the other hand, would be usual for Paul because he normally qualifies the designation "minister" with a genitive to indicate a connection with spiritual realities or with the other world. If authentic, the instance of διάκονος at 1 Thess. 3:2 would be the earliest literary evidence of the Christian use of the word and would thus be interesting for its overtly religious character and for its connections with both traditional usage, as just alluded to in connection with Col. 4:7, and with the language Paul would soon adopt in 1–2 Corinthians to explain and defend his role in the Christian mission to Corinth.

At Acts 19:22 Luke indicates that Paul had a pool of such emissaries at his disposal. On the occasion when he resolves to go to Macedonia and then decides to stay for a time in Asia—Luke makes little attempt to grapple with historical detail at this juncture, glossing over the complications that the collection for Jerusalem had introduced into Paul's mission—Paul sends ahead Timothy and Erastus, who are recorded as "two of them that ministered unto him" (Acts 19:22, RV, this translation reflecting more clearly the structure of the Greek: διακονούντων αὐτῷ. In this Luke is saying something more precise than that the two men were Paul's "assistants" *(NEB)* or "helpers" *(RSV, NJB)*. The phrase means that they were two of those who in the course of the mission were sent out in Paul's name; the Greek dative expressed here as "unto him" has the same grammatical function as that discussed at Rom. 15:25 and designates the sender. From a linguistic point of view the participial phrase would imply that Timothy and Erastus are to speak only to issues as instructed by Paul or to arrange affairs only after his plans,[6] but Luke was certainly envisaging that Timothy and Erastus would also encourage the churches in much the same way as we have seen Tychicus and Timothy instructed to do (Col. 4:7–9; 1 Thess. 3:2). The same kind of work is reported as having been done by Mark, who is said to be useful to Paul "for mission" (εἰς διακονίαν, 2 Tim. 4:11).[7]

Emissaries of the Community

Paul's commendation of the household of Stephanas occurs in a context of a mission from the Corinthian community to Paul in Ephesus (1 Cor. 16:15–18). The statement reads in part, "they have set themselves *[RV]* to the ministry of the saints *[AV]*" (εἰς διακονίαν τοῖς ἁγίοις, 1 Cor. 16:15). This "ministry" is associated immediately with words designating collaborators in the work of the gospel, "co-worker"and "labourer," and the Corinthians are urged to put themselves under the direction of people so involved (16:16). Paul next expresses delight at the arrival by him of Stephanas and his two companions (16:17), again urging that such men be given recognition (16:18): "they refreshed my spirit as well as yours." These people, then, would all seem to be directly engaged in sustaining the community of Corinth in the word of God, and the passage enables us to identify the particular "ministry" singled out by Paul as that which the household of Stephanas took upon itself. It was to constitute (perhaps also to organise and finance) a deputation to Paul or to liaise officially with him on behalf of the community. The expression εἰς διακονίαν has now been encountered five times in this sense of mission or delegation,[1] and at Rom. 15:25, Philem. 13, and Acts 19:22 we have also seen how the delegating authority is expressed by a dative, so that like Paul's mission for the churches of Macedonia and Achaia the mission of Stephanas's household is in the name of "the saints."[2]

The passage thus leaves no opening to interpret the standing of the household of Stephanas in the community as the result of their having "really worked hard to help the saints" *(JB)* or to elevate any such sense of service into a principle of early Christian community organisation. The passage provides us rather with a glimpse of a community well enough structured to be able to send to foreign parts a deputation of three men to take up the community's affairs with its founder. A degree of ambiguity exists as to whether Stephanas and his colleagues set up the mission of their own accord or whether, as is much more likely, they made themselves available for a delegation determined upon by the saints in the manner alluded to by Paul at 16:3.

Like Stephanas, Phoebe (Rom. 16:1) is travelling as a representative of her community. We learn this from Paul's commendatory note where among her credentials is "διάκονον of the church at Cenchreae."[3] She is the community's emissary—no doubt also the bearer of Paul's letter—but the precise nature of her business is unknown; to judge from the vagueness of the request to help her in whatever she might need, the business may not have been with the Roman community at all. As "helper" (16:2) she was also probably a woman of means who had been sent to Rome to use her influence, perhaps her wealth, in matters affecting her community.

The same usage appears in the letters of Ignatius.[4] The Philadelphians are urged to emulate neighbouring churches who have sent representatives to Antioch to celebrate the arrival of peace there, some sending bishops, others presbyters and deacons. Ignatius recommends that Philadelphia send "a διάκονον to perform there an embassy for God" (Phld. 10.1), adding that a blessing shall descend on the person worthy of such διακονία (10.2). The latter term means "mission"[5] and the former term designates the person who carries it out as emissary of the community;[6] the Greek is forceful. The word is not designating this officer a deacon, if only because Ignatius is not likely to urge the Philadelphians to limit their representation to the level of deacon when other churches are sending bishops or presbyters, especially as he is aiming to inspire the Philadelphians to make their own expression of solidarity with his own Antioch.

The writer has here closely associated his term for church delegate with diplomatic terms (πρεσβεῦσαι . . . θεοῦ πρεσβείαν). Because we have seen from chapter 8 that the terms do not mix in non-Christian writings, we recognise in this writer a novel development that arises from his conception of any act of a "church of God" being an act of God; if there is to be a messenger from one church of God to another it must perforce be through a messenger of God, the term for which he takes from the same religious background as his term for "mission." We have a clear view of how sensitive he was to such terminology in accounts of other delegations. In the delegation sent to himself at the port of Smyrna the bishops, presbyters, and deacons come in the name of their churches for the honour (IgEph. 2.1) and glory (Mg. 15) of God, by God's will and that of Jesus Christ (Tr. 1.1); of the delegation to Antioch he tells Polycarp, "This is the work of God and of yourselves" (Pol. 7.3), and the "godly" council, which is to be convened by Polycarp "blessed of God," is to appoint "God's courier" to go to Antioch "for the glory of God" (7.2). The word here is θεοδρόμος. In the letter to Smyrna the word is θεοπρεσβεύτης, "God's ambassador" (Smyr., 11.2), and the deed of celebrating Antioch's "godly" peace is "worthy of God" and "for the honour of God" (11.2–3). The sentiment in the instructions to the Philadelphians is exactly parallel, διάκονος being the equivalent of the other godly terms and designating a "godly courier" who shall travel for the "church of God" and "for the sake of the name of God" (Phld. 10.1–2).

The word is thus not an ecclesiastically determined term like the modern word "deacon"; indeed, given the alternative words used at Smyr. 11.2 and Pol. 7.2, we see that it is not even a technical term meaning "church delegate" but is applied by the writer ad hoc to an ecclesiastical situation because of its religious background in the area of message. In this it resembles usage in regard to Tychicus who is courier "in the Lord" (Col. 4:7), although there for Paul and not for a community, and usage in regard to Phoebe (Rom. 16:1), although there its religious character is not so expressly indicated.

The usage recurs in what Ignatius has to say in regard to Philo and Rheus. These two men, one from distant Cilicia and the other from Syria, had followed Ignatius via Philadelphia and Smyrna, joining him at last in Troas, from where he wrote to the Philadelphians and Smyrneans and thanked them, in passing, for the hospitality they had extended to this pair. In Phld. 11.1 he writes, "You received them as the Lord received you," which is an expression of the principle that the acts of a church are the acts of God; in Smyr. 10.1 he writes, "You did well to receive them as διακόνους θεοῦ; they give thanks to the Lord because of you," where the principle is the same, the gratitude being expressed to the Lord present in the church that manifests his love. The point for the interpretation of "ministers of God" (Smyr. 10.1) is that Philo and Rheus, who apparently had no commendatory letters to these communities from their own but were finding their way as best they could to Ignatius himself, were nonetheless received as if they had possessed them: they were received as coming from communities of God. That Philo was himself a deacon (Phld. 11.1) is incidental to the sense of the word διάκονος at Smyr. 10.1; like Rheus, who is commended simply as "an elect man," Philo is called "a man of good repute." If they had been officially accredited by their respective churches in Cilicia and Syria to attend on Ignatius, it is perhaps unlikely that some Philadelphians would have shown them disrespect (Phld. 11.1).

Burrhus is another deacon of whom Ignatius uses this terminology. As a deacon he was originally a member with his bishop Onesimus and others of the Ephesian delegation to Ignatius at Smyrna (IgEph. 2.1). Ignatius then asked to retain him (ibid.), and subsequently refers to him as "sent" in an official capacity by the Ephesians and Smyrneans (Phld. 11.2). Writing to the Smyrneans Ignatius says that Burrhus is the exemplar of the "ministry of God" (θεοῦ διακονίας, Smyr. 12.1), by which he means that Burrhus is the model of representation from a church of God. The fact that Burrhus also represents the Smyrneans supposes no more than that by a mutual arrangement with Ephesus they sponsored him, probably financially as well as officially; it has no bearing on the nature and functions of the early diaconate.

12

Commissions under God, Church, and Spirit

We have been looking at the words as designations for missions and for roles within missions beyond the confines of early Christian communities. The same words also signify commissions within communities. Of course the idea of a divine commission underlies the idea of the apostolic mission, but whereas in that context the commission was conceived as coming directly from God, as being directed principally towards the transmission of a heavenly message, and as being carried out in a community which was envisaged broadly as all those who are to be saved, the commission is now received by one Christian from another or by all Christians as members of the Lord's church, is directed towards traditional objectives rather than those directly revealed by God, and is carried out within the community where the commissioned person resides. We will begin, however, with three instances in Paul relating to both Christ and the imperial power as under mandate from the other world.

Christ and Rome

In the midst of instructions to the Roman community about the need of mutual support Paul draws a moral lesson from the way that the Gentiles have been received by Christ; although salvation belongs by the right of promise to Judaism, he reminds them, it has been extended to the Gentiles by mercy, such divine benevolence being then the standard of fraternal love. This scheme of salvation necessitated a strictly circumscribed role for Christ, who had to officiate according to the terms of God's promise to Israel. He became, accordingly, $\delta\iota\acute{\alpha}\kappa o\nu o\nu$ $\pi\epsilon\rho\iota\tau o\mu\hat{\eta}s$ (Rom. 15:8, *AV; RV:* "a minister of the circumcision"). The phrase does not mean "a servant of the Jewish people" *(NEB)* but "a minister of God who carried out his charge within the sphere of Judaism"; the genitive "of circumcision" is descriptive like the genitives describing the "angels of light" and "ministers of righteousness" (2 Cor. 11:15), and Christ is "minister" not because of a relationship to the Jewish people but because of a relationship with God: he is God's agent.[1] The sense recurs at Gal. 2:17: "Is Christ then an agent of sin [ἁμαρτίας διάκονος]?" *(RSV)*. Here, in asserting that Christ did not effect justification for Jews by virtue of the law but by his death, Paul also asserts that justification is not attained by observance of the law but by faith in the one who died; if however the pursuit of justification in this way by faith alone were an offence, Jewish Christians like him would be sinners like the Gentiles, and Christ would be accountable for the sin. The genitive "of sin", unlike that at Rom. 15:8 but like that at 2 Cor. 3:6, is objective. A third instance of the use occurs at Rom. 13:4 *(NEB)*:

> You wish to have no fear of the authorities? Then continue to do right and you will have their approval, for they are God's agents [διάκονος] working for your good. But if you are doing wrong, then you will have cause to fear them; it is not for nothing that they hold the power of the sword, for they are God's agents [διάκονος] of punishment, for retribution on the offender.

Local Office

The leading instruction to Timothy is "do the work of an evangelist, fulfil your ministry" (2 Tim. 4:5); this "ministry" is the sum of his responsibilities as an evangelist which are put to him in the preceding passage (4:1–4) and in instructions about his relationship with the apostle (2:2) and about his role as the appointed guardian of the faith of his community (1:6; 2:15). The sense of tradition and of structured congregation that is evident here is not so expressly stated in the brief exhortation that the Colossians are to pass on from Paul to Archippus—"See that you filfil the ministry which you have received in the Lord" (Col. 4:17)—but the indications are that as a fellow soldier of Paul and with a congregation centred on his house (Philem. 2) he has taken over the position vacated by Epaphras, the original teacher of the community. In these cases, then, the abstract noun is a comprehensive designation of the role of the chief evangeliser, and on both occasions the concept is of an office under the overriding authority of the apostle. Similarity to the usage of Paul at Rom. 11:13 and of Luke at Acts 20:24 is clear to the extent that work for the gospel is stated as a sacred commission, but in 2 Tim. 4:5 and Col. 4:17 the commission is by the hand of the apostle and is to a local community of believers instead of to regions abroad where the gospel is not heard.

The episcopate is such a "ministry" in Ignatius, that is, an office for sustaining the local church in unity (Phld. 1.1). For the Shepherd of Hermas, on the other hand, bishops are to concern themselves particularly with care of the poor and the widows, sheltering them "by their ministry" (Sim. 9.27.2) as a tree shelters the sheep and thereby fulfilling their duty to the Lord. To be noted in this instance is that the "ministry" is a ministry under the Lord and not a ministry or service to the poor and the widows. The same perception of care of the needy as a sacred commission or διακονία appears in exhortations to deacons: those deacons will die who "perform the duties of their office [διακονήσαντες] badly" by robbing widows and orphans in order to enrich themselves "through the charge [διακονίας] which they received to carry out [διακονῆσαι]"; they will live, however, if they turn "to their Lord" and "fulfil their charge" (τελειώσωσι τὴν διακονίαν αὐτῶν, Sim. 9.26.2). Again the "ministry" is a service to the Lord and an office to be "fulfilled" which has been received in the church but held ultimately under the authority of the Lord.[1]

In some passages the nature of the sacred commission cannot be determined. When we read of Onesiphorus, "you well know all the service he rendered [ὅσα . . . διηκόνησεν] at Ephesus" (2 Tim. 1:18), we do not know what kind of "ministering" the author is referring to. A reference to Onesiphorus's travels to find Paul (1:17) should not be taken to suggest that he is another courier for the gospel like Tychicus because in this instance the verb "ministered" has a direct object—"he ministered" or "effected so many things"—which strongly favours a statement about deeds done in fulfilment of an office or at the behest of a leader in Ephesus.

More obscure is a statement from Hebrews. We read: "God is not so unjust as to overlook your work and the love which you showed for his sake in serving the saints, as you still do [διακονήσαντες τοῖς ἁγίοις καὶ διακονοῦντες.]" (Heb. 6:10). The interpretation of this phrase really depends on the identity of the persons to whom the assurance is addressed. The passage occurs in a strong exhortation (5:11–6:20) where 5:11–6:8 warns about immaturity in Christian understanding and about the finality of apostasy, with 6.9–20 then offering the assurance of the great promise. The whole passage aims to dispose the readers for the doctrinal exposition on the theme of Melchisedek which, after being alluded to at the beginning and end (5:10; 6:20), begins to be developed in 7:1. The emphasis on maturity (5:12–6:2) shows that the instructions are directed at Christians who have had considerable experience of their calling—we have a view of how grim the experience could have been (10:32–34)—and suggests that those addressed might be a group of the mature within a larger community that also contained "children." They are addressed as "holy brethren" (3:1) or "brethren" (10:19; 13:22), are distinguished from the "leaders" (13:7, 17, 24), and, so it would seem, from the rest of the "saints" (13:4). At 6:9 the writer expresses confidence in their command of "the better things that belong to salvation," and in 6:10 makes a slightly more detailed reference to what they have actually done before expressing in 6:11 the hope that the situation will remain unchanged for the future. They are credited with "work" and a "love" manifested in some form of "ministering." For this author "work" is likely to be good works in the moral sense (cf."dead works," 6:1; 9:14), and it could be surmised from that the author would then wish to mention some public activity of a Christian nature. The problem is to identify what public activity the writer had in mind. His participial phrase with a dependent personal dative "to the saints" is syntactically the same as phrases in Rom. 15:25 and Acts 19:22, which speak of journeys undertaken in interests of the mission, but here there is no context of travelling or of being sent. We probably have to think of acts undertaken within the community itself, the kind of tasks that are the responsibility of the more senior members. The phrase says nothing explicitly about the mandate for such activity, unless the "saints" are to be understood as the leaders of the group; this is hardly likely but it would explain the dative, which otherwise is unusual.[2] The phrase, like the exhortation itself, does presuppose the existence of an authority, but beyond that perhaps the most we can recognise is roles filled that had a bearing on the Christian life of the congregation and have been prescribed for these members. By whom? Certainly not by "the saints" if that term means the general body of the congregation.

The rhetorical and didactic style of Heb. 6:9–12 does not obscure the fact that the passage is cast in thoroughly traditional Christian terms reminiscent of passages like 1 Thess. 1:2–10 or 1 Pet. 1:13–2:3, and similar terms litter the messages to the churches in Revelation. The message to the church in Thyatira mentions "work, love, faith, ministry and endurance" (Rev. 2:19). If "ministry" can be related to anything in particular in the tradition, it is perhaps to the idea of endeavour for the gospel to which it has given a large measure of expression; as such it might be the equivalent of the "labour" which is listed with the "works and endurance" of the church in Ephesus (2:2), and could be broadly paraphrased as "your dedication to the charge which I have laid upon you as church."

Extending Ministry

From what we have seen of Acts, Luke has displayed a nuanced use of the verb and abstract noun that matches at several points the usage of Paul. Thus both use the verb of co-workers in the mission (Acts 19:22; Philem. 13), while the abstract noun is used of the apostolic commission (Acts 1:17, 25; 20:24; Rom. 11:13), of mediating divine revelations (Acts 6:4; 21:19; 2 Cor. 4:1; 5:18; 6:3), and of delegations between churches (Acts 11:29; 12:25; Rom. 15:31; 2 Cor. 8:4; 9:1, 13). On the other hand Luke declined to use the common noun for the reason, we suspect, that he found its connotations inappropriate for the kind of religious message the gospel was in his eyes. In other words he was fully aware of the uses and colour of these words, as we would expect of a writer of his sensitivity and skill. Accordingly when we come to the passage Acts 6:1–6, where the words are played off against one another at what has all the appearance of being a critical moment in his early history, we can anticipate that he is not being off-hand in his usage but that it is integral to the important statement he is framing about development and change. His earlier uses of the word "ministry" in Acts 1 had a grave and sacred import which, as we have seen earlier, repeats itself in the solemn rededication of the Twelve to their commission at 6:4. In introducing his new theme of the Twelve and the Seven, where "ministry" and "ministering" recur at 6:1 and 6:2, there is room for surprise if we are to think that Luke is writing only about "the daily distribution of food" *(NIV)*.

Translations in terms of "distribution" *(RSV, NEB, NJB)* at 6:1 and of "to give out food" *(NJB)* or "to wait at table" *(NEB)* at 6:2 not only preclude the consideration of other intentions on the part of the author that might suit better with talk of "ministry of the word" at 6:4 but they are putting meanings in English that Greek is not likely to bear. "Distribution" is not really one of the attested meanings of the abstract noun; in contexts like this, where widows are undoubtedly being provided with something, the abstract noun is known to mean attendance on people or the arrangement of attendance even to the point of ministration as ritual.[1] With "to serve tables" *(RSV)* a stranger situation exists in relation to the Greek because the Greek attendant does not wait on tables but on people at tables; tables in fact are very rarely mentioned because the concept of διακονία is to fetch for a person.[2] Those who would wish to read the phrase as an economic activity—"to handle finances" *(GN)*—have no precedent to go by at all; to the Greek reader this verb with this noun "tables" would say that the operators are coming and going to the table whereas our eternal image is of the money dealer seated at his table.[3] The impersonal dative is likely to be indicating, in fact, that the verb has a meaning other than attending on people at all, and in the context of a crisis in ministry that meaning is going to be ministry itself—that is, the public and even ritual performance of duties in the community. Such an intent on the part of the author gives weight to the formal pronouncement of the Twelve to the full assembly of the disciples, "It is not right . . .," with the solemn implication of that phrase that "it is not right before God,"[4] and the Twelve proceed to publish the ordinance about a new ministerial duty in the same terms and on the same level of language as they use in regard to their own divinely commissioned ministry. They are saying that their appointed ministry is not to be at tables—the dative is of respect (as at Priscus, *HGM* 276)[5]—when they are at the command of the ministry of the word. Thus the only members of the sacred community who thus far have been instituted in an office, and by the Lord himself, choose to speak of "ministering" at the moment when they are authorising the first church-made office. Whether the duties to be performed occur at tables where people eat or at those where people receive financial assistance is not said.

The Twelve call the task "duty," which is a further indication of how much the passage is about commissioned roles. The duty is the διακονία of 6:1. So named, it is a public function under someone's direction already existing in the community; it is a "daily" ministry in contrast with the ministry of Acts 1, which is to take the Twelve "to the end of the earth." As ministry it is indeterminate, and we have to make our own estimates of the kind of care extended to the widows; the word says more about Luke's conception of a community with official structures than about how the community cared for its widows.[6]

Moral Duties

The *Mandata* of Hermas sees the twelve commandments as a "charge" under God or διακονία which each Christian has to "fulfil" (Mand. 12.3.3) as it does in particular the commandment of liberality (Mand. 2.6); in carrying out this charge from the Lord (διακονῶν) a person will have life under God because he is living within God's mandate. So the first Similitude reminds "the slaves of God" that they belong to an unworldly city with a master and laws of its own; among these is the law of liberality imposed on those whom the master has enriched "so that they will fulfil their charges to him" (ἵνα ταύτας τὰς διακονίας τελέσητε αὐτῷ,

Sim. 1.9). The dative "to him" shows the orientation of these "ministries"; they came from God and are performed as a responsibility to him. Whereas the poor man is endowed with the power of prayer and does his particular duty by exercising that power, the rich man will carry out "the ministry of the Lord" or mandate from God by dispensing the Lord's gifts (Sim. 2.7).

Gifts

The group in 1 Peter is instructed about its responsibilities in regard to the diverse gifts within the congregation. Like "good stewards of varied grace" they must be "communicating each gift" among themselves (αὐτὸ διακονοῦντες, 1 Pet. 4:10). The expression denotes Christians open to the reception of gift and in the process of transmitting the gift to others;[1] the community is greatly enriched because gifts to individuals become part of the varied grace of all. The preacher instances a gift from the wisdom of God, which is to be given expression in speech within the congregation, and then a gift from the strength of God—such as supported Paul in ministry (2 Cor. 12:9–10)—which is to be given expression in "ministering" (1 Pet. 4:11). The "ministering" will comprise public duties within the congregation but, as at Heb. 6:10, they are designated in only this indeterminate way.

In enumerating in 1 Cor. 12 the range of activities in which the members of the Corinthian community engaged, Paul was chiefly illustrating diversity. Enlivening the diversity was one divine principle (1 Cor. 12:11), and on this basis the "gifts" are the "ministries" (1 Cor. 12:5) which are the "operations," and Paul differentiates among them only on an artificial framework in aligning gifts with the spirit, ministries with the Lord, and operations with "the same God who inspires them all in every one." One might surmise that ministries are aligned with the Lord on the ground that the church, his body, is the extension of his own commission in time; Christ is the only part of the divine that came under a mandate of "ministry" of God, and as Lord remains that part of the divine into which all are baptised to become his body for the work of spirit, Lord, and God. These "ministries," then, are as diverse as the parts of the body, and whether viewed globally or as they actually exist and operate in their individuality and diversity they are given in order to be carried out, every gift of God being necessarily a charge or commission from him to put it to work. Paul is writing of the diversity from three points of view: first, from the viewpoint of its origin in the liberality of God, who enriches each in a particular way; second, from the point of view of the responsibility under God which the gifts have now become for all believers; and third, from the point of view of the diversity manifesting itself in the life of the community, this last being the aspect that Paul finds necessary to explore further and exemplify in the image of the interdependence of diverse parts of the body. The "ministries" are thus not viewed as oriented towards the community as "services" but, as the *Living Bible* paraphrases, they are "different kinds of service to God," much as Hermas later expressed moral duties as ministries under God (*Sim.* 1.9). Finally, the diversity is presented as "ministries" not with the idea of mutual benefit in mind but with that of a common mandate; this body is obliged to use all its parts, and each part is obliged to work within and for the body, its listening faculty as well as its capacity to teach and instruct.

If Paul speaks of "ministries" broadly in the passage about gifts in 1 Corinthians as the sacred obligations that make for the life of the spirit, in Rom. 12 he speaks of "ministry" narrowly among the gifts as the apostolic mandate already celebrated by him at Rom. 11:13. In fact, as he enters this passage of exhortation, he invokes that "grace given to me" (Rom. 12:3) just as he concludes the section with a comprehensive statement about it (15:15–21). In his listing of gifts, then, we are not surprised to find the same apostolic mandate, which is not his alone but belongs to all who take part in the mission of the gospel, prominent among the gifts which sustain the body: prophecy, "ministry" (12:7), teaching, exhortation. It is these other gifts, all to do with delivering the word, that determine the apostle's meaning in his use of διακονία here; ideas about "practical service" which translators provide at this point (*NJB*, Moffatt) or "administration" (*JB*, *NEB*) are an intrusion in a reflection on those gifts of the word which before any other gift make and sustain the body of the church. The curious element in this listing, however, is the priority given to prophecy over the apostolic mandate when "first apostles, second prophets, third teachers" (1 Cor. 12:28) is the norm. Perhaps Paul is acknowledging not only that prophecy is the gift most cherished within an individual congregation (1 Cor. 14) but that the Roman community was indebted to it more than to his own apostleship.

The author of Ephesians also celebrates gifts in the church but is speaking only of gifts upon teachers ("pastors" being taken as part of these; Eph. 4:11). This is in accord with the emphasis that the epistle has earlier placed on the communication of God's mystery, the process where Paul had the leading role as διάκονος (3:7); the emphasis on teaching is also in accord with the outcomes held up for emulation in this passage, namely, unity in faith, solidarity in doctrine, and maturity in truth (4:13–16). With teaching then the overriding theme and teachers the only figures mentioned, the "work of ministry" (ἔργον διακονίας, 4:12) can only be understood as part of this teaching process within the church so that it signifies here, against the background of the heavenly Christ dispensing his word through teachers, the work done by the kind of "minister" who dispenses heavenly knowledge (Eph. 3:7; Col. 1:7, 23, 25); the usage is close in meaning to instances at 2 Cor. 3:7–9; 5:18.

This interpretation is saying of course that "ministry" in 4:12 is not a work in which the saints, understood as a group distinct from the teachers, can participate, and as a reading of 4:12 also runs counter to contemporary English translations, if not to most older and traditional views of these phrases. Most notably we see the import of the meaning of this word in this passage from the revision of the translation undertaken by the Revised Standard Version (*RSV*). The original (1946), "And his gifts were that some should be apostles, . . . for the equipment of the saints, for the work of ministry, for building up the body of Christ," which is in accord with the meaning the context requires for "ministry" as a work of the apostles and other teachers, was revised (1971) in favour of "to equip the saints for the work of ministry . . . ," which would be to transfer the responsibility for purveying the sustaining mystery to believers, if the same sense of "ministry" is retained. But that would do violence to a chosen terminology because only apostles, prophets, and their collaborators are gifted by heaven with this kind of "ministry."

The correlation of the three phrases in 4:12 has long been a matter of dispute, and our purposes do not require a solution to it, but we can briefly consider what light is offered in reading "ministry" as exclusively a work of teachers. The last of the phrases, "for building up the body of Christ," expresses the overall purpose of the gifts and is achieved through doctrinal solidarity. The first two phrases express either two separate objectives—"for the equipment of the saints, for the work of ministry"—which effectively and naturally accommodate "ministry" as the work of the teachers, or only one objective—"to equip the saints for the work of ministry." The latter division is now widely preferred (whether for reasons of syntax, style, or theology we need not pursue) and can only accommodate "ministry" as a teaching mandate if "saints" include the teachers; if that is understood to be what the author intended, however, he has dangerously obscured his exposition at the very stage where he is trying to state the fundamental reality of growth in the church's life through sound teaching; he would be leaving room for the interpretation that all saints have access to "ministry." The interpretation of "ministry" as a function necessarily exclusive to teachers thus requires us to read the first two phrases as two separate objectives. Those who find this scheme makes for an inadequate or too passive life for "saints" in the church are underestimating the role attributed by the author to sound doctrine; it assimilates the whole church into the mystery where growth into the fulness of Christ occurs. No one is left out. No one has more experience of the mystery than anyone else.

13

Deacons

Christian literature has so far furnished no indication that Christians created a terminology based on the cognates we are examining. Passages in the New Testament dealing with the collection for Jerusalem and with the apostle as διάκονος are good illustrations of this because their interpretation requires us to draw on non-Christian usage, and in the comparison we realise that non-Christians and Christians are speaking the same language and projecting the same values. For writers in both traditions usage was fluid and applications were varied. Thus among Christians Ignatius speaks of the διακονία of a bishop without prejudice to that of a deacon, whose very title is a cognate of the word; he also applies the word διάκονος to a deacon in a sense other than "deacon"; the authors of Hebrews and 1 Peter apply the words to Christians and non-Christians, just as Paul applies them to the world of evil and to the political sphere and yet takes from them his own preferred title. With this background to the usage in mind, we will consider the title of the Christian deacon.

Philippi

One phrase in the prescript to Paul's letter to the Philippians is translated in many ways, from "the bishops and deacons" *(RSV)* to "church leaders and helpers" *(GN)*, the first word being ἐπίσκοποι and the second διάκονοι; interestingly both words have an etymological connection respectively with the English words "bishop" and "deacon." Commentators are understandably reluctant, however, to see in the people greeted by Paul in this young community of Philippi officials resembling the later familiar figures of church order. They do generally appreciate, nonetheless, that the only reliable guide to the meaning of διάκονοι here is the fact that it is bracketed with the other word, called for convenience "overseers." They conclude that the word διάκονοι points to the existence at Philippi of office bearers subordinate to the "overseers." This approach is supported by the present study, which suggests that in conjunction with a word meaning "overseers" or something nearly equivalent διάκονοι will indicate agents of these people; it also suggests that the term is an honorific, if not a technical, title.

An understanding of the term in the line of agent will restrict the field in which it can be compared with the use of the same term in the pagan guilds where minor officials at religious festivals assumed the term as a title meaning "waiters." The currency of the term at these festivals is nonetheless evidence that it retained the religious character, which is reflected in fact in much general usage,[1] so that the choice of such a term over one like ὑπηρέτης, which is broadly similar but which has a background in largely bureaucratic language, indicates where their preferences lay. When this religious aspect is overlooked in favour of what is considered a merely honorific value of the title in the inscriptions, the conclusion tends to be drawn that Christians adopted the term because it was current as a title in the contemporary guilds and associations, and when the meaning of the word there ("waiter") is also taken into consideration the way is open to explain the Christian title by reference to the table of the Eucharist or to the succouring of the needy rather than to the "overseer" himself. We have already seen in a later Christian writer (Ignatius, Phld. 10.1) how choice of the religious term reflected an awareness that what one did as a member of a Christian community was a godly activity.

If the term implies subordination, which in this case is to the "overseers," the same officer of an organisation is not going to be designated by both terms. Georgi has explained the Christian provenance of both terms from his view that they were synonymous designations of Cynic-Stoic preachers[2] but we have seen in chapter 8 that for one of the terms this cannot be. Lemaire on the other hand has proposed that the two terms together form a Hellenistic alternative to the Jewish term "elders," and that although the double term became quickly obsolete in the face of developing technical senses for each term the double phrase in Phil. 1:1 marks an important stage in "progressive specialisation" of ministerial vocabulary.[3] This hypothesis is advanced largely in order to avoid reading later technical senses back into Paul, and with little attempt to explore the possibility of less technical meanings there, but it skirts the fact that the words have no history of synonymity and raises the further problem of how the word διάκονος came to be used in the entirely novel sense of ministers of the community.

The role of the διάκονος could only be known if that of the "overseer" was known. The suggestion that this may have included at Philippi responsibility for the offering made to Paul does not necessarily suppose that the "overseers" existed only for this purpose or were so called because of this activity, nor is it seriously prejudiced by the fact that they are not mentioned outside the prescript. The suggestion cannot be substantiated, however, and the nature of the title given to their assistants cannot contribute to the elucidation of the issue.

Relevant perhaps to an understanding of both terms is the possibility that the "overseers" were a "local idiosyncracy"[4] among Pauline communities, and that in view of the presumably restricted character of their role their assistants were not people who had been appointed by the "overseers" or by the community to their title but were actually graced with the title for the first time by Paul himself. The structure of the address itself hardly encourages this suggestion, and the problem is not such as to require it, but given the honourable mention of "overseers" it would be typical of Paul, especially in a letter so encouraging as this, not to overlook those who had co-operated with them. The choice of term would be appropriate because Paul's prescripts are particularly redolent of his sense of God at work in the church, and by bestowing a distinct dignity on those engaged in what were perhaps ephemeral activities he would remind the community that even such persons were doing a godly work. The religious connotation of the term is even such as to suggest that it would not be linked in this way to the word ἐπίσκοποι unless that also was basically a religious term.

The Church of Timothy

Whether or not Paul introduced the title in the prescript of the Philippians as an honorific designation or takes up from the Philippians a title for officers of the "overseers" within their Hellenistic community, there is no doubt that in 1 Tim. 3 we meet such officers who with some certainty we can call "deacons." In attempting to identify the function of the deacon, however, we need to distinguish usage in this passage, where the deacon is principally characterised by association with the "overseer," from usage elsewhere in the Pastoral letters where it refers variously to the special apostolic ministry of the apostle (1 Tim. 1:12), to participation in the mission by Mark (2 Tim. 4:11), to Timothy's office (2 Tim. 4:5) and function (1 Tim. 4:6) as head and teacher of a local church, and to Onesiphorus carrying out indeterminate churchly functions (2 Tim. 1:18). With the possible exception of the last instance, these uses refer to activity in the direct interest of the gospel by those with a commission of a missionary or local nature. To infer that 1 Tim. 3:8–13 must also be understood against this background is to overlook the guiding principle that each instance of these cognate words is to be interpreted as far as possible within the limits of its own context.

No single item in this section necessarily points to duties of deacons in an evangelising or in any other recognisable capacity. Thus the code of conduct imposes no more than an elemental decency, "holding the mystery of faith in a clear conscience" might be expected of any sincere Christian, being "tested" may be simply an assessment of character, "good standing" no more than public esteem, and "confidence" personal assurance. Some of these points and others in the passage are open to more specialised interpretation and can then, according to the approach taken to διάκονος and its cognates, be advanced as grounds for seeing the deacons as either preachers or social workers, but our own survey of the word and its cognates leaves room for neither of these misunderstandings. The most instructive fact would be that the passage about deacons follows one about the "overseer"; even if this indicates little more than that the two offices are in some way coordinated, it would at least suggest that the deacon is the assistant of the other. Because his function cannot be satisfactorily determined, despite the general emphasis in the Pastoral letters on teaching—an emphasis that may have arisen for historical reasons and not because teaching characterised the "overseer's" office, it is unlikely that this section of 1 Timothy can provide any more precise idea of the diaconate.[1]

Didache

The situation is clearer in the Didache (15.1): "appoint yourselves overseers and deacons." These are men of upright character, like those of 1 Tim. 3, whom the community is to appoint and respect because they carry on the work of the itinerant prophets and teachers; the statement to this effect is particularly emphatic (15.1): "for they are carrying on for you the sacred work of the prophets and teachers." This, taken in conjunction with the instruction about the Eucharist (14.1–3), leaves little doubt that the author is writing in particular about the cultic role of these local ministers. Of its nature the title again suggests that the deacons are assistants to the overseers, who would at the least be supervisors of the eucharistic assemblies. This would occasion assistance at the eucharistic table by the deacons.

Clement of Rome

The passage in which the two titles occur in 1 Clement (42.4) is even more strikingly liturgical, falling within the detailed exposition of Levitical models for church order. The word διακονίαι itself sets the pattern for this, being applied to the ministrations of the Levites as part of the broader priestly liturgy (40.5); this is the only instance of a cultic sense in early Christian literature and exemplifies Hellenistic as against Septuagintal usage, which otherwise is influential here.[1] The precise divine ordinances by which the ancient priestly liturgy had been governed were preserved for the Christians by the apostles who appointed their first converts "to be bishops and deacons of the future believers" (42.4). That these designations are presented by Clement as cultic titles becomes clear from the developing discussion. Clement passes by way of Isa. 60.17 (which in his version exposes the root of the two Christian ministries in a divine ordinance)[2] to a consideration of Num. 17 where Moses established Aaron's divine entitlement in the face of jealous rivals "to carry out the sacred service of the priest" (43.4). That earlier dispute about "the glorious title of priesthood" (cf. 43.2) is presented as an analogy of the disputes in Corinth about "the title of the office of bishop [ἐπισκοπή]" (44.1), and Clement asserts that the apostles attempted to forestall these by making appointments to this role and by making provision for other approved men to continue "service" (λειτουργία, 44.2). That the term "service" and its Greek cognate words continue to refer exclusively to cult is seen at 44.4 ("those who have blamelessly and holily presented the gifts"), so that "the office of bishop" (ἐπισκοπή) which is under dispute is referring to the central function within Christian cult. Clement has thus changed from the term "priesthood" of Levitical cult to one meaning something like "presidency" in Christian assemblies, and if the arrangements as he reports them suppose more than one person in the presidency, then of co-presidency; the role was filled by presbyters (cf. 44.5) who had been appointed and approved (44.3).

No mention is made in this chapter of the title "bishop" itself or of the title "deacon," Clement referring only to "the aforementioned" (44.2, cf. 43.4–5). Because the former of the two titles is a cognate of the word signifying the office or function under contention, the expression "aforementioned" presumably refers only to the bearers of that title; to suppose that it is inclusive of "deacons" would mean that these also are presbyters, which, as well as being unprecedented, makes light of the differences between one word implying authority or preeminence and another implying some kind of subordination.[3] Clement's vagueness here is not seriously misleading, however, once we realise that "deacons" are mentioned at 42.4–5 only because at that point he is putting a case for the divine ordering of Christian cult and has judged it necessary to give the full complement of liturgical personnel; in addressing himself subsequently to "the dispute about the title of the office of bishop" (44.1), the deacons are irrelevant.

The deacons of this document, accordingly, are the nonpresbyteral liturgical assistants of presbyters in the presbyters' capacity of bishop. Because the liturgy included a sacred meal, the deacons presumably acted as ritual waiters, but they would have done this not on a title of being waiters for the assembly but in their capacity as attendants to those responsible for the conduct of the service; their role would also have included functions associated with other aspects of this service like readings and offerings.[4]

Ignatius and Polycarp

In Ignatius the two titles lack this rigid liturgical orientation, and the two officers fit well in the traditional church order of bishops and deacons. The deacons are clearly a third subsidiary group within a local organisation headed by a bishop and his presbyters. Ignatius frequently exhorts the various churches to be subject to the bishop,[1] who is practically a monarchical figure,[2] or to the bishop and presbyters.[3] Occasionally he includes the deacons with the two latter (Phld. 7.1; Pol. 6.1), although no particular significance attaches to this because the expression "give heed to the bishops, presbyters, and deacons" is less a juridical formula than an exhortation to observe that unity which is the mark of the church (e.g., Mag. 6.1–2; cf. 13.1–2). More typical of the status of the deacons is the exhortation to the congregation to hold them in respect (Tr. 3.1; Smyr. 8.1). They are nonetheless integral to the church (Tr. 3.1; 7.2; Phld. Int.).

To delineate the role of the deacon in Ignatius accurately it is necessary, as in the Pastoral epistles, to distinguish usage in respect of deacons from occasional other usage denoting the bishop's godly commission as head of the church (Phld. 1.1) or signifying that certain individuals, whether deacons (cf. Phld. 10.2) or other members of the community, are godly representatives of one church to another (Phld. 10.1; Smyr. 10.1) on a godly mission (Phld. 10.2; Smyr. 12.1).[4] Failure to observe this distinction has led Lemaire to discern a homogeneous development of the meaning of the common noun in Christian sources from "messenger" to "itinerant preacher" or "liaison officer" (deacon), the latter figure being gradually restrained for more or less political reasons from pursuing such a roving commission.[5] In the documents we have so far examined, however, there is no sign that the deacon was ever anything other than a personal assistant to the overseer, and any function of the deacon we have been able to detect has been liturgical in character. In the writings attributed to Ignatius the deacon is a member of a local hierarchy who is almost inseparable from the bishop and presbyters;[6] interestingly the existence of this hierarchical terminology has not inhibited the writer from drawing on the broader usage already referred to.

Several passages indicate how deep-rooted is the connection between bishop and deacon. Tr. 3.1 urges all to "respect the deacons as Jesus Christ," just as they respect the bishop for being "the type of the Father" and the presbyters for being "the council of God and the college of Apostles"; the axis of the analogy here is Father-Jesus, which expresses the idea that as Jesus did the will of the Father the deacon does the will of the bishop. Jesus could as well have been called the "deacon" of the Father, and Mag. 6.1, where the same parallels are drawn, comes close to saying this, for the deacons are said to be entrusted with the διακονία or sacred commission of Jesus Christ—that is, deacons have a sacred commission like that of Jesus to carry out another's will. The point of comparison being in the field of agency, there is no ground for the surprise expressed by some writers that deacons rather than apostles are compared with Jesus Christ.[7] In a variation on this theme, Smyr. 8.1 exhorts all to "follow the bishop as Jesus Christ follows the Father and the presbytery the apostles," while deacons are to be reverenced "as the command of God," the thought being that they are the embodiment of God's will as expressed by the bishop.[8]

On one occasion the deacon is shown to be responsible also to the presbytery, although the relationship is less well defined, presumably because this is not the essential mark of the office. Thus the deacon Zotion is commended because he is subject to the bishop "as to the grace of God" and to the presbytery "as to the law of Jesus Christ" (Mag. 2). Here the basic relationship is again to the father bishop, who is the source of God's blessings for the church, but this is complemented by a relationship to the presbytery which, as the council of the apostles, maintains for the bishop and for the church the law or norm of its existence.

Doubtless the intimate relationship between deacons and bishop is the reason why the bishop Ignatius refers to deacons as "co-slaves" (Eph. 2.1; Mag. 2; Phld. 4; Smyr. 12.2), meaning that he and they are both working out the mind of God for the church, and in Tr. 2.3 we are reminded to eliminate any suggestion that the deacon is named after his duties at the eucharistic table: "they are not διάκονοι of food and drink but are ὑπηρέται of the church of God"; the latter Greek term is used in its public service sense of "officer," so that although the deacons may attend at tables they do so as "godly officers" of the bishop. In other words, they are named after their association with the bishop rather than after their role at a table, and the term διάκονος rather than ὑπηρέτης was chosen to designate this officer within the church because of its religious character. That character is evidenced in the earlier phrase from the same passage, "being deacons of the mysteries of Jesus Christ" (Tr. 2.3). The word "mysteries" here is not likely to refer to the Eucharist, Ignatius using it elsewhere only to designate Christian verities (Eph. 19.1; Mag. 9.1); the phrase does not mean "purveyors" of mystery (like Paul and Apollos as "stewards of mysteries," 1 Cor. 4:1; cf. 2 Cor. 3:6), but describes the sacred locale of religious knowledge within which the deacons operate (like the descriptive genitive "of righteousness" in Paul's phrase, 2 Cor. 11:15).

In two passages the Ignatian church order is set against a background of cult. Thus Tr. 7.2: "He who is within the sanctuary is pure, but he who is outside is not pure; that is, whoever does anything apart from the bishop and the presbytery and the deacons is not pure in his conscience." Although this passage is a literary figure designed to urge unity in the church as a safeguard against the contamination of heresy (cf. 6.1–2), its terms imply that the church is most surely itself and immune from harm when it sees itself as a group organised for worship. The same correlation of bishop, presbytery, and deacons with cult and unity is more direct in Phld. 4, again after a passage that warns against heresy:[9]

> Be careful therefore to use the Eucharist (for there is one flesh of our Lord Jesus Christ, and one cup for union with his blood, one altar, as there is one bishop with the presbytery and the deacons my fellow servants), in order that whatever you do you may do it according unto God.

Thus, although the threefold ministry may superficially appear in Ignatius as a more or less technical instrument of the church's unity, he may reveal in this statement an understanding after the manner perhaps of Clement and the Didache that the ministry is essentially expressed in the liturgy. Certainly his recurrent image of the bishop as the type of the Father places all the action of the church on a godly basis and leaves no doubt that for him $\delta\iota\acute{\alpha}\kappa o\nu o\varsigma$ is in itself a sacral word, whether applied as a title to the deacon in the sense of the bishop's agent or as a designation of the godly community's representatives.

Little of these basic characteristics are explicit in the exhortation addressed by Polycarp to the deacons of Philippi (Pol. 5.2), largely because, like the exhortation to the wives and widows, it is cast in general terms. There is no mention of a bishop, and the deacons are linked to the presbyters as the group to which the young men are to subject themselves (5.3). No specific diaconal function is to be deduced from the list of virtues that the deacons are urged to practice, compassion in particular being a quality that the presbyters also are expected to exhibit (6.1). They are, however, said to be "deacons of God and of Christ and not of men" (5.2), and we note that these genitives indicate the person in whose name the deacons act, not the person towards whom they provide a service, service to God being expressed by $\delta o\upsilon\lambda\epsilon\acute{\upsilon}\epsilon\iota\nu$ and $\lambda\alpha\tau\rho\epsilon\acute{\upsilon}\epsilon\iota\nu$ (2.1; 6.3). They are instructed to "walk according to the truth of the Lord" (5.2) and are compared with Christ (when the presbyters are compared with God 5.3), a comparison enforced by saying that Christ became $\delta\iota\acute{\alpha}\kappa o\nu o\varsigma$ "of all" (5.2). The designation of Christ as $\delta\iota\acute{\alpha}\kappa o\nu o\varsigma$ here is of course occasioned by the discussion about deacons, and the phrase is not illustrating the role of deacons—as if Polycarp were presenting Christ as physically ministering to people—but illustrates the disposition of Christ in that his commitment to his role under God was absolute ("of all"), this being then the disposition that deacons should take to their duties.[10]

Hermas and Justin

In Hermas the character of ministry is almost completely shrouded behind the author's concern that the church in all its parts should grow to perfection in virtue. A prophet himself, he views the church neither in time nor in any place but as an arena for renewal of the spirit (cf. Vis. 3.8.9–11), in L. Pernveden's phrase, "an entity that is essentially celestial and independent of any local confinement."[1] Consequently if he speaks of "overseers" only in the plural, this is not necessarily to imply that he did not know of a local bishop, and if the liturgy is not mentioned this could be because the eye of the reforming prophet is focused on morality. Thus, from Sim. 3.18 he begins a review of all the church's members to reveal the good and the evil to the end that the church might be purified and become one in faith and love. After examining in this way seven groups of the faithful, he arrives at "apostles and teachers who preached through the whole world and taught the word of the Lord" (Sim. 9.25.2), and then considers διάκονοι (9.26.2), "overseers" (9.27.2), the martyrs (9.28.2), and undefiled infants (9.29.1). This arrangement, moving through the last living members of the church to the blessed ones who have gone before, itself suggests that 9.26–27 speaks of deacons and bishops as ministers in the church and that they are at the sacred apex of the church in this world. The fact that the only aspect of their ministry mentioned is works of charity—an area where deacons but strangely not bishops[2] are found to be deficient—need not surprise us because the administering of funds for the support of the needy is of course an area where the church's ministers would most naturally manifest tendencies towards good or evil; because morality is the prophet's immediate concern, his concentration on an area where the moral worth of ministers was conspicuous does not mean that the ministers did not also have other official ministerial duties.

The two titles appear in Vis. 3.5.1, "the apostles and bishops and teachers and deacons," where the order is unusual in that apostles and teachers form a pair elsewhere (Sim. 9.15.4; 9.16.5; 9.25.2). If the passage is presenting an idealised picture of the past founders of the church (as in the case of the apostles and teachers of 9.25), the broken order could, as Zahn suggested, serve the author's purpose better than a list in which the final pair, "bishops and deacons," might merely remind the reader of his contemporaries.[3] The apostles and teachers are elsewhere regularly spoken of as preachers, but not the bishops and deacons, who are said either—in the idiom of 1 Tim. 3:10 and 13—to "carry out their office" (διακονήσαντες, ibid.) or to engage in charity (9.26; 9.27). It is difficult to know anything more of them.

If Hermas is thus silent about a role in liturgy for deacons, that role by contrast is the only one recorded by Justin. The Eucharist is distributed to the assembly by "those among us called deacons" (Apol. 1.65.5), who later take it to the members of the assembly who might be absent (1.67.5.).

Origins of the Title

While the picture of the deacon in Justin is a familiar one, on the basis of documents prior to him it is nonetheless not possible to argue with certainty that a liturgical role is the essential mark of the diaconate. 1 Clement can be adduced in support, and would add significance to some statements in Ignatius and the Didache, but the question remains obscure because none of these documents is expounding a theory of the office. Hermas speaks only of care for the needy, but there are no clear echoes of this elsewhere. Attempts to introduce the idea of a preaching or missionary deacon are unconvincing, as we noted of Lemaire's approach to the terminology of Ignatius. Our own survey, centering on linguistic rather than on broadly literary or historical aspects of the literature concerned, seems to show that the notion underlying the relationship between bishops and deacons is that of the agent. So far as the common noun is concerned, the notion of agency is not widely represented in Christian usage of the period,[1] but we do know that the word always speaks of a correlation, which in other areas of usage has been seen to be with God, Christ, or the community (e.g., 2 Cor. 6:4; 11:23; Rom. 16:1; IgnPhld. 10.1); in the usage respecting deacons the only consistent correlation is with the figure called "overseer," and this, as a title implying preeminence, unavoidably suggests that διάκονος is designating the overseer's personal attendant. As the overseer emerges more clearly as the bishop, the attendant takes on the characteristics of his agent, an idea made explicit in Ignatius's correlations of Father and bishop with Jesus and deacon (Mag. 6.1; Tr. 3.1; cf. Smyr. 8.1 and Polycarp 5.3).

In the later church orders the analogy of deacon and Jesus continues to have a profound influence on thinking about the diaconate while that of bishop and Father gives way to ideas of pastor and priest, and expressions of the analogy are to be read as the understanding on the part of a Greek-speaking church that the title "deacon" designates an agent. Thus, although Hippolytus speaks of the deacon's care for the community and of his liturgical functions, he relates both intimately to his standing vis-à-vis the bishop, who alone imposes hands—without the presbyters—because the deacon is ordained "in the service of the bishop [in ministerio episcopi] to do what is ordered by him," and who in the prayer over the deacon invokes only the identical aspect of the mission of Jesus, "whom you sent to carry out thy will [ministrare tuam voluntatem] and make known to us thy desire" (*Trad. ap.* 8). The *Apostolic Constitutions*, whose view of the diaconate we observed at the end of chapter 3, presents the deacon as the "mind and soul" of the bishop (3.19.7; cf. *Didascalia* 2.26.5) and provides the following theological comment on the idea of deacon as agent (some of its phrases serving also as a precise note on the meaning of the word διάκονος; 2.26.5):

> Let the deacon present himself to the bishop after the manner of Christ to the Father, and he is to serve him [λειτουργείτω] blamelessly in all matters, just as

Christ, doing nothing of his own will [ἀφ' ἑαυτοῦ ποιῶν οὐδέν], constantly did everything perfectly according to his Father's will.

This later tradition thus reflects what a semantic study of the earlier material has suggested in regard to the meaning of the common noun within church order. At the same time the later tradition can hardly spell out the reason why the early communities should have chosen this word in the first place as the title for one of their officers. If the word signifies the bishop's functionary, the reason it was selected for this signification will not lie in the nature of any particular task undertaken for the bishop, whether that be the distribution of bread and wine, the handling of his financial affairs, the care of his destitute, or the preaching of his word. Of these activities only the first and last have any real place in the field of meaning covered by the word διάκονος, and of these the preaching of the word has no place in the history of the diaconate as preserved in the earliest documents. On the other hand there are clear statements later (ὡς διακονούμενος ἱερεῦσιν, *Const. Ap.* 8.24.4: "ministering for the priests") and some indication earlier (ἐκκλησίας ὑπηρέται, Ign Tr. 2.3: "officers of the church") that in the distribution of the Eucharist the deacon is acting for the other clergy. Thus the Christians would not have chosen the term simply because it was current in pagan associations as the title for ritual waiters;[2] the relevance of that title to the title of deacon lies instead in its religious character, which it shares with many other uses of the word and its cognates in areas of religious ritual, message, and agency. What this religious character in turn points to is that from the beginning the diaconate was considered to be an arena of the sacral. This implies of course that the title with which the title "deacon" stands in correlation, namely, "overseer," itself designates an officer within the same sphere of the sacral, which makes improbable the view that the bishop was originally a mere administrator of affairs. The most likely field in which the two sacred titles might originate is that of cult, and if this can be taken back to Phil. 1:1 "deacons" perhaps already existed at Philippi even if "bishops" as such did not.[3]

14

The Gospels

In the gospels the words under discussion mainly designate menial attendance of one kind or another. The parable of the royal marriage feast includes "table attendants" (Matt. 22:13)—like those at the village marriage feast (John 2:5, 9)—who are distinguished from the "slaves" sent to bring the guests (22:3, etc.) The parable of the judgment of the peoples uses the verb in the wider sense of personal attendance on a master, again a royal person (Matt. 25:44).[1] In parables in Luke the verb occurs in the sense of table attendance on the part of a slave in regard to his master (Luke 17:8) and on the part of a master returning from a marriage feast in regard to his slaves (12:37).

In regard to Jesus the verb occurs in the same sense when angels ministered to him (Mark 1:13; Matt. 4:11), when Peter's mother-in-law ministered to him (Matt. 8:15; "to them," Mark 1:31; Luke 4:39), and when the women ministered to him (Mark 15:41; Matt. 27:55; "to them," Luke 8:3); in Martha's ministering to Jesus the verb is used absolutely (John 12:2; Luke 10:40), the latter passage including also the statement that she was preoccupied with much "waiting."[2]

Jesus as Waiter

Curiously Luke seems to have avoided stating that ministrations at table were directed exclusively at the Lord. Thus he omits the ministrations of angels at Luke 4:1–13, follows Mark 1:31 in writing "to them" at Luke 4:39, and substitutes "to them" (the better reading) at Luke 8:3 for "to him" of Mark 15:41; in this passage he has also moved the statement from the scene of crucifixion in Mark to the middle of Jesus' ministry, removed any reference to discipleship by omitting "who followed him," and added "out of their means," thus giving the phrase "they ministered to them" a purely profane sense in a historical note. There is thus no scene in Luke where the master is the centre of such attentions at table—

the attentions of the woman of the city (Luke 7:37–38) are not of this kind; indeed Martha's own special attentions (Luke 10:40) are directed away from his person to make room for the attention given to his teaching by Mary. On the other hand, in the image of faithful coming from the four quarters to sit at table in the kingdom of God (13:29) the focus shifts to the general joy of the eschatological feast at which, as we see in the sayings about watchful servants (12:37), the master "will gird himself and have them sit at table, and he will come and serve them." This feast reemerges as a theme of the discourse at Luke's last supper (22:16, 24–30), and his treatment of this and the earlier related material inclines one to speculate that within the theme he is also posting another motif in that his intention has been not to blur the portrait of the Lord "as one who serves" (22:27). A final glimpse of him is provided in the scene at Emmaus, which epitomises so much of Luke's presentation, when illumination and satiety come to disciples—who had invited him for refreshment—from the hand of the Lord at table (24:29–32).

We cannot fail to be aware that of the evangelists it is the Hellenistic author who has included in his account of the supper a factor reminiscent of the Hellenistic festive custom of masters waiting on their slaves.[1] The Hellenistic character of the language in 22:24–27 is recognised even by scholars who maintain that the passage enshrines a tradition independent of or older than Mark 10:42–45.[2] This character shows through in the participial expression of verse 27, ὁ διακονῶν ("the one attending") not because it instances a Hellenistic preference for a participle but because it is the preferred Greek designation in terms of these words for the waiter in action;[3] as such it refers to the activity of bringing food or drink, and no other meaning or connotation—such as of the washing of feet—can justifiably be read into it here. Both the form of the word and the general context are thus telling us that we are presented with an image of Jesus as a waiter; this understanding is confirmed by the contrast between the one sitting at table and "the one attending" in verse 27, and is no doubt supported by the parallel of the youngest and "the one attending" in verse 26, for from Homeric times it was the Greek ideal that youths should honour their betters in age by waiting on them.[4] The meal motif then runs on to verse 30, where Jesus holds out an assurance for the disciples of a place in his kingdom at his table.

The ethical teaching of verse 26 is made by way of a simile (ὡς) that anticipates the simile of verse 27, "I am among you as one who serves"; we are thereby alerted to the fact that the point of the teaching is not going to be found in its literal reading. The statement "I am among you as one who serves" comes at the end of verses 25–27 but expresses the controlling idea of the whole passage, which begins with the grand word "kings" and ends just as effectively with the word "waiter": verse 25 depicts the pride and honours of kings and officials; verse 26, which implies that great ones, leaders, and banquets exist also among disciples, urges the disciples to adopt the attitude of the young man who waits on older dignitaries; and verse 27 asserts that Jesus, himself the "greater one" on this occasion, does just this, reversing the convention and exemplifying the attitude of the young man. Because the teaching is cast in terms of the waiter—not merely of the servant, which as an image could have a broader reference to many aspects of life—and has been included, differently from the other synoptics, in the supper account, the simile refers only to events there at table; the image of waiter would be an unnatural figure by which to allude beyond the supper to situations like Jesus' care for the disciples or for the sick. Essentially what transpired at this supper is that Jesus passed the cup to the disciples, broke and gave them bread; reclining in their midst, however, he is not "going about fetching" things, which is what his Greek word for the waiter connotes, and he is therefore only "like" (ὡς) the waiter. The disciples are not being urged to repeat his kind of service or to carry out practical service in other situations of daily life—the author later reports that the Twelve had indeed to decline involvement in such roles in order to devote themselves to the ministry of the word (Acts 6:1–6)—but they are being instructed, even as they hold positions of authority, to adopt the attitude that Jesus' actions symbolise. Any other meaning the Greek reader may have taken from this scene, which has been so carefully composed by the Hellenistic Luke, we will consider at a later stage.

Servant of All

Similar ethical teaching is put before the Twelve on several other occasions but nowhere with such a background of service at table. In the earlier dispute about who was the greater (Luke 9:46–48) Jesus discounts the relevance to the kingdom of grades among the disciples and states that the least is already great; this teaching is tied in to the incident with the child. Here, but without any reference to a child, Mark 9:35 states, "If any one would be first, he must be last of all and servant of all." This statement is omitted at Matt. 18:1–5, which resolves the question of greatness in terms of the child's lowliness, but a similar statement occurs in that gospel's instruction against seeking the titles of rabbi or master (Matt. 23:11): "He who is greatest among you shall be your servant." In a different situation again, when ten of the disciples are indignant at the sons of Zebedee for seeking powerful positions in the kingdom, Jesus first sets the teaching, as at Luke 22:25, against a background of earthly power and then adds (Mark 10:43–44; cf. Matt. 20:26–27): "Whoever would be great among you must be your servant, and whoever would be first among you must be slave of all."

From the variety of contexts in which this kind of teaching occurs we can see that the teaching has an uncertain home in the tradition, although the diversity could itself be a sign of the teaching's importance. In two of the contexts where there is dispute among the disciples, the saying about the servant or διάκονος presupposes ambition for greatness (Mark 9:35; 10:43; cf. Matt. 20:26); in the third of such contexts the saying supposes merely that some disciples hold positions of authority or preeminence (Luke 22:26). This is the case also in Matt. 23:11, which is not set in a context of dispute. Luke 9:46–48, which is a reworking of Mark 9:33–35 and 36–37 but which does not take up from there the term διάκονος, expresses the idea of Mark 9:35, namely, that disciples must be "least" or "lowly" (Mark, "last"), which is what the same author is expressing in the statement about "the one attending" at Luke 22:26. Whereas there the designation is "waiter," in the other sayings the term διάκονος does not refer to service at table; instead at Mark 10:43 (Matt. 20:26) it is sustaining the idea of kingdom introduced by the disciples and embellished by Jesus in order to show that in a kingdom the least or last position, the one furthest from the position of power or honour, is to be the slave (Mark 10:44) of the one exercising power or his διάκονος (10:43), this being a recognisable term for a household attendant and slave in royal courts and grand houses.[1] The position is the equivalent to that of the waiter at a banquet or of a child among adults, which are other images the same teaching draws on. A dispute about roles in a kingdom is presumably the occasion of a reference to the household slave again at Mark 9:35, although the formal character of the common noun would itself suffice to explain its occurrence in ethical teaching.[2] The same might apply to the instance at Matt. 23:11, but here the word is the antonym of terms meaning "teacher" and is possibly meant to suggest the disciple of the teacher; we see that connotation working out more clearly at John 12:26 in the combination of "serving," "servant," and "following."[3]

No special significance, consequently, attaches to the occurrences of the term διάκονος in these sayings, and there is no call to explain its presence by way of a connection with its currency in the language of church order or in any other part of a presumed Christian lexicon. In this ethical teaching, with the possible exceptions of instances at Matt. 23:11 and John 12:26, the term is merely a synonym of words meaning "slave" or "last," and the three terms form one pole in a simple moral axiom. None of them is a summons to an ethical response in a particular line of activity, but they all illustrate in the same way and to the same degree an ethical value or attitude without which one cannot be a member of the kingdom; the qualification "of you" or "of all" merely localises the import of the teaching and indicates that the standard is absolute. The teaching is part of teaching on lowliness and childlikeness, and asserts that only by assuming the most lowly position, namely, the one antithetically placed to the exercise of ordinary power, will one be on the level at which the kingdom operates. On that level will leaders arise.

The Son of Man

In Mark 10:45 this teaching moves into a statement about the Son of man himself: "For the Son of man also came not to be served [διακονηθῆναι] but to serve [διακονῆσαι], and to give his life as a ransom for many." These infinitives echo the cognate noun in verse 43, suggesting that what is expressed in respect of the Son of man conforms to the situation depicted in respect of the disciples. This appears to work well enough in regard to "not to be served," which would indicate that the Son of man stands at the opposite extreme from power and honour and that this is to be the station of the disciples. When the statement adds however (in the usual translation) "but to serve," problems begin to emerge of the kind outlined in chapter 2; basically these are that it is not clear whom the Son of man might be serving or what kind of service he proffers but as well that the verb is an unusual word in the circumstances, unusual enough to have suggested to many a connection with Luke 22:27. In this way the verb has occasioned much speculation as to the tradition and redaction of the verse. Further, the second half of the verse (45b: "and to give his life as a ransom for many"), instead of clarifying the context for an understanding of "to serve," has tended to accentuate the word's singularity here because a theology of ransom would not seem to have a natural correlation with the idea of service, especially in a context that is basically ethical.

Moreover if we consider these two verbs in the light of general usage a further difficulty arises in that the rendering "serve," which forms the basis of current interpretations of the verse, has shown itself to be an unsatisfactory expression of what the Greek verb is saying in almost all instances, exceptions being possible in some aspects of attendance in a household or at table. Such a marginally satisfactory meaning should not be presumed in Mark 10:45. What the Son of man's "to serve" consists in should be determined against the background of specific uses of the Greek verb and in the light of the context. Part of the context is the passive infinitive "to be served," which is itself one of the most unusual instances of the verb because it is not only a comparatively rare passive but is predicated directly of a person in a way that no other passive is.[1] From this we are probably right to infer that the passive has been occasioned by the need to produce an antithesis to the active infinitive "to serve"; if so, the antithesis further emphasises the central position which these words hold here, and suggests that interpretations based on predetermined meanings might upset the balance that this pair of words is apparently meant to maintain between what precedes and what follows them. This is to imply that whatever the earlier history of verses 42–45—that is, of whatever independent sayings the passage may have been composed—the redactor is setting the two infinitives against the background of the passage as a whole. This could be important since, as has been amply illustrated, context is more than normally important for determining the sense of words from this cognate group. Meanings are often elusive and are sufficiently varied as not to be comprehensible within a single English word group like "ministry" or "service"; indeed within each of the areas of meaning (message, agency, attendance) more than one English word group may be needed to catch the sense, and we have seen that simply to transpose a meaning attested (or assumed to be attested) in one passage to another passage which might be more obscure can lead to misinterpretation and even to unnecessary textual conjecture.[2] Against this background we should be able to venture some judgments as to what meanings the redactor could or could not have intended.

To take the latter first, we can eliminate any interpretation of the active infinitive that is advanced on the grounds that the New Testament provides evidence of a meaning that is inherently Christian and distinct from meanings in literature generally. Those who discern such a meaning here describe it variously as care for others (Brandt, Beyer), lowly service (Schweizer), service at table (Roloff, Boulton), or liturgical service (Reicke).[3] Of these ideas, however, care and lowly service are not expressed by the words in any literature; liturgical service is denoted in a Christian author only at 1 Clem. 40.5, a use which in fact has antecedents in non-Christian sources; and service at table, which is expressed in Christian writings of the era only in the gospels and at Sim. 8.4.1, 2, is not the so-called fundamental meaning of the words in either Christian or other literature but is one expression of the notion of go-between. The usages of Christians and non-Christians are in fact identical, as are the kinds of problems they present for interpretation, and even the one technically Christian term ("deacon") is to be understood in the light of general—not specifically Christian—usage.

Several other commentators, while not advancing comprehensive theories about this group of words, maintain that the sense "to serve" (with the connotation of "at table") is due in one way or another to a tradition that associated this saying with the Eucharist, Luke 22:27 usually being cited as evidence of this.[4] With regard to Luke there is no doubt that service at table is what is meant, and it is possible that a saying like Mark 10:45 enshrining this sense could have had an independent existence, but it is not possible to show that Mark redacted any such saying to produce what now exists in its present context, and it is hardly conceivable that any writer of Greek could have done so. If the connotation of service at table is wholly unnatural in Mark 10:45, the fact that Luke places equivalent ethical teaching in the context of the supper and phrases it in terms of the same verb (Luke 22:27: "the one attending") becomes irrelevant to the meaning of the infinitive in Mark 10:45, and the Markan infinitive in its turn becomes very doubtful ground on which to argue the greater originality of Luke 22:27. While however it is difficult to understand from a linguistic point of view how Mark could have transformed a traditional saying about waiting at table into one expressing directly something about ethics or soteriology, and impossible to discern the signs of any such process, the proposal that Luke 22:27, if it is not an unrelated saying, is a redaction of the ethical aspects of Mark 10:42–45 does not create the same kind of linguistic problem and is compatible with Luke's interests. His Hellenising tendency is apparent in the whole passage, the participial ὁ διακονῶν, as we have seen, itself being one sign of this, and the importance he attaches to eschatological banquets (12:36–37; 14:15–24; 24:13–27) is such that he could conceivably have taken up the infinitive διακονῆσαι from a statement where it expressed a more abstract idea ("to carry out a charge") and have given it a concrete sense ("to serve at table") in what is his most important banquet scene in order to meet the interests of those readers in whose native Greek tradition a banquet with the Lord was itself the redemption of which Mark speaks.[5] The "benefactors" (Luke 22:25) of this world might on occasion engage in a similar ritual to indulge their dependants in the freedom of a festive meal, but theirs is but a temporary relaxation of the inexorable law that men and women live in subjection to the rule of the world and occasional largesse is but an understated display of the lordship to whose oppressive forms the Gentiles quickly revert. Jesus by contrast remains among his own, for whom it is important to observe, under Luke's guidance, not an act of service on Jesus' part but the level at which God works, the occasion on which God works (cf. 12:37; 24:30–31), and on which salvation comes. Somewhere here is a statement by Luke about Jesus' redemptive act, a subject in regard to which Luke is not so much reticent as breaking with a Markan formulation not to the Hellenistic taste. If the question arises as to whether Luke depends on Mark or Mark on Luke, it would at least appear that redactional processes are more evident in Luke.

A separate approach taken by commentators to Mark's "to serve" yields the idea of service to the brethren in the course of daily life, a Christian philanthropy, the absolute expression of which is the death of Jesus for the sake of many.[6] Such a sense, however, is unparalleled in other Christian sources and unprecedented in non-Christian sources, where the verb designates specific types of undertaking in the areas of message, agency, and attendance but not the philanthropic or charitable service envisaged in this interpretation.

Another understanding of the infinitive is in the sense of serving God. In Swete's sense of "submission to God"[7] this is unacceptable because the verb in whatever application always speaks of activity; serving God however can also mean doing things for God or at his command, and this has suggested to some that the verse places Jesus in line with the great servant prophets[8] or identifies him with the Suffering Servant of Isaiah.[9] This idea is helpful but, as it stands, imprecise. Certainly Mark's verb is not the link between his statement about the Son of man and the Septuagint's statement about the Suffering Servant at Isa. 53:11 because there the verb is different;[10] since one of the uses of Mark's word, however, is to express the idea of someone carrying out a charge, the word would apply very well to a servant of God carrying out the charge laid upon him, even though in such a case the word would not be expressing directly the idea of servanthood but the activity of the servant in respect of his commission. The commission itself can be to any kind of activity, and for this reason the word is predicated in literature of persons of diverse personal relationships with the commissioning authority. The word διακονῆσαι is thus not a sign in itself that the Son of man belongs to the category of servant even if verse 45b is taken as speaking of the Suffering Servant.

In Christian writings, outside the table usage of the gospels and Sim. 8.4.1, 2, the verb always signifies carrying out a task established either by God, by the terms of an ecclesiastical office, or by the authority of an apostle or by an authority within the community,[11] in all cases with that special connotation of the sacred that characterises so much of its use in all senses and that of its cognates in non-Christian sources, and which leads Paul to designate both his own apostolic task and the spiritual functions of all Christians as "ministries" or διακονίαι. The statement at 1 Cor. 12:5 shows in fact Paul's perception that the church is that sphere of action and grace where interlocking and complementary "ministries" are carried on in the name of the Lord. In the light of such usage the terms of the Son of man's statement in Mark take on a clearer definition. The infinitive itself, no less than the title "the Son of man" and the prophetic verb "came," speaks of a particular personal commission under God, and from this point of view the statement is at once more theological than ethical. Because the infinitive does not indicate what the commission entails, verse 45a could hardly have existed independently,[12] and needs to be complemented by verse 45b, which then defines the sacred role as one of ransom for many through the death of the officebearer. We ought to take note that while the other statements in this section of Mark about the Son of man's career have likewise emphasised humiliation and death, they include as well the rising after three days (8:31; 10:34); this is omitted at 10:45 because it is not relevant to the teaching on how the kingdom is entered and might even be seen as stimulating the ambition for power and glory, which is distracting the disciples. In this omission verse 45 establishes that the Son of man's διακονῆσαι leads to the opposite of all that is powerful and glorious so that he becomes the absolute standard for disciples who would belong to the kingdom. The ethical lesson is pointed not by the infinitive as itself a term designating this kind of humiliation, but by the death that the commission to effect the ransom entails for the Son of man.

The relationship between this active infinitive and the passive "to be served" needs to be clarified. Although this translation gives the general sense, it does not reproduce the particular connotation that enables the active infinitive with its meaning of carrying out a sacred task to stand in antithesis. The passive belongs to the part of usage described as domestic and personal attendance; analysis of that usage has shown that, as in other areas of meaning, the verb looks to the activity rather than to the status of persons, who in this instance are "attendants," and are perhaps best designated as "those who come and go at the behest of another." This is the aspect of meaning that makes the passive unusual. The situation envisaged by the statement is that the Son of man is not one who holds such a position in the world as to have attendants—the διάκονοι of the rich and powerful[13]—coming up to him and being despatched by him about various tasks of his own choosing; he has his own task to go to, and it is for the purpose of setting the profane grandeur of one way of life against the prophetic dedication of the other that Mark has brought these oddly fitting infinitives together.

Afterword

The reader who has followed the argument of this book will be looking for responses to questions raised in the opening chapter. These questions arose from the impact of a notion called "diakonia" on what is understood to constitute ministry in Christian communities. Unreasonable as it may appear at first, my feeling is that this afterword is not the place for answers. Certainly one aim of the study has been to clear the decks for yet one more close consideration of just who does what in a church: of why some are ordained and others are not, of whether ministry pertains only to the ordained, of whether the ministry of one communion is rightly shared by another—indeed, whether exclusive claims can be laid upon ministry by a church, of whether ministry is sex-determined, of whether ministry is authenticated by grace, of whether churches can modify their order yet continue to claim a tradition of ministry, and so on. Would there be no end to an afterword on ministry? And because this study has aimed mainly to work towards a more accurate view of what the first practitioners of Christian ministry meant when they spoke of διακονία, and has attempted to correct what it has presented as misconceptions current in this area for the last fifty years, its implications need to be worked through in more detail than is appropriate at the end of an already long book and with a finesse beyond the capacity of one writer. At the same time, when one reads a recent writer's discouraging view that in regard to first Christian perceptions of ministry it is "almost impossible to reconcile all the facts in a single coherent and convincing account,"[1] one must hope that with the perceptions of διακονία proffered in the preceding chapters we might edge closer to a coherent account.

The few observations that follow merely take a lead, accordingly, from indications arising in the study on approaches that seem appropriate to some issues concerning Christian ministry today. The first approach is to the first question raised in the book, which was whether "diakonia"—this being the modern loanword we began with—is a suitable designation for the churches' responsibility to redeem the oppressed and the dispossessed. Related to that by reason of today's prevailing view of the diaconate is the question of which direction the study ought to take us in if we are to arrive at defining more satisfactorily the character and role of ordained diaconate today. Next, on the broader scene of ministry, are some reflections on the relationship between ministry and charism. Closer to the nub are the succeeding considerations as to where ecumenism now stands on its already long journey towards an agreed understanding of ministry. This discussion will take us into the ultimate question of just what early Christians considered that ministry to be without which a group of Christians did not constitute a church. Finally we will need to ask whether today, according to that early Christian view, women can take up that kind of ministry.

In July 1984, on the occasion of the fortieth anniversary of the founding of what is now the Commission on Inter-Church Aid, Refugee and World Service (CICARWS), the Central Committee of the World Council of Churches approved a statement on "The Diaconal Task of the Churches Today," which included the following assessment of "diakonia": "Diakonia as the church's ministry of sharing, healing and reconciliation is of the very nature of the church."[2] When we ask, in the light of our study, whether such language accords with early Christian idiom, we can only say in brief that it does not. Further, we need to add that early Christians would not understand why their modern counterparts have so restricted the ambit of the ancient word that it now applies only to the caring sphere of the Christian life. The projected extensions of CICARWS's programme into the next century under the name of *Diakonia 2000* is, accordingly, on the face of it, a misnomer,[3] as is the nomenclature so long adopted by the German Evangelical church for its social work: the Diakonische Werk. The only reason an early Christian would have designated social work as a $\delta\iota\alpha\kappa o\nu\iota\alpha$ would have been—after the manner of the rare but plainly intelligible application of *The Shepherd of Hermas*—that social work was understood to be one more among the set of responsibilities that the gospel lays on the Christian;[4] of itself, as applied to social work, the Greek word would not be expressing social concern. Care, concern, and love—those elements of meaning introduced into the interpretation of this word and its cognates by Wilhelm Brandt—are just not part of their field of meaning.

At this juncture an associated problem arises in connection with the diaconates of nineteenth-century Germany, for these were founded expressly, as we saw in chapter 1, to be an office of care for the needy. In the case of the permanent diaconate instituted in the Roman Catholic church in the course of the Second Vatican Council, the problem is much less acute because theoretically care of the needy is only one dimension of the new clerical office—although, to judge from their activities and the style of their spirituality as reflected in their journals, care of the needy and of the marginalised would seem to be the most notable dimension for by far the majority of deacons. As regards status within their respective churches, the situations of these re-instituted Protestant and Roman Catholic diaconates are of course again different; not only does the Roman Catholic permanent diaconate specifically include within the diaconal tasks "the service of the liturgy" and "of the Gospel," but its deacons also receive sacramental ordination and thereby enter the ranks of clergy. Allied with this Roman Catholic innovation of the 1960s are subsequent reviews of the diaconate's nature and role that are continuing—not without a level of controversy—in some Protestant churches. What has our study to say in the face of these developments? Of the evangelical diaconates it can only say that the titles "deacon" and "deaconess" were adopted in the nineteenth century on the mistaken understanding that the apostolic diaconate was essentially for works of mercy. At the same time, of those and of the newer institutions one must also note with respect how much hope and endeavour have been invested in them over a century and a half. Mainly, however, we note that, while all of them invoke the ancient name, ironically the closest to the reality of diaconate as suggested by this study would seem to be a function that has long existed within the Reformed Church of Scotland, which is not in fact called diaconate, but which an astute observer of that church would like to see so called. This is the eldership. Writing of the relationship of elders to the diaconate, T. F. Torrance has had this to say:

> It seems to be clear that in the New Testament there is no evidence that can stand up to objective criticism for the title "elder" used in our way. This does not mean that there are no biblical grounds at all for the kind of office which we refer to under the eldership. It would seem to be entirely consistent with biblical teaching that there should be associated with those specifically ordained to the ministry of the Word and Sacrament others who are "ordained" to a complementary ministry within the congregational life and activity of God's people. . . . While ministers are ordained to dispense the Word and sacraments *to* the people, elders are set apart to help the people in their reception of the Word and in their participation in the Sacraments, and to seek the fruit of the Gospel in the faith and life of the community. . . . It is imperative that we set about once again to reform our church polity in accordance with the revealed Word of God. That should not entail a rejection of the rich tradition of service fulfilled by the Reformed eldership which has often been admired in other Communions, but rather that the eldership, assimilated to the biblical and early Christian diaconate, would recover something of its wholeness as an essentially spiritual and evangelical *diakonia*, taking its distinctive character from the Gospel of the Lord Jesus Christ which it is meant to serve. So far as the Church of Scotland is concerned, this would have the much needed effect of deepening mutuality and complementarity between the *presbyteral* ministry of the Word and Sacrament and the *diaconal* ministry of shared obedience to Christ. . . . Such a reformation in the office of the elder-deacon within the Reformed Church might then even play a modest ecumenical role in prompting fuller recovery of the biblical and early Christian pattern of *diakonia* in other Churches as well.[5]

There is much else relevant to diaconate in this learned and succinct pamphlet on *The Eldership in the Reformed Church,* but the line of thought represented in these excerpts coincides so nearly with that emerging from our study of διακονία that for the moment nothing more need be ventured here.

The next substantial matter requiring brief comment is the relationship of ministry to charism. In the light of currently dominant perceptions of how ministry is meant to work in the churches, we can hardly avoid looking back to Paul's statement "concerning spiritual gifts" (1 Cor. 12:1). We know that these gifts (χαρίσματα) come in "varieties" and that they are variously called "services" (διακονίαι) and "workings" (ἐνεργήματα). Designated in such general terms, they are impossible to identify as specific activities within the community, and in translation the terminology is vague, confusing and, without doubt, liable to mislead. Two things, however, are helpfully said, and the first concerns the meaning of the Greek expression διακονίαι in this passage. Firstly, in the light of our study, these cannot be services in the sense of initiatives that a member of a congregation might take upon himself or herself for the benefit of the community. The reason we must say this is that διακονία is not constituted by a rush of blood to the head, as if a Christian community were a market to be targeted by some spiritual entrepreneur; the "ministry" is real only when it is imposed on the minister, and is called "ministry" by very reason of the imposition, by force of the imposing, by virtue of the requisitioning. Ministry is a summons, a duty. We would thus say that in a Christian community we are in the midst of "different kinds of gracings, sacred duties, and spiritual deeds." These three may well be identical, and what they have in common is their divine source; this divine origin is the ground of Christians seeing them as "grace," their divine origin is the ground too of their strength as "deed," and is also, of course, the ground of early Christians proclaiming them to be "sacred obligations" or διακονίαι.

In designating these activities of the Christian community at this point Paul is purposely general; his paragraph is not a programme of action for a church but a recognition of what its origins are. A church is an other-worldly reality, operating on energies that are not homegrown; its impact might be on people, even on societies, but its force is God's: the names Paul gives to the specifically Christian things people do in community—"gracings," "summonses," "performances"— are not perfunctory generalisations but are each a recognition of the divine at work.

If that is something about the meaning and spirit of Paul's terms, the second thing to say about them concerns the context in which he uses them. If 1 Cor. 12:1–11 is not a programme, 1 Cor. 12:28 is. The whole descriptive exercise of chapter 12 has been to remind people that the community is indeed replete with divine capacities under the divine guidance but that on this side of heaven the operation is headed by those who hold the authority of the word. The variety of life is not the principle of life; life remains the outcome of the ministry of the word. It could well be—and Paul's admonitions in some passages exemplify this— that the minister of the word has to take it upon himself or herself to declare what is and what is not an authentic expression of the divine.

In its journey to arrive at an authentic expression of the nature of Christian ministry, the Faith and Order Commission of the World Council of Churches in conjunction with churches that have pursued bilateral or other similar paths has travelled further than any individual church or any theology of a particular tradition. Its present position is stated in the Lima document (*Baptism, Eucharist, and Ministry* [BEM]) alluded to in part I. In the official responses of the churches to this document, however, we are in a position to assess just how far the journey has really taken us. In some six volumes to date, we read much in appreciation of the endeavour, but are constantly confronted with the seemingly irreducible dichotomy that divided thinking on ministry in the 1930s between Protestant and

Orthodox—the Roman Catholic of course being aligned theoretically with the latter although at that time not directly involved in the ecumenical process. In general, Protestant churches welcome what BEM has to say of ministry, requiring mainly only that the reality of ministry be given outlets throughout the whole church and not just through the ordained or official ministries. It is precisely here that the Roman Catholic and Orthodox responses point to an underlying conceptualisation of ministry in BEM which is unsatisfactory to the point of producing an unacceptable theology of ministry. Briefly, we can put their problem this way: Protestants were once content to see the common character of Christians in priesthood, the order of the church being then characterised by ministry. The modern position in the Lima statement says, in contrast, that, since all Christians are called into ministry, we now have both priesthood and ministry being predicated of all Christians, with the consequence that the order of the church has lost its traditional identity. One Protestant church has even commented in response that "one must be careful not to applaud such a development merely because one is committed to a one-sided Protestant understanding of the ministry."[6]

The official Roman Catholic response, while appreciative that the Lima document "goes in the direction of the major lines of what we recognise 'as the faith of the Church through the ages,'"[7] contains several thinly veiled references to major theological obstacles that prevent full reception of the document, and these come together, for example, in a brief statement replete with scholastic and Tridentine terminology like "in our view, ordained ministry requires sacramental ordination in the apostolic succession."[8]

The official response of some Orthodox churches has already been alluded to in part I, and in an Inter-Orthodox Symposium preparatory to these responses we see, along with the Roman Catholic concern about sacramentality, a particular concern that at root there is a problem about how we are called upon in the Lima statement to conceptualise ministry. In the plain words of Archbishop Kirill of Smolensk and Vysma, "The Lima document does not make a sufficiently clear distinction between the ministry of the people of God and the ordained ministry."[9] And Metropolitan Antonie Plămădeală of Transylvania in the Romanian Orthodox church, whose book introducing notions covered by the contemporary loanword "diakonia" has been noted in part I, spoke forcefully and at some length on that occasion about the need for clarification of the meaning of "ministry":

> What does it mean for the Orthodox and what does it mean for the Protestants? Does it mean the same thing? Does it have the same content? At a superficial and very general look the answer would be: yes. But when one proceeds to define concretely its content, he finds that the Orthodox give it one meaning, while the Protestants another. . . . If we acknowledge *Ministry* to be equal with *Slujire* (Service) we can agree with the text and can easily achieve convergence with the Protestants, but then we would not speak about *the sacrament of priesthood* but about something totally different. Such convergences would be false convergences.[10]

The implications of such thinking on the part of two such major Christian groupings within the ecumenical movement are clear. We can view them in the light of the journey all these churches of such disparate traditions have joined.

Although the journeying has been good, and the travelling surface much less uncomfortable than in the past, and though the company has enjoyed the rich interplay of its characters, the destination would not seem to be much closer; one might indeed be tempted to ask if the company has not returned to the ecumenical impasse from which it set out, having in the meantime, in the course of its circuit, succeeded only in familiarising itself with the mountains and gorges of the surrounding terrain.

What in the preceding paragraphs has been called the Protestant view of ministry is the dominant view in the Lima statement. Our study has earlier shown that the modern formulation of this view has been grounded in the inadequate linguistic work by Brandt and Beyer, and that their work has nonetheless legitimised a new low theology of official or ordained ministry that is different not only from the high Christian theologies but in fact from the theology of ministry dominant in the earlier Protestant tradition. In their authentically modern way churches are increasingly speaking of a common call to ministry that does not depend on ordination: "for everybody"—to use the words of an Anglican Working Party on the theology of laity—"bishops, priests and laity together, the great sacrament of our common calling is our baptism";[11] so that the churches then speak (to continue with phrases from the same Working Party) of *"our churchly ministries . . . of both clergy and laity"* (these extending to "those who cook, type and clean"), or *"our ministries with family, friends and neighbours,"* of *"our 'Monday morning' ministries . . .* within the structures of the secular world—political, industrial," of *"our 'Saturday night' ministries . . .* in leisure and hospitality and entertainment and sports and holidays."[12] In speaking like this, they are speaking of ministry as a range of activities that some earlier Christians would instead have spoken of in their familiarly figurative way as the upbuilding of the body; in regard to activities that are as broad as our contemporary, first-world lives, we still need to make sense out of that earlier Christian vision of fostering, nurturing, maturing, and healing the body, because in our activities we need to see ourselves as giving expression to the communion we have received in Christ—the blessed *koinonia* that is our *fellowship* within the Spirit and within which all are knitted up in a cross-weave of responsibilities.

These responsibilities are works of love and service of immense variety; all together and singly they are, indeed, a "sacred obligation" of today's Christians and can thereby rightly be called, in the language of the earlier Christians, διακονία or (in the plural) διακονίαι. This is how Paul expressed it when he was putting forth that broad view (1 Cor. 12:5). Saying this, however, is not the same as saying that all share in ministry, as if ministry were a global or churchwide reality that Christians are called to give expression to in their various activities and at different levels of the experience of life, or as if Christians were baptised into a ministerial condition. Ministry, in the earlier thinking and by force of the Greek language, was much more particular. A ministry is a charge put upon someone, and those who gave us the language took it originally from a broadly religious background because they wanted to speak of churchly charges: for most, their ministries were commissions handed out by church authorities; for a few—very few—there was a sense of a ministry received instead directly from God. The "gifts" that Paul acknowledged in 1 Cor. 12:4 may well have been "ministries" in 1 Cor. 12:5 only by virtue of having been recognised and endorsed by the community or its leaders; he was certainly not implying that anything any Christian undertook or did was (or should have been recognised as) ministry. To adopt, by contrast, the modern turn of phrase and call everything ministry is to trivialise a tradition of language and to hide from us the reality of the διακονία which Paul knew sustained church. This is "the ministry [*RSV:* dispensation] of righteousness" and "of reconciliation" (2 Cor. 3:9; 5:18), which is the ministry of "first apostles, second prophets, third teachers" (1 Cor. 12:28), and of whoever has been entrusted with the word and commissioned to proclaim it. Between these ministers and those Christians who share in the word through them there is no rivalry, incompatibility, or opposition because, as one of the more adventurous theologians of today's church points out, "the object of the life of the people is the revealed mystery and the object of the teaching, worship and government by the hierarchy is this same mystery, the limits of hierarchical power being the limits of revelation."[13] If today an impasse has been reached in the theology of ministry, to what way out from there does the ancient conceptualisation of ministry point us?

The challenge to the churches is to examine their traditions to discover in what measure they have given expression to the ancient ministry of the word. Ministry of the word is the ministry by which Christian churches first came to be, and it is possessed throughout the generations by those churches that recognise themselves as part of the apostolic tradition. Put another way, in the words of George Tavard, "the first thing that needs to be recognised is not ministry, but Churchhood; and the way to such a recognition is the analysis of the awareness of being the Church. . . . Once Church has been seen, ministry has also been seen at work."[14] In their search to identify this ministry in their churchhood, churches may have to subject to close scrutiny an array of social, cultural, and artistic expressions of faith characterised by their particular traditions; they will scrutinise their terminology (has it become merely a hard-edged jargon, a carapace?), their in-house debates as well as their extramural controversies, such things also as their architecture and vestments, and of course rituals and even sacrament. Among all these are historical overlays of a sometimes merely cultural kind, but sometimes also profound, if now muted, expressions of theological conviction about ministry of the word.

In what, for brevity's sake, we can call the broadly Protestant traditions, an awareness of the word has never been far from their centre of experience: church furnishings of an earlier era bear it in upon us so forcefully. (On a recent visit to a country town in Victoria, where the attractive external features of the one-time Presbyterian and now Uniting Church building suggested that an inspection of the interior would be rewarding, I was instructed by a passing member of the congregation in how the furnishings were at that moment about to be modified for the purpose, ironically, of freeing up the focus on word and making them more consonant with liturgies centering on the table—a development, surely, under some influence of ecumenism.) In the Orthodox and Roman Catholic traditions—indeed, in much of the Anglican—the task of establishing the priority of word in the richness of their rituals is more complex, but at root their very insistence on priesthood as sacrament is a statement about the word in their midst. For in their priesthoods they are upholding above all else the role of the eucharistic minister, the one whose characteristic and essential task within the church is to proclaim the mystery of another whose death and resurrection is commemorated at tables. It was for this central proclamation that their priests were initiated into their roles through sacrament. From within the traditionally cultic or sacerdotal tradition of Roman Catholic theology of priesthood, this perception was forcefully articulated by Karl Rahner in the aftermath of the Second Vatican Council; eschewing a theology based on "sacramental powers" as "totally inadequate" and with "no directly perceptible basis in scripture," he wrote: *"The priest is the proclaimer of the word of God. . . . He is this in the supreme mode in which this word can be realised, that namely of the anamnesis of the Death and Resurrection of Christ which is achieved in the celebration of the Eucharist."* [15] What unites ministries across ecclesial communities, whether their theologies and rituals are high or low, is this word to which individuals feel called, are called, and to which they are faithful. In the one baptism in the one faith they share that ministry of the word in which Paul glorified and which Pope Paul VI acclaimed as the bearer of a "divine power." [16]

If ministry in the churches today comprises more than was embraced in the διακονία of Paul, its additional components cannot be essential to the formation and maintenance of a Christian church. That is to say that Paul's own conception of ministry suffices for church, and a major idea like priesthood in the sacrificial or cultic sense is extraneous to the character of ministry. In the light of Paul's perception of ministry, some churches today could well undertake a critical review of some of the processes they have traditionally chosen to adopt in the selection and preparation of their future ministers. If ministry is of the word, and is not a cultic role, what likelihood is there—to take a telling example—that young men of some twenty-five years of age will be effective bearers of the word, especially when they live the artificial life of theological colleges and may even be required to be celibate? Would a Paul be calling these junior members of a sophisticated adult world to proclaim the word to their parents' and grandparents' generations, among whom are certainly those many people with a more profound experience of the Christian calling? More likely Paul would still be nourishing the young men with the milk of doctrine. Around himself, in fact, he gathered in his mission men and women of maturity, and they in their turn, as he moved on, recognised and summoned those among their own communities on whom they judged the burden of ministry would best and safely lie. Only subsequently, with the formalising of this process through ritual, was the intention behind the process gradually obscured, and entry into ministry then increasingly occurred less by virtue of accreditation in the word than by canonical regulation.

An early regulation, of course, was the exclusion of women from ministry. On an authentic view of ministry, the exclusion was simply wrong and has become unjust. As explained in part I, many of those who support women's inclusion in official ministry have insistently appealed to a theology of ministry as service, arguing that the serving role is characteristically feminine and that early Christian women were in fact preeminent in this role. The appeal collapses, of course, in the light of the present study. Interestingly, a leading protagonist of women's rights in the church has come to shun any appeal on such grounds. Any theology of ministry as service, Schüssler Fiorenza argues, serves "to internalise and legitimate the patriarchal-hierarchical status quo" and is taking advantage of the "cultural socialisation of women to self-less femininity and altruistic behavior,"[17] and is in fact "disempowering to women."[18] Instead of deepening their capacity for service, accordingly, women should, on this view, throw off the chains of servitude and join the other dispossessed forces in society in the struggle to achieve rights and equality even within the churches. The present study should put a stop, however, to this revisionist line against the received theology of service because, as is the case with the "diakonia" Schüssler Fiorenza is attempting to discredit, the early Christian terms for service are misunderstood. It is in fact precisely because early Christians chose a noble designation for the essential ministry of church that women have a right of entry into it. That is, early Christians chose to base essential ministry on the conviction that the word of God is communicated to men and women with equal force and clarity, and is then to be communicated to others by those among them, whether male or female, who are faithful hearers and honourable bearers of its import.

The ministry we have described in its sources in the earliest Christian writings—having viewed it there in the light of mainly pre-Christian sources—is one we could recognise as late as 1951 in a typical statement of the period: "The Ministry is that of the Word of God, and the minister's authority is plain."[19] At the same time, we have come to accept in the intervening decades the fact, as put by another writer as early as 1903, that ministry in the early church "was entirely unlike anything which has existed in any part of the Christian Church from the beginning of the third century downwards."[20] The core of that earliest conceptualisation of ministry is what we have sought to recover as it is encapsulated in the term $\delta\iota\alpha\kappa o\nu\iota\alpha$, which the first Christian writers took up from the language of religious rhetoric of centuries. If the attempt to set that term and the words associated with it within their cultural context has required a wide-ranging critique of mid-twentieth-century theological writings, the exercise reminds us of one striking phenomenon: the vast number of writings that the subject of ministry has generated over the past forty years.[21] Of itself this far-flung enquiry along many fronts is clear evidence of an enormous enthusiasm for reform of ministry, a reform, I am happy to see, that embraces in its objectives the mutual recognition of ministry between churches.

Probably more pressing than mutual recognition of ministries is the need for each church to embrace opportunities to reform its own ministry. I have already alluded to the priority of word in ministry and to women's competence and rights therein, and in the passing comment about the widespread and age-old practice of instituting only young men for a lifetime in ministry—most often as celibates— we encounter a phenomenon that has undoubtedly become an impediment to the flourishing of official ministries in many of today's otherwise enlivened communities of faith. In an age of such sociological acumen as our own, however, why should there be hesitation to acknowledge that careers in ministry—gifts of ministry, if you like—might be of limited duration and undertaken at any appropriate time of life? What might such mobility of ministry and interchangeability of role contribute to the enrichment of life both in formal Christian communities and in the broader community? How many calls have there been from small communities in areas virtually denuded of traditional clergy for the recognition of Christian community leaders as presiding ministers? In the Roman Catholic church the next Synod of Bishops is to review what is called priestly formation; will the synod be looking at what a correspondent to a journal calls her "vision of our clergy for the future: a highly motivated, diverse and dedicated body of people, moving into the Church's service from other areas of life, giving of their very best for a peak period, and when appropriate moving on again to enrich other areas of life with the experience they have gained"?[22]

What the present study might contribute to reform and ecumenism is, as I have indicated earlier in this afterword, more likely to be the fruit of collaboration between theologians than the result of one person's reflections. At the same time it seems to me there is a clear direction in which this study points, and that is to the higher ground of thinking about ministry. Accordingly, as I have wished to give this direction a strong emphasis, I have chosen some words to finish from a churchman who wrote in a very different era but with a feeling and a vision appropriate to our theme. Edwin Hatch concluded his eight Bampton Lectures at Oxford in 1880 on *The Organisation of the Early Christian Churches* with the following passage. In weighing it, we may need to abstract from the vigor of his social and political comment (but not entirely?—at what stage of history do sages not express themselves in some such way?) and even more carefully we will need to disassociate ourselves from his assumption that ministry is for men. In Hatch's view of ministry, nonetheless, and of its place in the church and among the churches, is very much what a consideration of ancient διακονία inclines me to. That word is, after all, the main evidence of ministry in what Hatch would have happily recognised as the Apostolic Age; all other evidences like priesthood, apostolic succession, and "gathering" peter out. Within the Christian tradition the reality of this ministry is as catholic as faith.

> Whatever be the form in which they are destined to be shaped, the work which the Christian societies, as societies, have to do, in the days that are to come, is not inferior to any work which has lain before them at any epoch of their history. For the air is charged with thunder, and the times that are coming may be times of storm. There are phenomena beneath the surface of society of which it would be hardly possible to overrate the significance. There is a widening separation of class from class: there is a growing social strain: there is a disturbance of the political equilibrium: there is the rise of an educated proletariat. To the problems which these phenomena suggest Christianity has the key. Its unaccomplished mission is to reconstruct society on the basis of brotherhood. What it has to do it does, and will do, in and through organisation. At once profoundly individual and profoundly socialistic, its tendency to association is not so much an incident of its history as an essential element of its character. It spiritualises that ineradicable instinct which draws man to man and makes society not a convention but a necessity. But the framing of its organization is left to human hands. To you and me and men like ourselves is committed, in these anxious days, that which is at

once an awful responsibility and a splendid destiny—to transform this modern world into a Christian society, to change the socialism which is based on the assumption of clashing interests into the socialism which is based on the sense of spiritual union, and to gather together the scattered forces of a divided Christendom into a confederation in which organisation will be of less account than fellowship with one Spirit and faith in one Lord—into a communion wide as human life and deep as human need—into a Church which shall outshine even the golden glory of its dawn by the splendour of its eternal noon.

Notes

PART I

1. *Boswell's Life of Johnson*, vol. 1 (London 1927), p. 387.
2. M. Löhrer, "La hiérarchie au service du peuple chrétien," in *L'Eglise de Vatican II*, vol. 3, by P. Th. Camelot and others (Paris 1966), p. 723 ("l'idée de service ecclésial, et d'ailleurs général").

CHAPTER 1

Word Studies

1. *Theologisches Wörterbuch zum Neuen Testament*, ed. G. Kittel, 2 (1935) 81–93; Eng. trans., *Theological Dictionary of the New Testament* 2 (1964) 81–93.
2. *Dienst und Dienen im Neuen Testament* (Gütersloh 1931). The original dissertation was entitled "Diakonie und das *Neue Testament*" (University of Münster 1923); see the biographical notes on Brandt in the index to *Evangelisches Kirchenlexikon* 4 (1961).
3. "Anfänge der soteriologischen Deutung des Todes Jesu (Mk. x.45 und Luk. xxii.27)," *New Testament Studies* 19 (1971–1972) 38–64. See also Roloff's earlier *Apostolat-Verkündigung-Kirche* (Gütersloh 1965), where his observations on the word group are frequent (see index) but of a more general nature.
4. *La fonction diaconale aux origines de l'église* (Bruges 1960). See also Colson's *Les fonctions ecclésiales aux deux premiers siècles* (Bruges 1956), pp. 163–74; *Ministre de Jésus-Christ ou Le sacerdoce de l'évangile* (Paris 1966), chap. 6; "Der Diakonat im Neuen Testament," in *Diaconia in Christo*, ed. K. Rahner and H. Vorgrimler (Basel/Wien 1962), pp. 3–22.
5. *Church Order in the New Testament*, Eng. trans. (London 1961).
6. *Les ministères aux origines de l'Eglise* (Paris 1971); "From Services to Ministries: 'Diakoniae' in the First Two Centuries," *Concilium* 10, no. 8 (1972) 35–49; "L'Eglise apostolique et les ministères," *Revue de droit canonique* 23 (1973) 19–46; "The Min-

istries in the New Testament, Recent Research," *Biblical Theology Bulletin* 3 (1973) 133–66; *Ministry in the Church*, Eng. trans. (London 1977).
7. "Diáconos helénicos y bíblicos," *Burgense* 4 (1963) 9–143. See also Guerra's "Cambio de terminología de 'servicio' por 'honor-dignidad' jerárquicos en Tertuliano y San Cipriano," in *Teología del sacerdocio*, vol. 4, by I. Oñatibia and others (Burgos 1972), pp. 295–313.
8. *Die Gegner des Paulus im 2. Korintherbrief* (Neukirchen-Vluyn 1964), pp. 31–38; Eng. trans., *The Opponents of Paul in Second Corinthians* (Philadelphia 1986), pp. 27–32.
9. In an informative critique of several major contributions to the topic of "diakonia" (including those by Colson, Brandt, and Beyer, but not Georgi's), C. Tatton expresses dissatisfaction with "the basic issue of providing a satisfactory definition of *diakonia*"; he concludes, nonetheless, that "diakonia" is "a concept indispensable to a proper understanding of the whole Gospel." See "Some Studies of New Testament *Diakonia*," *Scottish Journal of Theology* 25 (1972) 423–34, citations from pp. 426 and 434. More basically Avery Dulles rejects the concept of "diakonia" as "the Church's service to the world" on grounds—only generally indicated—of biblical usage; see *Models of the Church* (Garden City, N.Y., 1978), pp. 105–6, and *The Resilient Church* (Garden City, N.Y., 1977), p. 17 (cited here).

"Diakonia" and the German Evangelical Diaconate

1. "Das Amt der Diakonissen . . ." (1842), in *Quellen zur Geschichte der Diakonie*, ed. H. Krimm, vol. 2 (Stuttgart 1963), p. 359.
2. *Die Diakonisse, oder: Leben und Wirken der Dienerinnen der Kirche für Lehre, Erziehung und Krankenpflege* (Düsselthal 1835).
3. See C. Woodham-Smith, *Florence Nightingale* (London 1950), pp. 89–92. She recorded her emotion at reaching Kaiserswerth for a fortnight's visit in 1850 on a return journey from a tour of Egypt: "With the feeling with which a pilgrim first looks on the Kedron I saw the Rhine dearer to me than the Nile" (p. 82). She was admired by Fliedner and was godmother to one of his children, whom she later educated at her own expense.
4. Letter to Amalie Sieveking (8 February 1837), in *Friederike Fliedner und die Anfänge der Frauendiakonie*, by A. Sticker (Neukirchen 1961), p. 334.
5. Sticker, *Fliedner*, p. 356.
6. "Kurze Geschichte der Entstehung der ersten evangelischen Liebes-Anstalten zu Kaiserswerth" (1856), in Krimm, *Quellen*, vol. 2, pp. 211–17.
7. *De captivitate babylonica ecclesiae* (1520; Weimarer Ausgabe, vol. 6, p. 566). The translation of Luther's "diaconia" by "Diakonie" in Krimm, *Quellen*, vol. 2, p. 27, is misleading; ; "diaconia" means "diaconate" ("Diakonat" in German), as in the parallel "diaconia—Episcopatus" (Weimarer Ausgabe, vol. 26, p. 63).
8. In evidence of 1856 before "Der Evangelische Oberkirchenrat," in Krimm, *Quellen*, vol. 2, p. 366.
9. See extracts from the proceedings in Krimm, *Quellen*, vol. 2, pp. 361–73.
10. Letter of November 1835, in Krimm, *Quellen*, vol. 2, p. 350.
11. In evidence concerning "Die Diakonie und den Diakonat" before the High Consistory (1856), in Krimm, *Quellen*, vol. 2, p. 371. On "the deeds of God" as the counterpart to "the word of God" in the life of the individual Christian, see Wichern's address to the Wittenberg assembly of 1848 in Krimm, *Quellen*, vol. 2, pp. 241–45.
12. From his collected writings in Krimm, *Quellen*, vol. 2, p. 386.
13. *Von der Barmherzigkeit* (1869) in Krimm, *Quellen*, vol. 2, p. 380.

14. *Die christliche Liebesthätigkeit,* vol. 1 (Stuttgart, 1882), pp. 67–72. Uhlhorn's subsequent observations on "diakonia" in the New Testament fully support contemporary attitudes to the idea. Thus, p. 73: "Not every Christian is a presbyter but each one is truly a deacon, a servant of all."
15. See the documents from the synod of March 1957, in Krimm, *Quellen,* vol. 3 (Stuttgart 1966), pp. 296–306. The synod's theme was "The diakonia of the Church in a changed world." For details of the modern organization, see R. Leudesdorff, "Diakonie," in G. Otto, *Praktisch Theologisches Handbuch* (Hamburg 1970), pp. 102–30. The main library collections on "diakonia" are held by the head office of the Diakonisches Werk in Stuttgart and by the Diakoniewissenschaftliches Institut of the University of Heidelberg.
16. Foreword to *Quellen,* vol. 2, p. 11.
17. Sticker, *Fliedner,* pp. 79, 97. The origins of the modern evangelical diaconate are recounted by P. Philippi, *Die Vorstufen des modernen Diakonissenamts (1789–1848) also Elemente für dessen Verständnis und Kritik* (Neukirchen 1966). For an early account of its development and for some sources, see T. Schäfer, "Diakonen- und Diakonissenhäuser," in *Realencycklopädie für protestantische Theologie und Kirche,* 3d ed. (1896–1913), vol. 4 (repr. 1969), pp. 604–16. Outside of more recent encyclopaedias, resumes are provided by H. Krimm in *Diaconia in Christo,* ed. K. Rahner and H. Vorgrimler (Freiburg/Basel/Wien 1962), pp. 190–200; World Council Studies, No. 2, *The Ministry of Deacons* (Geneva 1965); *Le Diacre dans l'église et le monde d'aujourd'hui,* ed. P. Winniger and Y. Congar (Paris 1966); and in the Karl Barth Festschrift, *Service in Christ,* ed. J. I. McCord and T. H. L. Parker (London 1966); see among other writers, K. Bliss, *The Service and Status of Women in the Church* (London 1952), pp. 79–89; P. Philippi in World Council Studies no. 4, *The Deaconess* (Geneva 1966); C. Bridel, *Aux seuils de l'espérance, le diaconat en notre temps* (Neuchatel 1971), pp. 39–42. For Sweden in particular, see G. Elmund, *Den kvinnliga diakonin i Sverige 1849–1861* (Lund 1973). For eleven European foundations, see the articles by V. Vinay in *Dizionario degli istituti di perfezione,* vol. 3 (1976), cols. 476–89; his general article on modern "Diaconisse, Diaconi," cols. 470–76, has a full bibliography.
18. See the Panel Presentation by K. H. Neukamm of the Federal Republic of Germany in *Diakonia 2000: Called to be Neighbours,* Official Report, WCC World Consultation, Inter-Church Aid, Refugee and World Service, Larnaca, 1986, ed. K. Poser (Geneva 1987), p. 62; see a further description of the operation in J. Degen, "Diakonia as an Agency in the Welfare State," *Concilium* 198 (1988) 102–9. For a concise survey (with statistics) of German deacons and deaconesses see *Deacons in the Ministry of the Church,* A Report to the House of Bishops of the General Synod of the Church of England (London 1988), pp. 30–33.
19. Vol. 3 (1965). See the entry "Social Welfare in the Lutheran Church."
20. P. Nordhues in *Theologie und Glaube* 65 (1975) 237, reviewing *Caritas und Diakonie,* a booklet issued by the Conference of German-speaking Pastoral Theologians (Mainz 1974). "Caritas" is also the name of the Roman Catholic relief and welfare agency, a fact that accentuates the difference from the Evangelical tradition in which, as mentioned previously, the corresponding agency is called "Das Diakonische Werk."

A Theological Conception

1. *Cyclopaedia of Biblical, Theological and Ecclesiastical Literature,* ed. J. M'Clintock and J. Strong, 2 (1880) 704–7.

2. "Deacon and Deaconess," in *Encyclopaedia Biblica*, ed. T. K. Cheyne and J. S. Black, 1 (London 1899) col. 1038.
3. See the discussion in T. Klauser, "Diakon," *Reallexikon für Antike und Christentum* 3 (1957), col. 903.
4. *Bibliographie internationale sur le sacerdoce et le ministère*, ed. A. Guitard and M. G. Bulteau (Montreal 1971); *Index to Religious Periodical Literature; Internationale Oekumenische Bibliographie; Répertoire bibliographique des institutions chrétiens.*
5. "Serve" (K. Hess), in *The New International Dictionary of New Testament Theology* 3 (1978), being the translation of the article "Dienen," *Theologisches Begriffslexikon* 1 (1967); other articles or comment on "diakonia" occur in such works of reference as the following: *Evangelisches Kirchenlexikon* 1 (1956), "Amt, geistliches, I. Biblisch" (H. D. Wendland), and "Diakonie, I" (W. Schütz); *Die Religion in Geschichte und Gegenwart* 2 (1958), "Diakonie" (H. Wagner); *Lexikon für Theologie und Kirche* 3 (1959), "Amt, III" (K. H. Schelke), and "Diakon" (J. Gewiess); *The Interpreter's Dictionary of the Bible* 3 (1962), "Ministry, Christian" (M. H. Shepherd); *Biblisch-historisches Handwörterbuch* 1 (1962), "Dienen, Dienst" (J. Müller-Bardorff); *Das Wort in den Wörtern* (1965), "Dienen" (F. Melzer); *The Encyclopedia of the Lutheran Church* 3 (1965), "Social Welfare in the Lutheran Church, I, II" (H. J. Whiting); *Bibel-Lexikon* (1968), "Amt, II" (F. Mussner); "Diakon" (A. van den Born), and "Dienst, II" (H. Gross); *Sacramentum Mundi*, Eng. trans. 2 (1968), "Deacon" (N. Jubany); *Vocabulaire de théologie biblique* (1970), "Autorité" (F. Amiot and P. Grelot); *Conceptos fundamentales en la Doctrina social de la iglesia* 4 (1971), "Servicio" (J. L. Gutierrez Garcia); *Evangelisches Staatslexikon* (1975), "Diakonie" (H. C. von Hase); *Exegetisches Wörterbuch zum Neuen Testament* 1 (1980), "διακονέω κτλ." (A. Weiser).
6. "Diakon," in *Die Religion in Geschichte und Gegenwart* 2 (1958), col. 160.
7. *The Role of the "Diakonia" of the Church in Contemporary Society* (Geneva 1966), p. 13.

The Second Vatican Council

1. G. A. Lindbeck, "A Protestant Point of View," in *Vatican II, An Interfaith Appraisal*, ed. J. H. Miller (New York/London 1966), p. 227.
2. Y. Congar, "The Laity," in *Vatican II*, pp. 244–45.
3. Dogmatic Constitution on the Church, article 10; see *Vatican Council II, The Conciliar and Post Conciliar Documents*, ed. A. Flannery (Dublin 1977), p. 361.
4. "A Protestant Point of View," p. 219, and see pp. 221–22.
5. R. Spiazzi, "The Main Lines of Priestly Spirituality," *L'Osservatore Romano* (English weekly edition), 29 January 1976.
6. J. F. Hotchkin, "The Christian Priesthood: Episcopate, Presbyterate and People in the Light of Vatican II," in *Lutherans and Catholics in Dialogue*, vol. 4, ed. P. C. Empie and T. A. Murphy (Washington/New York 1970), pp. 205 and 208.
7. P. J. LeBlanc, "A Survey of Recent Writings on Ministry and Order," *Worship* 49 (1975) 47.
8. The commission's report was published as *Le Ministère sacerdotal* (Paris 1972).
9. The synod did not publish its document on the subject but see reports on its discussions and difficulties in *Informations catholiques internationales*, nos. 394, 397 (1971), also on preparatory work, no. 382 (1971); for analysis of some speeches and an evaluation,

see E. Schillebeeckx, *The Church with a Human Face*, Eng. trans. (London 1985), pp. 209–36.
10. Article 29: "in diaconia liturgiae, verbi et caritatis Populo Dei . . . inserviunt," in *Vatican Council II*, p. 387, translated: "they are dedicated to the People of God . . . in the service of the liturgy, of the Gospel and of works of charity."
11. Article 24 (trans. in *Vatican Council II*, p. 378).
12. *We Who Serve: A Basic Council Theme and Its Biblical Foundations*, Eng. trans. (London 1969), p. 11.
13. *We Who Serve*, p. 12; for statistics, p. 13 n. 2. The theme can be followed up by way of P. Delhaye and others, *Concilium Vaticanum II, Concordance, Index, Listes de fréquence, Tables comparatives* (Louvain 1974). See also the article "Servizio" (T. Federici), in *Dizionario del Concilio Ecumenico Vaticano Secondo*, ed. S. Garofalo and T. Federici (Rome 1969), cols. 1827–30.
14. *The Runaway Church* (London 1975), p. 61.
15. J. McCulloch in *The Times* of London, 11 March 1976.
16. *Your Faith* (Chawton, Hants 1976), p. 23.
17. Cited from *The Tablet* (London), 17–24 April 1976.
18. Cited from *The Catholic Herald* (London), 24 September 1976.
19. *Commentary on the Documents of Vatican II*, ed. H. Vorgrimler, Eng. trans., vol. 3 (New York 1969), p. 309.
20. Commentary, pp. 313–14. See his whole discussion of articles 2 and 3, pp. 308–18. Also K. Morsdorf on art. 17 of the Decree on the Bishops' Pastoral Office in the Church in *Commentary*, Eng. trans., vol. 2 (1968), pp. 234–35.
21. *Le Ministère sacerdotal*, p. 74.
22. "History of Lumen Gentium's Structure and Ideas," in Vatican II, p. 128.
23. See Congar's observation on the deficiency of Roman Catholic theology at this point, in "The Laity," p. 248.
24. M. Löhrer, "La hiérarchie au service du peuple chrétien," in *L'Eglise de Vatican II*, vol. 3, by P. Th. Camelot and others (Paris 1966), p. 725. The article concludes with a bibliography reflecting largely Roman Catholic interest at that time in the relationship between hierarchy and service.
25. *Ministerial Consciousness: A Biblical-Spiritual Study* (Rome 1975), pp. 20–33.
26. Ibid., p. 24, in reference to art. 11 of the Declaration on Religious Freedom.
27. Ibid., p. 208.
28. Ibid., p. 240.
29. Ibid., p. 200.
30. *The Church 2000*, An Interim Report offered by the joint working party on pastoral strategy for study and consultation (Westminster 1972), para. 152.
31. *Minsterial Consciousness*, pp. 21 and 24.
32. *Diaconia in Christo*, ed. K. Rahner and H. Vorgrimler (Freiburg 1962).
33. "Der Diakonat im Neuen Testament," in *Diaconia in Christo*, pp. 3–22.
34. J. Lécuyer, "Les diacres dans le Nouveau Testament," in *Le Diacre dans l'église et le monde d'aujourd'hui*, ed. P. Winninger and Y. Congar (Paris, 1966), pp. 15–26.
35. "La hiérarchie comme service selon le Nouveau Testament et les documents de la tradition," in *L'épiscopat et l'église universelle*, ed. Y. Congar and B. D. Dupuy (Paris 1962), pp. 67–99 (see pp. 72, 81, 86); see there too Congar's exemplary study, "Quelques expressions traditionnelles du service chrétien," pp. 101–23.
36. *Jungerschaft und Apostelamt* (Freiburg 1957); the later English version was *Discipleship and Priesthood, A Biblical Interpretation* (London 1966).
37. M. Löhrer, "La hiérarchie au service," p. 727; J. Feiner, "Commentary on the De-

cree" (on Ecumenism), *Commentary*, vol. 2, p. 121; also there p. 99 and, in his history of the decree, pp. 13–14 (address by A. G. Vuccino), and 22.
38. "History of the Constitution," in *Commentary*, trans. vol. 5 (1969), p. 22; cf. p. 19.

The World Council of Churches

1. "Christian Councils—Instruments of Ecclesial Communion," *One in Christ* 8 (1972) 145.
2. For the report on service, see L. E. Cooke, "Preliminary Reflections on the Assembly Document on 'Service,' " *The Ecumenical Review* 13 (1960–1961) 234.
3. C. Simonson has written to the point in *The Christology of the Faith and Order Movement* (Leiden 1972).
4. *The Ecumenical Review* 13 (1960–1961) 69.
5. Cooke, "Preliminary Reflections," p. 243.
6. Ibid., p. 238.
7. S. M. Cavert, "The New Delhi Story," in *The New Delhi Report*, ed. W. A. Visser 't Hooft (London 1962), pp. 8–14. Takenaka's paper, "Called to Service," is printed in *The Ecumenical Review* 14 (1961–1962) 164–76.
8. "Called to Service," p. 169.
9. *The Ecumenical Review* 11 (1958–1959) 443. In the quotation in text, biblical references, which include Matt. 25:31–46 and 1 Pet. 4:11, are omitted.
10. See n. 5 above.
11. *The First Assembly of the World Council of Churches*, ed. W. A. Visser 't Hooft (London 1949), and see also the third volume of the preparatory studies, *The Church and the Disorder of Society* (London 1948); *The Evanston Report*, ed. W. A. Visser 't Hooft (New York 1955).
12. *Biblical Authority for Today*, ed. A. Richardson and W. Schweitzer (London 1953).
13. H. von Campenhausen, "Church and State in the Light of the New Testament," in *Biblical Authority*, pp. 293–309.
14. "Guiding Principles for the Interpretation of the Bible," in *Biblical Authority*, p. 243.
15. W. Schweitzer, "Biblical Theology and Ethics Today," in *Biblical Authority* pp. 151–54.
16. See the report of sec. II on "Evangelism," art. 3–5, in *The Evanston Report*, p. 161.
17. *The Evanston Report*, art. 17–19, p. 102.
18. Report of sec. VI on "The Laity," art. 2, in *The Evanston Report*, p. 161.
19. In *The Evanston Report*, art. 15, p. 170; cf. art. 12, p. 168.
20. Report of the Advisory Commission on the Main Theme of the Second Assembly, "Christ—The Hope of the World," published with the six preliminary Evanston Surveys in *The Christian Hope and the Task of the Church*, ed. H. P. Van Dusen and N. Ehrenstrom (New York 1954), pp. 42–43.
21. "Survey on Evangelism," in *The Christian Hope*, p. 59. (The more exact biblical reference is to 1 Cor. 9:19–23.)
22. Sect. III, "Social Questions: The Responsible Society in a World Perspective," in *The Evanston Report*, p. 126.
23. "Survey on Social Questions," in *The Christian Hope*, pp. 28–30.
24. *The New Delhi Report*, p. 93.
25. Art. 76 and 78, p. 111.
26. See for example the paragraphs referred to in the preceding note.
27. Art. 71, p. 110.
28. Cavert, "The New Delhi Story," pp. 13 and 33.

29. Art. 85, p. 113.
30. Art. 80, pp. 111–12.
31. *World Conference on Church and Society* (Geneva 1967), pp. 182–83.
32. For some confused reactions of theologians to their experience at the conference, see appendix XVI in the *Minutes and Reports of the Twentieth Meeting of the Central Committee of the World Council of Churches* (Geneva 1967).
33. See for example the concluding paragraphs (44–48) of the report on World Economic and Social Development in *The Uppsala Report 1968*, ed. N. Goodall (Geneva 1968), pp. 54–55; the report of the Moderator of the Central Committee in *Breaking Barriers, Nairobi 1975*, ed. D. M. Paton (London/Grand Rapids 1976), pp. 229–30. See further P. Abrecht, "The Development of Ecumenical Social Thought and Action," in *The Ecumenical Advance*, ed. H. E. Fey (London 1970), pp 233–59; and his "The Social Thinking of the World Council of Churches," *The Ecumenical Review* 17 (1965) 241–50. "Diakonia" remains, of course, a basic concept for the World Council; see, for example, "The Diaconal Task of the Churches Today, Statement approved by WCC Central Committee, Geneva, July 1984" in *Hope in the Desert, The Churches' United Response to Human Need, 1944–1984*, ed. K. Slack (Geneva 1986), pp. 133–35, which includes the statement from the Vancouver Assembly, "Diakonia as the church's ministry of sharing, healing and reconciliation is of the very nature of the church" (p. 134).
34. Cited from a working paper by P. Löffler, "The Sources of a Christian Theology of Development," in *In Search of a Theology of Development*, papers from the Sodepax consultation at Cartigny, 1969, p. 69; the companion volume of bibliography by G. Bauer, *Towards a Theology of Development*, shows no interest in service as such.

Counterpart to Witness

1. Report of sect. 2, "What is the service of God today?" art. 1, in *This Happened at Nyborg VI* (Geneva 1971), p. 72. The assembly's theme was "Servants of God, Servants of Men," which formed the title also of Bishop Werner Krusche's leading address printed as appendix 6 and in *The Ecumenical Review* 23 (1971) 205–21.
2. The assembly's "Message to the Churches," in *Nyborg VI*, p. 117.
3. M. Opočensky, "European Theology—Other Theologies: An East European Contribution," in *European Theology Challenged by the World-Wide Church*, Occasional Paper no. 8 of the Conference of European Churches (Geneva 1976), p. 94. Opočensky refers in particular to work in Hungary, in discussion E. H. Amberg refers to the German Democratic Republic (p. 102), and in discussion of a paper on the possibilities for an Asian theology G. Nagy cites the "theology of diakonia" as a relevant instance from socialist countries of "situation-related" or "contextual" theology (p. 28). See observations by Nagy also in a sectional address in *Nyborg VI*, pp. 200–202, and his "Sent into the World of Society," in *Unsere Sendung in der Welt/Our Commitment in the World*, Hungarian Lutheran Study Document for the Sixth Assembly of the Lutheran World Federation at Evian (Budapest 1970), pp. 106–10.
4. *Biserica Slujitoare* [The Servant Church] (Bucharest 1972), pp. 333–34 (from the conclusions in English; the book includes a detailed list of contents in French).
5. *The Orthodox Approach to Diaconia* (Geneva 1980), pp. 41–42.
6. T. Schober, "Mission und kirchliche Entwicklungshilfe heute unter politischen und diakonischen Aspekt," in *Die Verantwortung der Kirche in der Gesellschaft*, ed. J. Baur and others (Stuttgart 1973), p. 213.
7. Art. 35, in *The Ecumenical Review* 25 (1973) 358. On Leuenberg consultations, see

M. Lienhard, *Lutherisch-reformierte Kirchengemeinschaft heute* (Frankfurt 1972), and *Zeugnis und Dienst reformatorischer Kirchen im Europa der Gegenwart,* ed. M. Lienhard (Frankfurt 1977).
8. A theme in evidence at the International Congress on World Evangelism, Lausanne 1974.
9. Easter leader in *The Times* of London, 29 March 1975.
10. *The Role of the "Diakonia,"* p. 16.
11. In chap. 3, "Division of Inter-Church Aid, Refugee and World Service," of *Work Book for the Assembly Committees,* prepared for the Fourth Assembly at Uppsala (Geneva 1968), pp. 43–44.
11. *Uppsala to Nairobi,* ed. D. E. Johnson (New York/London 1975), p. 124.
12. See G. Murray, "Joint Service as an Instrument of Renewal," in *The Ecumenical Advance,* ed. H. E. Fey (London 1970), p. 221, citing the division's extended mandate after New Delhi: "to express the ecumenical solidarity of the churches through mutual aid in order to strengthen them in their life and mission and especially in their service to the world around them *(diakonia),* and to provide faculties by which the churches may serve men and women in acute human need."
13. *Ecumenical Terminology* (Geneva 1975), p. 258.
14. *Hope in the Desert,* The Churches' United Response to Human Need, 1944–1984, ed. K. Slack (Geneva 1986), p. 134.
15. *Diakonia 2000: Called to be Neighbours* ed. K. Poser (Geneva 1987), p. 124.
16. J. E. Fischer, "Inter-Church Aid and the Future," in *Hope in the Desert,* p. 131.
17. "Our Ecumenical Diakonia—Both Large and Small," in *Hope in the Desert,* pp. 103–4. And see the reservations at this point of A. Dulles discussed in n. 9 of the section "Word Studies" at the beginning of chapter 1.

The Doctrine of Ministry

1. *Ministry to Word and Sacraments* (Philadelphia 1976), pp. 343–44.
2. Major applications of "diakonia" are amply illustrated in the following pages; of interest might be a few of the less expected applications from an era when the notion was becoming established, especially in Roman Catholic circles: in reference to pastoral councils, E. Corecco, "Ecclesiastical Parliament or Synodal Diakonia?" *International Catholic Review* 1 (1972) 23–36; H. Schmitz, "Der Bischof und die konsiliare Diakonie," *Trierer Theologische Zeitschrift* 84 (1975) 236–39; in reference to church bureaucracy, G. Ceretti and L. Sartori, "La curie au service d'une papauté rénovée," *Concilium* 108 (1975) 129–41; and in reference to theology, C. Dagens, "Le ministère théologique et l'expérience spirituelle des chrétiens," *Nouvelle Revue Théologique* 98 (1976) 530; K. Barth, *Church Dogmatics,* trans. vol. 4, pt. 2 (Edinburgh 1958), p. 693.
3. As in the subtitle of *Tous responsables dans l'Eglise?* (Paris 1973).
4. *Ministry and Ordination,* A Statement on the Doctrine of the Ministry Agreed by the Anglican-Roman Catholic International Commission (Canterbury 1973), para 2.
5. *Lutherans and Catholics in Dialogue,* vol. 4, *Eucharist and Ministry* (Washington/New York 1970), para. 9.
6. *Ministry in Ecumenical Perspective* (Rome 1969), pp. 80 and 92.
7. *The Pioneer Ministry* (1961; repr. London 1971), p. 13.
8. "The Ministry in the New Testament," in *The Apostolic Ministry,* ed. K. E. Kirk (London 1946), p. 181.
9. *Eucharist and Ministry,* para. 9.

10. See the account of one of the earliest and greatest of these debates, the Westminster Assembly of the 1640s, in Robert S. Paul, *The Assembly of The Lord* (Edinburgh 1985).
11. Report of sec. V, "The Ministry of the Church," paras. 35–36, in a *Documentary History of the Faith and Order Movement 1927–63*, ed. L. Vischer (St Louis 1963), pp. 34–35. See further *Faith and Order: Proceedings of the World Conference, Lausanne, August 2–21, 1927*, ed. H. N. Bate (London 1927).
12. Notes appended to the report, para. 44, in *Documentary History*.
13. Notes appended to the report, para. 46, in *Documentary History*.
14. See the preparatory studies, *The Ministry and the Sacraments*, ed. R. Dunkerly (London 1937). For history and bibliographies, see R. Frieling, *Die Bewegung für Glauben und Kirchenverfassung 1910–1937* (Göttingen 1970); T. Tatlow, "The World Conference on Faith and Order," in *A History of the Ecumenical Movement 1517–1948*, ed. R. Rouse and S. C. Neil (London 1967), pp. 403–41.
15. Sec. V, "The Church of Christ: Ministry and Sacraments," para. 103, in Vischer, *Documentary History*, p. 60. On the points of difference see the whole subsection "Ministry," paras. 91–110. See further *The Second World Conference on Faith and Order Held at Edinburgh, August 3–18, 1937*, ed. L. Hodgson (London 1938).
16. Marc Boegner, a vice-president at the conference, recalls in his memoirs "violent clashes in full session"; see *The Long Road to Unity*, Eng. trans. (London 1970), p. 116. The subject had always generated warmth; see numerous evidences in the debate between Presbyterians and Independents in Paul, *Assembly*, e.g., pp. 184, 271 ("here was great heat," noted the contemporary diarist), 317. In concluding his monumental survey *Christian Unity: Its History and Challenge in All Communions in All Lands* (London 1929), G. J. Slosser noted that to his day "interdenominational animosities" continued to characterise the debate (p. 372).
17. On the latter see Frieling, *Die Bewegung*, pp. 261–62.
18. In Dunkerly, *The Ministry and the Sacraments*, pp. 343–67.
19. *The Fourth World Conference on Faith and Order*, ed. P. C. Rodger and L. Vischer (London 1964), p. 40.
20. D. M. Paton, "A Montreal Diary," in Rodger and Vischer, *Fourth World Conference*, pp. 26–27.
21. Sec. III, "The Redemptive Work of Christ and the Ministry of His Church," paras. 77, 80, 89, 98, 91, in Rodger and Vischer, *Fourth World Conference*.
22. G. Gassmann, "Die Entwicklung der ökumenischen Diskussion über das Amt," *Ökumenische Rundschau* 22 (1973) 460–61; H. J. Goertz, "Amt und Ordination in 'Glauben und Kirchenverfassung,' " *Una Sancta* 28 (1973) 291.
23. "The Ordained Ministry," in *Faith and Order, Louvain 1971, Study Reports and Documents* (Geneva 1971), p. 80.
24. "Un Compagnon de service au sein d'un peuple de serviteurs"; "Le Ministère ordonné dans une perspective oecuménique," *Istina* 18 (1973) 454.
25. "The Ordained Ministry in Ecumenical Perspective, An Agreed Statement of the Faith and Order Commission of the World Council of Churches," paras. 1–2, in *Modern Ecumenical Documents on the Ministry* (London 1975), p. 111.
26. "Taufe—Eucharistie—Amt," *Ökumenische Rundschau* 24 (1975) 203.
27. Paras. 25–26, in *Modern Ecumenical Documents*.
28. *Baptism, Eucharist and Ministry*, Faith and Order Paper no. 111 (Geneva 1982), pp. 19–32.
29. *Churches Respond to BEM*, vol. 1, ed. Max Thurian, Faith and Order Paper no. 129 (Geneva 1986), p. 96. See similar concerns in responses from the Lutheran Church in

America (p. 34) and the North Elbian Evangelical Lutheran Church of West Germany (p. 49). By contrast, see the grave problem created for some Orthodox churches by the broad view of ministry in vol. 2, ed. M. Thurian, Faith and Order Paper no. 132 (Geneva 1986), p. 9 (Russian); p. 21 (Bulgarian: "there is no word about the ministry as being a God-established blessed sacrament and . . . no difference is made between the hierarchical ministry and the various ministries of the lay people in the church"); pp. 27–28 (Finnish).
30. "Diakonia in Modern Times," in *Service in Christ,* ed. J. I. McCord and T. H. L. Parker (London 1966), p. 149.
31. "Stages in Questions and Development," in *Eucharist and Ministry,* p. 35.
32. *A Survey of Bilateral Conversations among World Confessional Families 1959–1974,* 3d ed. (Geneva 1975), p. 182. For similar judgments, see H. M. Legrand, "Le ministère ordonné dans la dialogue oecuménique," *Revue des Sciences philosophiques et théologiques* 60 (1976) 683; H. R. McAdoo (co-chairman of the Anglican/Roman Catholic International Commission I) in the introduction to *Modern Ecumenical Documents,* p. 6 (in respect of statements from St Louis, 1970; Canterbury, 1973; Les Dombes, 1973; and Accra, 1974). A. Dulles also reviewed a number of conversations in "Ministry and Intercommunion," *Theological Studies* 34 (1973) 634–78.
33. *Aux seuils de l'espérance* (Neuchatel 1971), p. 76: "la 'chaine diaconale.' "
34. *NJB* revises *JB* to "to knit God's holy people together for the work of service," noting that "holy people" seems to be teachers but could include all the faithful. Compare the Traduction oecuménique, "afin de mettre les saints en état d'accomplir la ministère" ("for the purpose of putting the saints in a position to perform ministry"). All three add notes on the ambiguity of the Greek.
35. Dogmatic Constitution on the Church, art. 7: "dona ministrationum . . . quibus . . . nobis invicem ad salutem servitia praestamus" ("gifts of ministries through which . . . we serve each other unto salvation," Eng. trans. *Vatican Council II: The Conciliar and Post Conciliar Documents,* ed. A. Flannery [Dublin 1977], p. 356). See the commentary in G. Philips *L'église et son mystère au IIe Concile du Vatican,* vol. 1 (Paris 1966), p. 111 (Eng. trans.: "we are to understand . . . the ministries which prepare the saints for the purpose of building up the body; in no way does this mean that the whole task should fall only on the shoulders of ministers, but each member must devote himself to the salvation of others."
36. M. Barth, *Ephesians* (Garden City, N.Y., 1974), p. 479: "All the saints (and among them, each saint) are enabled by the four or five types of servants enumerated in 4:11 to fulfill the ministry given to them, so that the whole church is taken into Christ's service and given missionary substance, purpose, and structure. This interpretation challenges both the aristocratic-clerical and the triumphalistic-ecclesiastical exposition of 4:11–12. It unmasks them as arbitrary distortions of the text. Are, therefore, the existence and function of a clergy simply dispensable? Indeed, the traditional distinction between clergy and laity does not belong in the church. Rather, the whole church, the community of all the saints together, is the clergy appointed by God for a ministry to and for the world."
37. H. Merklein, *Das kirchliche Amt nach dem Epheserbrief* (Munich 1973).
38. He translates: "das die Heiligen zugerichtet werden zum werck des Ampts da durch der leib Christi erbawet werde," adding a gloss on "Zugericht": "Das ist wol geruesst und allenthalben versorget und zubereit, das nichts feile zum Ampt der Christenheit"; see *Die gantze Heilige Schrifft Deutsch,* Wittenberg 1545, ed. H. Wolz (Munich 1972). Luther's thinking on ministry does not tie in consistently, however, with this approach to Eph. 4:11–12. Some scholars assert a shift in his thinking from an early position of

the essential equality of ordained and nonordained to a later position of ministerial competence exclusive to the ordained—so P. D. L. Avis, *The Church in the Theology of the Reformers* (Atlanta 1981), pp. 95–108. G. Haendler, by contrast, argues that later positions were consistent with the earlier; see *Luther on Ministerial Office and Congregational Function*, Eng. trans. (Philadelphia 1981). B. Cooke provides a summary and other literature, *Ministry to Word and Sacraments*, p. 595.
39. *Institutes of the Christian Religion*, iv.iii.2, trans. H. Beveridge, vol. 2 (London 1962), p. 317. See the discussion in A. Ganoczy, *Ecclesia Ministrans, Dienende Kirche und kirchlicher Dienst bei Calvin* (Freiburg 1968), pp. 180–84. On p. 366 Ganoczy draws attention to the difference at this point between Calvin and Vatican II.
40. *Theology of Ministry* (New York 1983), p. 193.
41. *Constitution and Regulations (Interim)* (Melbourne 1976), p. 22, para. 11.
42. Preface to the new edition of *The Church, Its Nature, Function and Ordering* (Melbourne 1984), pp. 1–2.
43. *Tous responsables*, p. 76.
44. Address to the Conference of European Churches, in *This Happened at Nyborg VI* (Geneva 1971), p. 216. In similar vein, D. von Allmen, "Amt—Ämter-Dienste—Ordination," in *Zeugnis und Dienst reformatorischer Kirchen im Europa der Gegenwart*, ed. M. Lienhard (Frankfurt 1977), p. 90.
45. "Quelques problèmes touchant les ministères," *Nouvelle Revue Théologique* 93 (1971) 785–800 (p. 790: "le sacerdoce ministériel se situe, non dans la ligne de l'ontologie constitutive du chrétien mais dans celle du ministère. C'est une participation *fonctionnelle*, qui comporte son fondement ontologique mais d'une ontologie de fonction ou de ministère"). Compare the interview, "Refléxion d'un théologien," reported in *Unité des chrétiens*, no. 7, 1972, pp. 16–20.
46. "The Redemptive Work of Christ," para. 80, in *The Fourth World Conference*, p. 62. The problem is not just with English. In *Dictionnaire alphabétique et analogique de la langue française*, vol. 4 (Paris 1973), Paul Robert notes that "ministériel" is not used in a general sense but in reference to office or function, citing in the first instance, ironically enough, "Chef ministériel" of the pope; J. Delorme writes of the tensions usage imposes on theological discourse, "Diversité et unité des ministéres d'après le Nouveau Testament," in *Ministère et les Ministères selon le Nouveau Testament*, ed. J. Delorme (Paris 1974), pp. 311–13. In *Why Priests?*, Eng. trans. (London 1972), p. 30, H. Küng notes the inconvenience caused by the lack of German words of the "minister" family. Writers in German prefer "Dienst" (service) to "Amt" (office, and in modern times the word has a particularly bureaucratic ring); the newer "Dienstamt" is sometimes favoured. Thinking on office as service has come full circle when a German theologian applies "Amt" to the "Dienst" of the nonordained, as H. J. Pottmeyer, "Thesen zur theologischen Konzeption der pastoralen Dienste und ihrer Zuordnung," *Theologie und Glaube* 66 (1976) 313–31 (315–18).
47. *Sons and Lovers* (1913; repr. Harmondsworth 1973), p. 16.
48. Cited in W. J. Bausch, *Traditions, Tensions, Transitions in Ministry* (Mystic, Conn., 1982), p. 98. This curious attitude of the bishops' committee towards "precise definition" was echoed in lectures at the Lay Ministry Conference in Sydney, December 1988, by the Director of the Office for Ministry Formation in Chicago, Lucien Roy, who, adverting to "the endless litany of polarities and dichotomies" in any discussion of ministry, averred he would "rather hang on to problems: if we only get the definition right we might miss the reality; in restricting the meaning of ministry, we are endangering a very fragile moment of the church" (from my notes). Referring to the 1987 Synod on the Laity in Rome, at which he was a consultant to the United States'

bishops, he reported that the reason for the delay in the issuing of John Paul II's letter on the synod was the difficulty of arriving at an understanding of ministry. (On the eventual papal response of 30 December of the same year see n. 63.) At the Synod itself, where bishops resisted a concept of "lay ministry," notable attempts to clarify the language were made by Cardinal Basil Hume of Westminster, who urged a tripartite terminology of "ordained ministry," "instituted ministries" (such as lector), and "commissioned ministries" (temporary in nature). See reports in *The Tablet*, London (17 October 1987). As he was to observe of the problem in a later review of the synod, "One danger is that if all activity is seen as ministry then the concept rapidly loses any shape or value." See *The Tablet*, London (7 November 1987). At a more academic level, in "A Theological, Pastoral Handbook" called *Ministry* (San Francisco 1987), Richard McBrien devised a terminology along the following lines to facilitate speech about ministry: "general/universal ministry"; "general/specific ministry"; "Christian/universal ministry"; "Christian/specific ministry" (pp. 11–12).
49. *Traditions*, p. 150.
50. *Towards an Ecumenical Consensus on Baptism, the Eucharist and the Ministry*, Faith and Order Paper no. 84 (Geneva 1977), p. 12.
51. *Churches respond to BEM*, vol. 1, p. 76.
52. J. Grabner, "Ordained and Lay: Them-Us or We?" *Worship* 54 (1980) 326.
53. Ibid., p. 327.
54. "Ministry and Ordination," para. 13, in *The Final Report, Windsor, September 1981* (London 1982), p. 36 One member of the commission, J. W. Charley, expressed more interest, nonetheless, in the "common ground" exposed by seeing ministry as service than in presenting ordained ministry as "isolated" and "determinative of the Church"; see his "Theological Commentary" in the Grove Booklet no. 22, *Agreement on the Doctrine of the Ministry* (Bramcote 1973), pp. 16–18.
55. "Elucidation 1979," para. 2, in *The Final Report*, p. 41.
56. "Response to the Final Report of ARCIC I," *One in Christ* 21 (1985) 173.
57. "Response of the French Episcopal Conference to the ARCIC Final Report," *One in Christ* 21 (1985) 338.
58. "Preparing for the Extraordinary Synod," *One in Christ* 21 (1985) 362.
59. Dogmatic Constitution on the Church, art. 10.
60. The interpretation of "character" in an ontological sense, influential but never universally current, is now virtually abandoned in Roman Catholic circles. See for example E. Ruffini, "Character as a Concrete Manifestation of the Sacrament in Relation to the Church," *Concilium* 1, no. 4 (1968) 52–58; P. Fransen, "Orders and Ordination," in *Sacramentum Mundi*, trans. vol. 4, ed. K. Rahner and others (New York/London 1969), pp. 324–25 ("a mythic theology of the priesthood . . . a metaphysical clericalism"). Official documents invoke the term, as the Declaration "Mysterium Ecclesiae" of the Sacred Congregation for the Doctrine of the Faith "In Defence of the Catholic Doctrine on the Church against Certain Errors of the Present Day," *Acta Apostolicae Sedis* 65 (1973) 396–408, trans. in *Clergy Review* 58 (1973) 950–62, but acknowledge the existence of various explanations ("Mysterium Ecclesiae," *Acta*, p. 406). Among these predictably is W. Kasper's in the sense of a summons "to serve others with every ounce of humanity," in "A New Dogmatic Outlook on the Priestly Ministry," *Concilium* 3, no. 5 (1969) 16; in similar vein Nathan Mitchell: "permanently commissioned as servants of Jesus' unique priesthood . . . ," in *Mission and Ministry* (Wilmington 1982), p. 312 (for the history of the concept, see pp. 235–39). The term would seem to serve little further purpose, as B. Cooke's pages on the topic perhaps illustrate, *Ministry to Word and Sacraments*, pp. 643–46. On the conflict of views on

character at the Roman Synod in 1971 see E. Schillebeeckx, *The Church with a Human Face,* Eng. trans. (London 1985), pp. 213–34.
61. See M. Richards, "Servants of the Word, Shepherds of the People: The Ordained Ministry after Trent and after Vatican II," *Clergy Review* 64 (1979) 244.
62. See J. M. R. Tillard's succinct outline of the rise of sacerdotal vocabulary in Grove Booklet no. 13, *What Priesthood Has the Ministry?* (Bramcote 1973). Priestly terminology is a sensitive issue in Catholic circles; see spirited complaints from a Roman Catholic bishop and two priests directed against D. Walker's treatment, "Is Ministry Essentially Priestly?" *Australian Catholic Record* 57 (1980) 107–16, with letters there, pp. 301–5, and also 58 (1981) 88–90; similarly a bishop's response to Duquoc (see n. 67 below), E. Marcus, "A propos d'une réflexion sur l'appel au presbytérat," *Etudes* 350 (1979) 415–23. R. E. Brown defended the retention of priestly terms on the reduced ground that ordained ministers are to exemplify sacrificial self-giving; see "The Challenge of New Testament Priesthood," *Emmanuel* 86 (1980) 314–22.
63. *Letter of the Supreme John Paul II to all the priests of the church on the occasion of Holy Thursday 1979,* Eng. trans., para. 4 (Melbourne 1979). His later Apostolic Exhortation, "Christifideles Laici" of 30 December 1988, a response to the 1987 Synod on the Laity, adopts the broader contemporary usage; it also reported the existence of a commission to provide "an in-depth study of the various . . . considerations which are associated with the great increase today of the ministries entrusted to the lay faithful" (para. 23, Eng. trans., *Origins:* NC Documentary Service, 9 February 1989, p. 572). The study's conclusions were still awaited at the time but were to reflect, according to this papal directive, "the essential difference between the ministerial priesthood and the common priesthood, and the difference between the ministries derived from the sacrament of orders and those derived from the sacraments of baptism and confirmation."
64. "Concepts of Ministry (2)," *The Tablet,* London (24 March 1979), p. 310 (and see n. 67 below).
65. J. K. McGowan introducing O. M. Liebard, *Clergy and Laity* (Wilmington 1978), p. xvii.
66. "Gospel, Church, Ministry," in *Minister? Pastor? Prophet?,* Eng. trans., ed. L. Grollenberg and others (London 1980), p. 24.
67. The phrases are those of *The Nottingham Statement,* being the official statement of the second National Evangelical Anglican Congress, April 1977 (London 1977), p. 33. Writers referred to are C. Duquoc, "Théologie de l'Eglise et crise du ministère," *Etudes* 350 (1979) 101–13; Eng. trans., "Concepts of Ministry," *The Tablet,* London (10 and 24 March 1979); "Situation of Ministries," *Pro Mundi Vita Bulletin* 78 (July 1979) 15–21; J. Moignt, "Services et lieux d'Eglise," *Etudes* 350 (1979) 835–49; ibid., 351, (1979) 103–19, 363–94; "Authority and Ministry," *Journal of Ecumenical Studies* 19 (1982) 202–25; W. R. Burrows, *New Ministries: The Global Context* (Melbourne 1980); E. Schillebeeckx, "A Creative Retrospect as Inspiration for the Ministry in the Future," in *Minister?,* pp. 57–84; *Ministry: A Case for Change,* Eng. trans. (London 1980); *The Church with a Human Face* (see n. 61); L. Boff, *Church: Charism and Power,* Eng. trans. (1981; repr., London 1985); *Ecclesiogenesis: The Base Communities Reinvent the Church,* Eng. trans. (1977; repr., London 1986); G. Tavard, *A Theology for Ministry* (Wilmington 1983); T. F. O'Meara, *Theology of Ministry* (New York 1983). The "interpretations" of Gabriel Marc are also relevant; see "The Institutional Church in the Future: Facts and Interpretations," *Pro Mundi Vita Bulletin* 82 (July 1980) 1–26.
68. Cited from *The Constitution of the United Presbyterian Church in the United States of*

America, pt. I, *Book of Confessions*, 2d ed. (New York), sec. 5.153; see also J. J. von Allmen, "Ministry and Ordination according to Reformed Theology," *Scottish Journal of Theology* 25 (1972) 75–88. For the Latin and French with von Allmen's commentary, see his *Le saint ministère selon la conviction et la volonté des Réformés du XVIe siècle* (Neuchâtel 1968), pp. 54–62. The teaching is reflected, despite introductory comments about "diakonia" as service to the brethren, in P. H. Menoud's "Ministry, N. T. 1. Service and Ministries," in the handbook edited by von Allmen, *Vocabulary of the Bible*, Eng. trans. (London 1958), pp. 262–63.

69. J. L. Ainslie, *The Doctrines of Ministerial Order in the Reformed Churches of the 16th and 17th Centuries* (Edinburgh 1940), p. 8. Ainslie recites some of these notions from the sources, pp. 6–11.
70. *The Second World Conference*, p. 245.
71. "The Lima document on 'Baptism, Eucharist and Ministry': The Event and Its Consequences," being the introduction to *Churches Respond to BEM*, vol. 1, p. 19.

The Language of Office

1. *Church Order in the New Testament*, Eng. trans. (London 1961), chap. 21, pp. 171–80. The German-language original *(Gemeinde und Gemeindeordnung im Neuen Testament)* appeared in 1959 and was a development of *Das Leben des Herrn in der Gemeinde und ihren Diensten* (Zurich 1946).
2. *Church Order*, p. 174.
3. T. W. Manson, *The Church's Ministry* (London 1948), pp. 24–27; W. D. Davies, *Christian Origins and Judaism* (London 1962), pp. 235–36 (both these writers were under the influence of Schweizer, *Das Leben des Herrn*); P. Grelot, *Le ministère de la nouvelle alliance* (Paris 1967), pp. 74–77; D. N. Power, *Ministers of Christ and His Church* (London 1969); A. Descamps, "Aux origines du ministère," *Revue théologique de Louvain* 3 (1972) 135–38; J. Coppens, "Le sacerdoce chrétien," *Nouvelle Revue Théologique* 102 (1970) 355; and, for strictures regarding attempts to make anything more of "diakonia," "Le caractère sacerdotal des ministères selon les écrits du Nouveau Testament," in *Teología del sacerdocio*, vol. 4, ed. I. Oñatibia (Burgos 1972), pp. 11–39.
4. "The Ministries in the New Testament, Recent Research," *Biblical Theology Bulletin* 3 (1973) 133–66.
5. "Structures du ministère dans le Nouveau Testament," *Istina* 16 (1971) 451.
6. J. P. Audet, *Structures of Christian Priesthood*, Eng. trans. (London 1967), p. 55, and "Priester und Laie in der christlichen Gemeinde," in *Der priesterliche Dienst*, vol. 1, ed. K. Rahner and H. Schlier (Freiburg 1970), p. 163; J. Blenkinsopp, *Celibacy, Ministry, Church* (London 1968), pp. 228–48, cf. p. 9; J. Blank, from a report on "Critical Questions about a Sacral Image of the Priest," *Concilium* 10, no. 8 (1972), 137–40.
7. J. Delorme, "Diversité et unité des ministères d'après le Nouveau Testament," in *Le ministère et les ministères selon le Nouveau Testament*, ed. J. Delorme (Paris 1974), pp. 315–16; A. T. Hanson, *Church, Sacraments and Unity* (London 1975), pp. 86–87, and *The Pioneer Ministry* (1961; repr., London 1975), pp. 104–6.
8. *Church Order*, p. 206.
9. Ibid.
10. H. Conzelmann, *History of Primitive Christianity*, Eng. trans. (London 1973), p. 106.
11. H. Schürmann, "The Church as an Open System," *International Catholic Review* 1 (1972) 212.

12. H. von Campenhausen, *Ecclesiastical Authority and Spiritual Power,* Eng. trans. (London 1969), p. 70.
13. E. Käsemann, "Ministry and Community in the New Testament," in *Essays on New Testament Themes,* Eng. trans. (London 1964), p. 81.
14. *Sexism and God-Talk* (London 1983), p. 197.
15. M. M. Simpson, "Vocational and Pastoral Aspects," in *Yes to Women Priests,* ed. H. Montefiore (Great Wakering/Oxford 1978), pp. 59–60.
16. *Ordination of Women in Ecumenical Perspective,* ed. C. F. Parvey, Faith and Order Paper no. 105 (Geneva 1980), p. 49.
17. "The Twelve," in *Women Priests,* ed. L. Swidler and A. Swidler (New York 1977), pp. 118–19.
18. "The Church in Process: Engendering the Future," in *Women and Catholic Priesthood: An Expanded Vision,* ed. A. M. Gardiner (New York 1976), pp. 82–83.
19. "Ordination: "What is the Problem?" in *Women and Catholic Priesthood,* p. 34.
20. *Women's Challenge: Ministry in the Flesh* (Denville, New Jersey 1977), p. 40.
21. *Women and Ministry in the New Testament* (New York 1980), pp. 78–79. Since the 1970s the theology of women in ministry has moved to a new level with E. Schüssler Fiorenza's *In Memory of Her* (London 1983); for this author, nonetheless, "diakonia" continues to add a significant dimension (see pp. 320–21). On this, see my afterword.
22. See respectively "Amt und Gemeinde," in *Glaube und Geschichte* (Mainz 1970), p. 397; "Ways of Validating Ministry," *Journal of Ecumenical Studies* 7 (1970) 260–61; "Amt und Ordination in der ökumenischen Diskussion," in *Der priesterliche Dienst,* vol. 5, ed. H. Vorgrimler (Freiburg 1973), p. 70.
23. B. Cooke, *Ministry to Word and Sacraments* (Philadelphia 1976), p. 401.
24. "Hierodiakonia" is from P. Philippi, "Uber den Begriff des kirchlichen Amtes," *Kerygma und Dogma* 16 (1970) 147; "syndiakonia" from C. Bridel, *Aux seuils de l'espérance* (Neuchatel 1971), p. 74.
25. von Campenhausen, *Ecclesiastical Authority,* p. 69.
26. *Baptism, Eucharist and Ministry,* Faith and Order Paper no. 111 (Geneva 1982), pp. 22–23, being paras. 15 and 16 of "Ministry." The views in the statement cited are recognisably coloured also by long-standing historical attitudes to religious authority in numerous Protestant traditions; of interest here is how these traditional attitudes are now cast in terms of "diakonia."
27. *Structures of Christian Priesthood,* p. 55.
28. *Why Priests?,* Eng. trans. (London 1972), pp. 26–27. This book was Küng's background study for the document that incorporated its ideas, *Reform und Anerkennung kirchlicher Ämter,* Ein Memorandum der Arbeitsgemeinschaft ökumenischer Universitätsinstute (Munich 1973); Eng. trans. in *Journal of Ecumenical Studies* 10 (1973) 390–401.
29. *The Church,* Eng. trans. (New York 1967), pp. 391–92. See similar passages in *On Being a Christian,* Eng. trans. (London 1978), pp. 486–87.
30. The latter, a Roman Catholic, is perhaps the major influence; see Küng's long footnote in *Structures of the Church,* Eng. trans. (New York 1964), p. 179.
31. See the extensive dossier, comprising reviews and reactions, in *Diskussion um Hans Küng "Die Kirche,"* ed. H. Häring and J. Nolte (Freiburg 1971).
32. The following is some representative literature from the period around 1965–1975 when "diakonia" was being built into writings on ministry: E. Käsemann, "Ministry and Community in the New Testament," in *Essays on New Testament Themes,* Eng. trans. (London 1964), p. 63 (from a paper originally delivered in 1949); P. Philippi, principally *Christozentrische Diakonie* (Stuttgart 1963), pp. 263–73, but also *Abendmahls-*

feier und Wirklichkeit der Gemeinde (Berlin 1960), pp. 104–5; "Über den Begriff des kirchlichen Amtes," *Kerygma und Dogma* 16 (1970) 144–52; " 'Amt' heisst 'Diakonia,' " *Männliche Diakonie* 55 (1975) 80–81; J. Gewiess, "Amt, I. Biblisch," in *Handbuch theologishcer Grundbegriffe,* ed. H. Fries, vol. 1 (Munich 1962), p. 33; K. H. Schelkle, *Discipleship and Priesthood*, Eng. trans. (New York 1965), pp. 34–39; "Ministry and Minister in the New Testament Church," *Concilium* 3, no. 5 (1969) 8–9; E. Molland, "Das kirchliche Amt im Neuen Testament und in der Alten Kirche," in his *Opuscula patristica* (Oslo 1970), pp. 209–11, originally in *Oecumenica* 3(1968); J. Hoffmann, "Ministère en théologie catholique," in *Vocabulaire oecuménique,* ed. Y. Congar (Paris 1970), pp. 367–68; W. Kasper, "Amt und Gemeinde," in *Glaube und Geschichte* (Mainz 1970), p. 397; "A New Dogmatic Outlook on the Priestly Ministry," *Concilium* 3, no. 5. (1969) 14; H. Fries and W. Pannenburg in a seminar, "Das Amt in der Kirche," *Una Sancta* 25 (1970) 107; L. Goppelt, *Apostolic and Post-Apostolic Times,* Eng. trans. (London 1970), pp. 177–78; R. Pesch, "Structures du ministère dans le Nouveau Testament," *Istina* 16 (1971) 438–43; K. Kertelge, *Gemeinde und Amt im Neuen Testament* (Munich 1972), pp. 16, 20, 154; H. Schürmann, "The Church as an Open System," *International Catholic Review* 1 (1972) 214; G. Moede, "Amt und Ordination in der ökumenischen Discussion," in *Der priesterliche Dienst,* vol. 5, ed. H. Vorgrimler (Freiburg 1973), pp. 19–20; J. Delorme, "Diversité et unité des ministères d'après le Nouveau Testament," in *Le ministère et les ministères selon le Nouveau Testament,* ed. J. Delorme (Paris 1974), p. 313; E. Nardoni, "Ministries in the New Testament," *Studia Canonica* 11 (1977) 5–36; E. A. Russell, "The Development of Ministerial Orders," in *Church Ministry,* ed. A. Mayes (Dublin 1977), pp. 10–11; E. Schillebeeckx, *Ministry,* Eng. trans. (London 1980), p. 21.

Deacons

1. *Presbytery and Not Prelacy* (Glasgow 1844), p. 209.
2. Dogmatic Constitution on the Church, art. 29, Eng. trans. *Vatican Council II, The Conciliar and Post Conciliar Documents,* ed. A. Flannery (Dublin 1977), p. 387.
3. Eng. trans., in *Vatican II,* pp. 433–36.
4. As illustrated in "Prime norme per l'attuazione del diaconato permanente nella diocesi di Roma," under the signature of the Vicar General Ugo Cardinal Poletti in *Il diaconato in Italia,* no. 23 (July 1976), pp. 68–70; Bishops' Conference of Latin America (CELAM), "Encuentro de Reflexion sobre el Diaconado Permanente," Bogota/Petaluma, Colombia, in *Diaconia XP* 9, no. 4 (1974) 3–26; "Arbeitspapier zum Berufsbild des Ständigen Diakons," by the Working Party of the Organisation for the Diaconate in the Federal Republic of Germany, 1975; "Der Ständige Diakonat," a statement of the Austrian Commission for pastoral Affairs, 1976.
5. See their journals *Diakonia Christi* (prior to 1987, *Diakonia XP;* published by the International Centre for the Diaconate in Freiburg, West Germany); *Communion et Diaconie* (France); *Diaconia* (Spain); *Deacon Digest* (United States); *Il diaconato in Italia* (Italy); *Infor-Diacres* (Belgium).
6. Final Document of the Third General Conference of the Latin American Episcopate, Puebla 1979, para. 697, *Puebla and Beyond,* Eng. trans., ed. J. Eagleson and P. Scharper (Maryknoll 1979), p. 219.
7. See Hornef's account, "Vom Werden und Wachsen des Anliegens," in *Diaconia in Christo,* ed. K. Rahner and H. Vorgrimler (Freiburg/Basel/Wien 1962), pp. 343–61 (this volume includes thirty-four items of his bibliography to that date); see also "Die

Anfänge der Diakonatserneuerung," in *Der Diakon,* ed. A. Fischer, H. Kramer, and H. Vorgrimler (Freiburg 1970), pp. 7–14; and his *The New Vocation,* Eng. trans. (Cork 1963), which includes a number of his earlier articles.
8. *Married Men as Ordained Deacons,* Eng. trans. (London 1955; German 1953).
9. M. D. Epangeul. "Role des diacres dans l'église d'aujourd'hui," *Nouvelle Revue Théologique* 79 (1957) 153–68; Pius XII, "Allocutio iis qui interfuerunt Conventui alteri catholicorum ex universo orbe, pro Laicorum apostolatu," *Acta Apostolicae Sedis* 49 (1957) 925; also see *Nouvelle Revue Théologique* 80 (1958) 79.
10. "Le renouveau du diaconat," *Nouvelle Revue Théologique* 93 (1961) 337–66 (citations from p. 339).
11. News item "20 Jahre Internationales Diakonatszentrum (IDZ)," *Diaconia XP* 20, no. 4 (1985) 35.
12. See Hornef, "Vom Werden," pp. 350–51; 361; H. Kramer, "Chronik: Beitrag der Caritas zur Erneuerung des Ständigen Diakonates," *Diakonia Christi* 21, nos. 2–3 (1986) 15–19.
13. Under the names of eighty-two theologians; its first publication in English, "A Functional Diaconate," appeared in *Worship* 37 (1963) 513–20; see also the untitled document circulated to the bishops at the council by Hannes Kramer and others of the "Original Deacon Circle" and printed as appendix 2, "Formal Request to Restore the Diaconate as a Permanent Order," in *Sacrament of Service,* by P. McCaslin and M. G. Lawler (New York/Mahwah 1986), pp. 141–50.
14. The issues of competence, especially in the light of the 1983 Code of Canon Law, is examined by J. W. Pokusa, "The Diaconate: A History of Law Following Practice," *The Jurist* 45 (1985) 95–135.
15. These approximate figures are based on "Anzahl der Ständigen Diakone im Januar 1988," *Diaconia Christi* 23, no. 1 (1988) 36.
16. See the series of reports in *Diaconia Christi* 21, nos. 2–3 (1986), from Brazil, France, Holland, Switzerland, South Africa (the only sanguine account), United States and the dioceses of Munster and Freiburg in West Germany.
17. *Deacons in the Church,* the report of a Working Party set up by the Advisory Council for the Church's Ministry (Church Information Office, Westminster 1974). Curiously, a still-later report overturned this view in favour of a permanent or "distinctive" diaconate that would model the church's ministry; see *Deacons in the Ministry of the Church,* A Report to the House of Bishops of the General Synod of the Church of England (London 1988). (The synod did not implement the recommendations of this report.) Within the same Anglican tradition, but from a North American perspective and then with strong influences from the Roman Catholic permanent diaconate there, James. M. Barnett projects a totally confident statement in *The Diaconate: A Full and Equal Order* (Minneapolis 1981). From the beginning he attempts to balance "diakonia" as service to others with the authority of a church office (pp. 17–22), and finally identifies deacons—largely on historical rather than primarily on theological or biblical grounds—not as ministers of the bishop but as "servants of the Church," ordained "to hold up *diakonia* as central to all Christian ministry" among those who make up the *laos* or people, namely, bishops, presbyters, and others (pp. 143 and 164).
18. *La fonction diaconale aux origines de l'église* (Bruges 1960). And see n. 4 in the section "Word Studies" at the beginning of chapter 1.
19. *La fonction,* p. 10.
20. The position was rejected, for example, by H. Braun, *Qumran und das Neue Testament,* vol. 2 (Tübingen 1966), p. 333.

21. According to *La fonction*, p. 37, διακονία itself is a "ministry whose object is the procuring of salvation"; see further *Les fonctions ecclésiales aux premiers siècles* (Bruges 1956), pp. 163–74; *Ministre de Jésus-Christ* (Paris 1966), chap. 6.
22. P. Lécuyer, "Les diacres dans le Nouveau Testament," in *Le diacre dans l'église et le monde d'aujourd'hui*, ed. P. Winninger and Y. Congar (Paris 1966), pp. 15–26.
23. "Diáconos helénicos y bíblicos," *Burgense* 4 (1963) 1–143, and see particular comments on pp. 104 and 135.
24. *The Opponents of Paul in Second Corinthians*, Eng. trans. (Philadelphia 1986), pp. 27–32 (the original German text appeared 1964); and see the previous discussion in the section "Word Studies."
25. *The Opponents*, p. 29.
26. *Les ministères aux origines de l'église* (Paris 1971), p. 197.
27. "Zur altchristlichen Verfassungsgeschichte," *Zeitschrift für Wissenschaftliche Theologie* 55 (1914) 149. Compare the comment about the literary flavour of the Greek in J. H. Moulton and G. Milligan, *The Vocabulary of the Greek Testament Illustrated from the Papyri and Other Non-Literary Sources* (London 1930).
28. *Aux seuils de l'espérance: Le diaconat en notre temps* (Neuchatel 1971), p. 62.

CHAPTER 2

1. *Kyrios Christos*, Eng. trans. (Nashville/New York 1970), p. 39. The German edition was first published in 1913.
2. H. Anderson, "The Gospel of Mark (London 1976), p. 256.
3. J. Bowman, *The Gospel of Mark* (London 1965), p. 218.

Service as a Saving Action

1. *Dienst und Dienen im Neuen Testament* (Gütersloh 1931).
2. Ibid., p. 79.
3. Ibid., p. 70.
4. Ibid., p. 71.
5. The passages mentioned by Brandt include, with parallels, Matt. 25:44; Mark 1:13, 31; 15:41; Luke 8:3; 10:40; 12:37; 17:8; John 12:2.
6. *Dienst und Dienen*, p. 71.
7. Ibid., p. 69.
8. Ibid., p. 70.
9. Ibid., p. 71.
10. Ibid., pp. 75–76.
11. Ibid., p. 77.
12. Ibid., p. 79.
13. Ibid., p. 80.
14. Ibid., p. 85.
15. "διακονέω, διακονία, διάκονος", *TWNT* 2 (1935), pp. 81–93, and in the English translation, *TDNT* 2 (1964), pp. 81–93.
16. *TDNT*, p. 84.
17. *Dienst und Dienen*, pp. 90–91.
18. *TWNT*, p. 84.
19. *TDNT*, p. 85.
20. *TDNT*, p. 86.

21. "Der Sinn des Leidens liegt in dem Dienst, der darin geschieht. So erst wird es zum Opfer" (*TWNT*, p. 85).
22. "Die soteriologischen Aussagen in der urchristlichen Abendmahlsüberlieferung und ihre Beziehung zum geschichtlichen Jesus," *Trierer Theologische Zeitung* 81 (1972) 202. See also K. Prümm, *Diakonia Pneumatos*, vol. 2, no. 1 (Rome n.d.), p. 130.
23. "Διακονέω and Its Cognates in the Four Gospels," *Texte und Untersuchungen* 73 (1959) 415–22.
24. Ibid., p. 415.
25. Ibid., p. 416.
26. Ibid., p. 421.
27. *Das Evangelium nach Markus* (Berlin 1959), p. 219.

Jesus' Service at the Supper

1. *Das Evangelium Marci* (Berlin 1903), p. 91. Wellhausen's phrase is "die Diakonie des Abendmahls."
2. *Jesus-Jeshua*, Eng. trans. (London 1929), p. 118.
3. *Jesu Abschiedsrede Lk 22, 21–38* (Münster 1957), p. 82. At Isa. 53:11, LXX reads: εὖ δουλεύοντα πολλοῖς.
4. *Jesu Abschiedsrede*, p. 89 n. 305.
5. Ibid., p. 91.
6. *Church Order in the New Testament*, Eng. trans. (London 1961), p. 178 with n. 646. See also *The Good News according to Mark*, Eng. trans. (London 1971), pp. 219–20. Similarly L. Goppelt, *Apostolic and Post-Apostolic Times*, Eng. trans. (London 1970), p. 178.
7. Less explicitly stated by Schweizer in *Lordship and Discipleship*, Eng. trans. (London 1960), p. 50.
8. *The Good News*, pp. 219–20; *Jesus*, Eng. trans. (London 1971), p. 93.
9. "Anfänge der soteriologischen Deutung des Todes Jesu (Mk. x.45 und Lk. xxii.27)," *New Testament Studies* 19 (1972–1973) 38–64.
10. Ibid., p. 53.
11. Ibid., p. 53 n. 6.
12. Ibid., p. 53.
13. Ibid., pp. 51–52. In support of this opinion Roloff makes brief allusions to classical literature, p. 52. n. 1.
14. Ibid., pp. 53–54. The judgement is based on a comparison of Mark 1:13, 31 and 15:41 (clear instances of waiting at table) with Mark 10:45. Instances approximating to the later specific meaning include Matt. 22:13; 25:44; Luke 10:40; 17:8.
15. "Anfänge," p. 55.
16. Ibid., pp. 56–57.
17. Ibid., p. 58.
18. Ibid., p. 59. Roloff's earlier considerations (pp. 38–43) had indicated that the soteriological meaning of Jesus' death had to be sought in the ὑπέρ- formula; this in turn is to be traced to the earliest eucharistic assemblies, as Mark 14:24 suggests (pp. 43–50); Mark 10:45 is then of particular interest because it includes both the theology of the ὑπέρ- formula and, in "to serve," the indication that the theology had developed on the basis of experience of the Eucharist.
19. "Anfänge," p. 54. This approach was outlined by Roloff in his earlier *Apostolat-Verkündigung-Kirche* (Gütersloh 1965); see there esp. p. 187. It has since been followed by X. Léon-Dufour, "Jésus devant sa mort a la lumière des textes de l'institu-

tion eucharistique et du discours d'adieu," in *Jésus aux origines de la christologie,* ed. J. Dupont (Gembloux/Leuven 1975), pp. 141–68.

The Community's Eucharist

1. *The Problem of History in Mark* (London 1957), pp. 82–83.
2. *The Son of Man in the Synoptic Tradition,* Eng. trans. (London 1965), pp. 208–11.
3. *Die theologische Bedeutung des Todes Jesu* (Dusseldorf 1970), pp. 282–85.
4. *The Son of Man,* p. 209.
5. *Christus Traditus* (Zurich/Stuttgart 1967), p. 171.
6. *Diakonie, Festfreude und Zelos* (Uppsala/Wiesbaden 1951), pp. 9–164, esp. 23–24, 30, 150–52.
7. "Deacons in the New Testament and in the Early Church," in *The Ministry of Deacons,* World Council Studies no. 2 (Geneva 1965), p. 8.

The Isaian Servant

1. *The Gospel according to Saint Mark* (Cambridge 1966), p. 486 (in a supplementary note to p. 342 of the first edition of 1959 where a less qualified statement of his opinion needed to be brought into line with a subsequent article by Barrett).
2. "Jésus et le Serviteur de Dieu," in *Jésus aux origines de le christologie,* ed. J. Dpuont (Gembloux/Leuven 1975), p. 128.
3. *Jesus and the Son of Man* (London 1954), pp. 41–42.
4. Ibid., p. 42 n. 1.
5. *The Christian and Gnostic Son of Man* (London 1970), p. 24 n. 86.
6. S. Lyonnet and L. Sabourin, *Sin, Redemption, and Sacrifice* (Rome 1970), p. 100.
7. *Jesus and the Son of Man,* p. 42.
8. *The Mission and Achievement of Jesus* (London 1954), p. 56. Similarly W. H. Allen, *The Gospel according to Saint Mark* (London 1915), p. 140.
9. *Primitive Christian Application of the Doctrine of the Servant* (Durham, N.C., 1929), p. 111.
10. *The Servant-Messiah* (Cambridge 1953), p. 73; *The Church's Ministry* (London 1948), p. 24.
11. *Das Evangelium nach Markus* (Göttingen 1960), p. 109.
12. *According to Mark* (Cambridge 1960), pp. 220–21.
13. *A Commentary on the Gospel according to Mark* (London 1960), p. 180.
14. *The Gospel According to St Mark* (London 1963), p. 165, with note on p. 167.
15. *Les ministères aux origines de l'Eglise* (Paris 1971), p. 41. n. 2.
16. *The Founder of Christianity* (London 1971), p. 104; the citation occurs p. 105 n. 12.
17. *The Son of Man in Myth and History* (London 1967), p. 324 and n. 4.
18. *Jesus the Messiah* (London 1948), p. 131. Mark's aorist establishes that the service is not merely "life long." For a contrary view on this grammatical point, see Swete in the section "In Fealty to God" below.
19. *The Gospel according to St. Mark* (London 1959), p. 446; also *Jesus and His Sacrifice* (London 1939), p. 102.
20. *Rédemption sacrificielle* (Montreal 1961), p. 147.
21. See "Das Lösegeld für Viele," in *Abba* (Göttingen 1966), p. 227 (the paper originally appeared in 1948), where he says simply, "διακονῆσαι refers to the Servant of Yahweh"; the statement is repeated verbatim in "παῖς (θεοῦ) im Neuen Testament," *Abba,* p. 209 (originally 1954).

22. *New Testament Theology*, vol. 1, Eng. trans. (London 1972), p. 293, and see p. 292 with n. 3. Thus also W. Hendriksen (although not in the context of the Isaian servant): "The service which it was the Son of man's purpose to render is described in the words: 'to give his life as a ransom in the place of many.' " See *The Gospel of Mark* (Edinburgh 1976), p. 415.
23. "Lösegeld," p. 225; see also *New Testament Theology*, p. 217.
24. "Lösegeld," pp. 224–25; *New Testament Theology*, p. 293.
25. *New Testament Theology*, p. 293.
26. Ibid., p. 292.
27. Cf. *The Central Message of the New Testament* (London 1965), p. 47.
28. *New Testament Theology*, p. 293. For a presentation similar to Jeremias, see W. J. Moulder, "The Old Testament Background and the Interpretation of Mark x.45," *New Testament Studies* 24 (1977–1978) 120–27.

The New Ethic

1. *Märtyrer und Gottesknecht* (Göttingen 1955), pp. 117–22.
2. *The Mystery of the Kingdom of God*, Eng. trans. (London 1925), pp. 73–80.
3. Ibid., p. 76.
4. Ibid., p. 77.
5. *Theologie des Vertrauens* (Hamburg 1967), p. 200; see also p. 240.
6. *The Seal of the Spirit* (London 1967), p. 40.
7. *The Temptation and the Passion: The Markan Soteriology* (Cambridge 1965), pp. 140–41; see also pp. 181–82. See further on Best in n. 3 of the next section.
8. *The Temptation*, p. 133, and see p. 151.
9. *The Gospel of Mark* (London 1976), p. 256.
10. Ibid., pp. 257–58.
11. *The Gospel of Mark* (London 1964), pp. 189–91.
12. *Jesus of Nazareth*, Eng. trans. (London 1960), p. 227.
13. *The Titles of Jesus in Christology*, Eng. trans. (London 1969), p. 56.
14. "The Use of ($\pi\alpha\rho\alpha$)$\delta\iota\delta\acute{o}\nu\alpha\iota$ in connection with the Passion of Jesus in the New Testament," in *Der Ruf Jesu und die Antwort der Gemeinde*, a Jeremias Festschrift, ed. E. Lohse (Göttingen 1970), p. 211.
15. "The Christology of Mark," in *L'Evangile selon Marc* by M. Sabbe and others (Leuven/Gembloux 1974), p. 483. This is a revised form of a paper from *Journal of Religion* 51 (1971) 173–87.
16. *Jesus and the Servant* (London 1959), pp. 74–78.
17. *The Son of Man in Mark* (London 1967), pp. 142–43. See also J. Schmid, *Das Evangelium nach Markus* (Regensburg 1963), p. 202.
18. "ὁ υἱὸς τοῦ ἀνθρώπου," *TDNT* 8 (1972), p. 455.
19. "Jesusgeschichte und Jüngergeschick nach Joh 12, 20–33 und Hebr 5,7–10," in *Der Ruf Jesu*, pp. 263–64. The same factor is admitted in tradition by A. Schulz, *Nachfolgen und Nachahmen* (Munich 1962), pp. 252–70; 332–33.
20. *Evangile selon Saint Marc* (Paris 1920), p. 264. See also E. P. Gould, *A Critical and Exegetical Commentary on the Gospel according to St. Mark* (Edinburgh 1932), p. 202; A. H. M'Neile, *The Gospel according to St. Matthew* (London 1952), p. 290.
21. *Ministerial Consciousness: A Biblical-Spiritual Study* (Rome 1975), pp. 37–94.
22. Ibid., p. 42.
23. Ibid., p. 64.
24. Ibid., p. 85.

25. Ibid., p. 77.
26. Ibid., p. 57.
27. Ibid., p. 94.
28. Ibid., p. 103.

Rule in the Community

1. *The History of the Synoptic Tradition*, Eng. trans. (Oxford 1963), pp. 93, 144, 407
2. Ibid., p. 163.
3. Ibid., p. 146, and see p. 88. E. Best sees this section originating in a rule devised in the Palestinian community but coming to Mark joined to verse 45 with "to serve" meaning "service of a personal nature freely offered," while v. 45a "relates the lifestyle of the disciple to that of his Lord"; see *Following Jesus: Discipleship in the Gospel of Mark* (Sheffield 1981), pp. 123–33, citing pp. 126 and 128.
4. *Das Markusevangelium* (Tübingen 1950), p. 108; and see E. Lohmeyer, *Das Evangelium des Markus* (Göttingen 1953), p. 223.
5. *Der Weg Jesu* (Berlin 1966), p. 367.

Functions in the Community

1. *Ältere Sammlungen im Markusevangelium* (Göttingen 1971), pp. 155–58.
2. *Markus—Lehrer der Gemeinde* (Stuttgart 1969), p. 166.
3. *The Gospel according to Mark* (Richmond, Va., 1969), p. 106.

Ecclesiastical Office

1. "Diáconos helénicos y bíblicos," *Burgense* 4 (1963) 65–66. Similarly in "Cambio de terminología de 'servicio' por 'honor-dignidad' jerárquicos en Tertuliano y San Cipriano," in I. Oñatibia and others, *Teología del sacerdocio*, vol. 4 (Burgos 1972), pp. 300–301.
2. "Light on the Ministry from the New Testament," in *Christian Origins and Judaism* (London 1962), p. 276. This essay originally appeared in 1952. For a different suggestion by Davies, see the following section.

Attendance on the Rabbi

1. *The Setting of the Sermon on the Mount* (Cambridge 1964), p. 422.
2. Ibid., p. 423.
3. "Die Diens van Jesus" (dissertation, Groningen 1968), p. 167.

In Fealty to God

1. *The Gospel according to St Mark* (London 1898), pp. 225–26. On the latter page Swete expresses a contrary opinion to W. Manson's on the significance of the aorist: "His life as a whole was a ministry ($\delta\iota\alpha\kappa\upsilon\nu\hat{\eta}\sigma\alpha\iota$, not $\delta\iota\alpha\kappa\upsilon\nu\epsilon\hat{\iota}\nu$)."
2. "Dienst," in *Bibel-Lexikon*, ed. H. Haag (Einsiedeln 1968), 336–37.
3. *Wie sprach Josephus von Gott?* (Gütersloh 1910), p. 13.
4. "The Background of Mark 10.45," in *New Testament Essays*, ed. A. J. B. Higgins (Manchester 1959), p. 9, and see p. 4.

5. "Mark 10.45: A Ransom for Many," in *New Testament Essays* (London 1972), p. 25.
6. *The Gospel according to St. John* (London 1955), p. 363.
7. "Gottes Sohn als Diener Gottes," *Studia Theologica* 27 (1973) 103–4.
8. Ibid., p. 85.
9. *The Gospel according to Mark* (Grand Rapids 1974), p. 383. M. Hengel writes of "Servant of God for the many," in *Studies in the Gospel of Mark,* Eng. trans. (London 1985), p. 44; see also pp. 33, 142 on context.

In Summary

1. "The Son of Man in the Synoptic Tradition," *Biblical Research* 13 (1968) 24.

CHAPTER 3

1. *Les ministères aux origines de l'église* (Paris 1971), p. 42.
2. For details, see the index of sources.
3. Including the variant at 1 Thess. 3:2 but not those at Luke 22:28 and Titus 1:9.
4. "Jewish and Christian Influence on New Testament Vocabulary," *Novum Testamentum* 16 (1974) 154.

Works of Mercy

1. *Enigmes de la deuxième épitre de Paul aux Corinthiens* (Cambridge 1972), p. 61. See also J. Roloff, *Apostolat-Verkündigung-Kirche* (Gütersloh 1965), p. 121; H. W. Beyer, *TDNT* 2 (1964), p. 89.
2. *Die Gegner des Paulus im 2. Korintherbrief* (Neukirchen Vluyn 1964), pp. 31–38; Eng. trans., *The Opponents of Paul in Second Corinthians* (Philadelphia 1986), pp. 27–32.
3. *Die Gegner,* p. 34; *The Opponents,* p. 29.
4. C. K. Barrett, *A Commentary on the Second Epistle to the Corinthians* (London 1973), p. 295.
5. *Die Geschichte der Kollekte des Paulus für Jerusalem* (Hamburg Bergstedt 1965), pp. 58–60.
6. "Diakonia in the New Testament," in *Service in Christ,* a Karl Barth Festschrift, ed. J. I. McCord and T. H. L. Parker (Grand Rapids 1966), pp. 37–48.
7. Ibid., p. 39.
8. Rom. 12:7; 16:1; 1 Cor. 16:15; Heb. 6:10; 1 Pet. 4:11.
9. *Les ministères aux origines de l'église* (Paris 1971), p. 198. Similarly Georgi, *The Opponents,* pp. 30–31.
10. *The Mission and Expansion of Christianity in the First Three Centuries,* Eng. trans. (London/New York 1908), p. 148.
11. *The Go-Between God* (London 1972), p. 141.
12. See for the early period E. Massaux, *Influence de l'Evangile de saint Matthieu sur la littérature chrétienne avant saint Irenée* (Louvain 1950). The closest is Polycarp 5.2 on deacons in reference to Mark 9:35.
13. *Conversion* (Oxford 1933), p. 210.
14. "Dakonia in the Early Church," in *Service in Christ,* pp. 49–64. Thus "diakonia" is

not a theme in E. Osborn, *Ethical Patterns in Early Christian Thought* (Cambridge 1976); Mark 10:45 is briefly mentioned as an ethical injunction (p. 23).
15. *Quellen zur Geschichte der Diakonia,* vol. 1 (Stuttgart 1960), pp. 18–65.
16. Ever since medieval times the passages have not always been easy of interpretation. See for example the embarrassment brought upon Origen for appearing to call the Logos a servant of sun, moon, and stars (*C. Cels.* 5.12.37), in M. Borret, *Contre Celse,* vol. 5 (Paris 1976), p. 45. n. 1.

"Diaconiae" at Rome

1. C. Baronius Soranus, *Martyrologium Romanum* (Antwerp 1613), n. b to 8 August. This was originally published 1586. In 1570 a similar opinion had been advanced by O. Panvinio in *Le sette chiese Romane* (so Kalsbach, n. 4 below). E. Hatch, *The Organisation of The Early Christian Churches* (London 1881), p. 53 n. 61.
2. "Les titres presbytéraux et les diaconies," *Mélanges d'archéologie et d'histoire* 7 (1887) 217–43; see also his edition of *Liber Pontificalis,* vol. 1, 2d ed. (Paris 1955), pp. 364–65, n. 7.
3. *Liber Pontificalis,* vol. 1, ed. L. Duchesne (Paris 1955), p. 364. The word for "monasteries" is actually in the dative ("Hic dimisit omni clero monasteriis diaconiae et mansionariis auri libras XXX"). Duchesne translates "the monasteries of diakonia" ("Les monastères de diaconie"). Instead of this genitive, the discussion that follows above might suggest that "diaconiae" should rather be understood as a loan word standing in apposition in the nominative to the dative "monasteries" for the purpose of identifying the type of monastery that had received benefactions. Thus: "to the monasteries [known as] diaconiae."
4. "L'origine orientale des diaconies romaines," *Mélanges d'archéologie et d'histoire* 57 (1940) 95–142. See also A. Frutaz, "Diaconia," *Enciclopedia Cattolica* 4 (1950), cols. 1521–35; A. Kalsbach, "Diakonie," *Reallexikon für Antike und Christentum* 3 (1957), pp. 909–17; earlier, H. Leclerq, "Diaconies," *Dictionnaire d'Archéologie chrétienne et de Liturgie* 4 (1920), cols. 735–38; and briefly, G. Barrois, "On Mediaeval Charities," in *Service in Christ,* a Karl Barth Festschrift, ed. J. I. McCord and T. H. L. Parker (Grand Rapids 1966), pp. 65–66.
5. Cassian, *Coll.* 18.7; 21.1,8,9,10; Gregory, *Ep.* 5.25; 10.8; 11.27 (Migne, *Patrologia Latina* 77.1080,1137).
6. R. Krautheimer, W. Frankl, and S. Corbett, *Corpus Basilicarum Christianarum Romae,* vol. 2, pt. 3 (Vatican City 1962), pp. 277–307. See also Krantheimer's *Rome, Profile of a City, 312–1308* (Princeton, New Jersey 1980), pp. 75–78 (S. Maria: "a foundation date around 600 or earlier," p. 77), 109–12.
7. *Corpus Basilicarum,* p. 305.
8. "L'origine orientale," pp. 137–42.
9. The relevant papyri are *VBP* 94 (fifth century CE); *P. Cairo Masp.* 67003; 67096; 67111; 67138; 67139 (all sixth century). Of occurrences at 67138.ir.2 and iiir.18 Maspero writes that the word designates "l'ensemble des terrains dépendant d'un de ces monastères, et, par suite, le monastère lui-meme" (vol. 2, p. 25).
10. W. E. Crum and H. G. Evelyn White, *The Monastery of Epiphanius at Thebes,* pt. 2 (New York 1973), pp. 201–02. This work was originally published 1926.
11. *The Monastery,* p. 52 n. 3.
12. *SB* 6009; 6010.
13. "Sur quelques objets coptes du Musée du Caire," *Annales du Service des Antiquités de l'Egypte* 10 (1910) 173–75.

14. A. H. M. Jones, "Inscriptions from Jerash," *Journal of Roman Studies* 18 (1928) 169–70, no. 36; J. W. Crowfoot, *Early Churches in Palestine* (London 1941), p. 139.
15. What was possibly another is listed in G. Valentini and G. Caronia, *Domus Ecclesiae* (Bologna 1969), no. 166. This identification is made with reservations and on architectural grounds.

Deacons of Old

1. 3.13.1; 14.2.3; 20.2; 4.6.7; 8.4; 9.2; 5.1.3; 7.31.5; 8.5.5; 6.4; 28.4.
2. *The Treatise on The Apostolic Tradition of St. Hippolytus of Rome*, 2d ed. (London 1968), p. xxxv.
3. *La tradition apostolique* (Paris 1946), p. 10.
4. Writing just prior to the council J. Colson turns even the full phrase to this sense: "Diakon und Bischof in den ersten drei Jahrhunderten der Kirche," in *Diaconia in Christo*, ed. K. Rahner and H. Vorgrimler (Freiburg 1962), p. 25.
5. *La tradition apostolique de saint Hippolyte* (Münster 1963), p. 22.

CONCLUSION TO PART I

1. J. B. Skemp, "Service to the Needy in the Graeco-Roman World," in *Service in Christ*, a Karl Barth Festschrift, ed. J. I. McCord and T. H. L. Parker (Grand Rapids 1966), p. 17.
2. "Die Problematik eines Theologischen Wörterbuchs zum Neuen Testament," *Texte und Untersuchungen* 73 (1959) 486. On a related but broader theme see F. F. Bruce, "The New Testament and Classical Studies," *New Testament Studies* 22 (1975–1976) 229–42.

PART II

1. On methods of referring to ancient literature see the note heading the index of sources.
2. See W. J. Burghardt, "Literature of Christian Antiquity," *Theological Studies* 37 (1976) 429–30.
3. Relevant sections are *Deipn.* 128–53; 170–73; 185–203; 234–61; 262–75; 290–94; 340–75; 376–83; 423–47; 658–62.
4. For example *Deipn.* 340d, e; 342c; 345c, d, e; 346e; 354d, e; 366a; 367f; 368a; 377c; 378c, f; 380d; 383e, f; 384a; 398b; 401b; 402c.
5. *Deipn.* 376e, f; 377a (Posidippus); 377d (Euphron).

CHAPTER 4

A Self-Sufficient Community

1. R. C. Cross and A. D. Woozley, *Plato's Republic* (London 1964), p. 80.

Trade

1. Including a member called "some other minister [θεραπευτήν] of bodily needs" (369d), who is not a servant because none yet exists, nor the medical practitioner because

medicine is not called for until the frugal community degenerates into luxurious ways (372e; 373d). He would perhaps be a preparer of foods, and the phrase reflects no interest in philanthropic service.

2. Καὶ μὴν κενὸς ἂν ἴη ὁ διάκονος, μηδὲν ἄγων ὧν ἐκεῖνοι δέονται παρ᾽ ὧν κομίζονται ὧν ἂν αὐτοῖς χρεία, κενὸς ἄπεισιν.

A Colourless Term

1. Translated by A. E. Taylor, in *Plato: The Sophist and the Statesman* (London 1961).
2. In the Greek, note the emphatic particles, also καί and ἄλλοι, which ensure continuity in the usage of διάκονος.
3. Guerra has rightly emphasised the notion of the state as a "political-juridical-religious unit," but has wrongly seen διακονία as one expression of the notion in Plato, "Diáconos helénicos y bíblicos," *Burgense* 4 (1963) 27–37.
4. In the first law διακονία occurs in conjunction with "retailer" and "merchant" (*Lg.* 919d, cf. 831e), as in *R.* 370–71. The passage is discussed later.

Functions of the Subordinate

1. J. B. Skemp disputes this in *Plato's Statesman* (London 1952), p. 185; he renders, "The most extensive class of servants." The discussion which continues in the text should make it clear that degree is the only possible sense.
2. The noun in the sense "attendants" (on priests) occurs incidentally at 291a.
3. So A. E. Taylor, *Plato: The Sophist and the Statesman* (London 1961), p. 307.
4. *Plato's Statesman*, pp. 187–88. He does not seem justified, however, in restricting its application here to "minor" civil servants, and his "underlings" is unnecessarily pejorative. See further S. Waszynski, *De servis atheniensibus publicis* (Berlin 1898), who divides magistrates into ἄρχοντες, ἐπιμεληταί and ὑπηρέται (p. 10).
5. See H. Kupiszewski and J. Modrzejewski, "ΥΠΗΡΕΤΑΙ," *Journal of Juristic Papyrology* 11–12 (1957–1958) 141–66. After reference to classical usage, they consider the many technical uses in the papyri. The question will be taken up again in chapter 9 on the papyri and in reference to the usage in Epictetus and Luke.

Functions of the Go-Between

1. Translated by L. Campbell, in *The Sophistes and Politicus of Plato* (Oxford 1867).
2. *Plato's Statesman* (London 1952), p. 187.
3. *Plato*, Eng. trans. vol. 3 (London 1969), p. 294.
4. Campbell, trans.
5. *The Sophistes and Politicus.*
6. Thus: strategy and hunting; painting and representative arts; carpentry and manufacture; husbandry and agriculture; horsebreeding and herdraising; draughts and arithmetic.

The Functions as Menial

1. H. N. Fowler, *Plato*, Loeb vol. 3 (London 1925); cf. A. Ammann, -ΙΚΟΣ *bei Platon* (Freiburg, Schweiz 1953), p. 54: "eine Kunst dienender Art."
2. A. E. Taylor, *Plato: The Sophist and the Statesman* (London 1961).
3. References to menial service are the rule in standard translations and commentaries

(Campbell, Fowler, Friedländer, Skemp, Taylor); see also G. Stallbaum, *Platonis Politicus* (Göttingen 1841). Campbell for example provides the not very helpful note that the diviner is "a species of servant."
4. *Sophistes and Politicus* (Oxford 1867) on *Plt.* 289d and p. vii.
5. *Plato's Statesman* (London 1952), pp. 185–86.
6. *The Sophist and the Statesman*, p. 225.
7. *Plato's Statesman*, p. 189.
8. *Plato*, Eng. trans. vol. 3 (London 1969), p. 294.
9. B. Jowett, *The Dialogues of Plato*, vol. 3 (Oxford 1897); I. A. Richards, *Plato's Republic* (Cambridge 1966). In their notes on the Greek text Jowett and Campbell fall back on "minister," again representing an attempt to avoid a connotation of simple meniality; see *Plato's Republic* (Oxford 1894).
10. J. L. Davies and D. J. Vaughan, *The Republic of Plato* (1866; repr., London 1925); T. H. Warren, *The Republic of Plato* (London 1897); F. M. Cornford, *The Republic of Plato* (Oxford 1955); H. D. P. Lee, *Plato, The Republic* (Harmondsworth 1955).
11. P. Shorey, *Plato, The Republic*, Loeb vol. 1 (London 1930).
12. E. Chambry, *La République*, Budé vol. 6 (Paris 1943).
13. *La République*, "comme intermédiaries pour l'achat et la vente."

Functions of the Attendant

1. Skemp's claim for "a higher sense" is therefore rejected; see *Plato's Statesman* (London 1952), pp. 185–86. It has been customary for theologians also to see in the usage of *Gorgias* an expression of exemplary public service; see Brandt, *Dienst und Dienen im Neuen Testament* (Gütersloh 1931), p. 22 ("a positive value"); Beyer, *TDNT* 2(1964), p. 82 ("the service . . . which consists supremely in the education of good citizens"); Guerra, "Diáconos helénicos y bíblicos," *Burgense* 4 (1963) 33 (identifying political "diakonia" with the making of better citizens); Lemaire *Les ministères aux origines de l'église* (Paris 1971), p. 32 (the noun designates "those who exercise a political responsibility for the good of the city, because they especially must have the spirit of service").

Speed

1. T. H. Warren, *The Republic of Plato* (London 1897), p. 209.
2. H. Frisk, *Griechisches Etymologisches Wörterbuch*, vol. 1 (Heidelberg 1960). Frisk relates ἐγκονέω to Latin "conor" ("undertake," "endeavour"). His opinions are largely repeated in P. Chantraine, *Dictionnaire étymologique de la langue grecque*, vol. 1 (Paris 1968), but Chantraine also reports M. Lejeune, *Bulletin de la Societé de Linguistique de Paris* 55 (1960) 24–26: Mycenean *kasikono*, "worker," "associate." A suggestion that seems to have been disregarded is that of O. Nazari reported in *Glotta* 6 (1913) 306: δια + ακονος, Sabine *ancus* ("slave").
3. On the Latin, see C. Mohrmann's several studies in *Etudes sur le latin des Chrétiens*, esp. vol. 3 (Rome 1965), pp. 79–81, originally *Vigiliae Christianae* 3 (1949).
4. *Lexilogus*, Eng. trans. (London 1861), pp. 230–35. This was the fifth English edition since 1835; the German appeared 1818. The citation in text occurs on p. 233.
5. *Early Middle English Verse and Prose*, ed. J. A. W. Bennett and G. V. Smithers (Oxford 1966), p. 203.
6. H. Lietzmann, "Zur altchristlichen Verfassungsgeschichte," *Zeitschrift zum Neuen Testament* 55 (1914) 140 (the word belonged to "more elevated language"); J. H.

Moulton and G. Milligan, *The Vocabulary of the Greek New Testament Illustrated from the Papyri and Other Non-Literary Sources* (London 1930). The latter are in only partial agreement, observing of διακονία, "like διάκονος itself and the verb it seems to have been somewhat literary except in an almost technical sense which brought it into common speech."

7. *Lexilogus zu Homer und den Homeriden* (1878–1880; repr., Amsterdam 1967), vol. 1, pp. 95–96; vol. 2, p. 155.

Hermes

1. *Lexilogus*, Eng. trans. (London 1861), p. 233.
2. For sources and other discussion see H. Estienne, *Thesaurus graecae linguae*, ed. C. B. Hase and others, vol. 2 (Paris 1833), cols. 1200–1201; O. Jessen in Pauly-Wissowa 5 (1905), cols. 318–19; E. Boisacq, *Dictionnaire étymologique de la langue grecque* (Heidelberg/Paris 1923), pp. 184–85; and the works of Frisk, Chantraine, and Goebel mentioned in the preceding section.
3. A reference to the sense "messenger" underlies Plutarch's observation on the then current dictum about two kinds of "words" (*Mor.* 777b): ὁ μὲν ἐνδιάθετος ἡγεμόνος 'Ερμοῦ δῶρον, ὁ δ' ἐν προφορᾷ διάκτορος καὶ ὀργανικός (which might be translated, "one resides in the mind, gift of Hermes the initiator, the other is on the tip of the tongue and is for communication and exposition"). Plutarch considers that the opinion is outdated. The thought is later reflected nonetheless in Aelian's comment on the ibis (*De natura animalium* 10.29): "The ibis is said to be beloved of Hermes the father of speech because its appearance resembles the nature of speech: thus, the black wing-feathers might be compared to speech suppressed and turned inwards, the white to speech brought out, now audible, the servant and the messenger of what is within, so to say." Aelian, trans. A. F. Scholfield, Loeb vol. 2 (Cambridge, Mass., 1959). Servant/messenger here: ὑπηρέτης/ἄγγελος, an instance of interplay, as often elsewhere, rather than of contrast with διάκονος; see the index of other Greek terms.
4. Stephanus, *Thesaurus*. The eagle also is messenger of Zeus, Antipater of Sidon, *AP1.* 7.161; of this A. S. F. Gow and D. L. Page observe, "A. no doubt understood it as equivalent to διάκονος," *The Greek Anthology, Hellinistic Epigrams*, vol. 2 (Cambridge 1965), p. 49. According to one of the oracles of Lucian's Alexander, Homer is πολέμων διάκτορον ("purveyor of wars," *Alex.* 33). For the Christian Hellenist Synesios, the morning is "the day's messenger," ὁ διάκτορος ἀμέρας (hymn 8.42); see C. Lacombrade, *Synésios de Cyréne*, vol. 1, *Hymnes*, Budé (Paris 1978).
5. LSJ records the instance once under διάκονος as "messenger," and under διάκτορος as "minister."
6. Accepting διάκτορον in Epictetus's citation of *Od.* a.37–39 on grounds adduced by W. A. Oldfather, "Further Attestation of a Reading of Zenodotus and Aristophanes in the Odyssey," *Classical Philology* 22 (1927) 99–100.
7. So also Stephanus, Thesaurus, vol. 2, col. 1201a. Of commentators, only F. H. M. Blaydes appears to make anything of an allusion to an "office" of Hermes; see *Aristophanis Plutus* (Halle 1886), p. 380.
8. G. W. Elderkin, "Two Curse Inscriptions," *Hesperia* 6 (1937) 389, table 3, line 8.
9. LSJ records otherwise only a verb so compounded in the comic poet Timocles (noting a variant and a conjecture, the latter being συνδιακονεῖ).
10. The allusion is noted also by Buttmann, *Lexilogus*, p. 231.

Greek Slaves

1. "Studien zur griechischen Terminologie der Sklaverei," in *Akademie der Wissenschaften und der Literatur im Mainz, Abhandlungen der Geistes- und Sozialwissenschaftlichen Klasse* (1964), pp. 1281–1310. Gschnitzer examines the terms δοῦλος, ἀνδράποδον, οἰκέτης, θεράπων, παῖς, ἀκόλουθος, ὑπηρέτης, but does not include διάκονος. The citation in text occurs on p. 1307. For a collection of ancient sources in translation, see T. Wiedemann, *Greek and Roman Slavery* (London 1981).
2. "Studien," p. 1284.
3. Ibid., pp. 1306–8.
4. *Synonymik der griechischen Sprache*, vol. 4 (1886; repr., Amsterdam 1969), pp. 124–29.
5. Ibid., pp. 145–48.
6. Ibid., p. 138.
7. Ibid., p. 143.
8. Ibid., pp. 135–36.

A Crossroads in Lexicography

1. H. Estienne, *Thesaurus graecae linguae*, vol. 2, ed. C. B. Hase and others (Paris 1833), col. 1181. On the usage designated "de Legatione," see chapter 8, "A Question of Diplomacy"; on the sense "instrumentum," see on 1 Mac. 11:58.
2. LSJ, p. 398.
3. *Synonyms of the New Testament* (London 1870), p. 30.
4. For example, J. Strong, "Deacon," in J. M'Clintock and J. Strong, *Cyclopaedia of Biblical, Theological and Ecclesiastical Literature* 2 (1880), p. 704; H. M. Gwatkin, "Deacon," in *A Dictionary of the Bible*, ed. J. Hastings 1 (1900), p. 574; E. P. Gould, *St. Mark* (Edinburgh 1896), p. 202 ("the performer of the services, without indicating his exact relation to the person served").
5. H. Cremer, *Biblico-Theological Lexicon of New Testament Greek*, 3rd English ed., trans. 2nd German ed. (Edinburgh 1880), p. 177.
6. *Biblico-Theological Lexicon*, p. 179. Other lexicons give meanings mainly in terms of service. Thus J. H. Thayer, *A Greek-English Lexicon of the New Testament being Grimm's Wilke's Clavis Novi Testamenti* (Edinburgh 1886), on 2 Cor. 3:3, "an epistle written, as it were, by our serving as amanuensis"; F. Zorell, *Lexicon Graecum Novi Testamenti* (Paris 1931), on 2 Cor. 8:4, "relief by almsgiving of the needs of the poor"; W. F. Arndt and F. W. Gingrich, *A Greek-English Lexicon of the New Testament and Other Early Christian Literature*, a translation and adaptation of W. Bauer's *Wörterbuch* (Cambridge/Chicago 1957), on Matt. 25:44, "when did we not help you?"
7. *TDNT* 2 (1964), p. 81; cf. *TWNT* 2 (1935), p. 81.
8. *The Semantics of Biblical Language* (Oxford 1961), p. 223.
9. "Semantics and Biblical Theology," in *Supplements to Vetus Testamentum* 22 (1972), p. 18.
10. In respect at least of the word "deacon," Bo Reicke is a modern theologian who would see the virtue of these earlier presentations; using Frisk's etymology, he concludes, "a deacon means someone who undertakes and accomplishes a difficult task. . . . In the word 'deacon' it is not inferiority but usefulness which is stressed." See his "Deacons in the New Testament and in the Early Church," in *The Ministry of Deacons,* World Council Studies no. 2 (Geneva 1965), p. 9. Outside the field of the

diaconate, however, he finds that the word group suggests "humility, sense of service and modesty" (p. 9). Here and in his more detailed *Diakonie, Festfreude und Zelos* (Uppsala/Wiesbaden 1951), pp. 23–24, 30, 150–5, he also proposes that usage in respect of "service to one's neighbour" had a background in Jewish Christian liturgy.

CHAPTER 5

A Servant Girl and a Scholar

1. W. R. M. Lamb, *Lysias,* Loeb (London/New York 1930), p. 11.
2. *Indices Graecitatis quos in singulos oratores atticos confecit J. J. Reiskius,* vol. 1 (Oxford 1828), s. v. Mitchell's phrase is appended in English to Reiske's "de ancilla a pedibus, quae adhibeatur allegationibus."
3. S. Jebb, *Aeilii Aristidis Adrianensis Opera Omnia,* vol. 1 (Oxford 1722), p. 82 ("non tamen se ad historias scribendas, vel tale quid contulerit, sed antiquitati Graecae subservire maluerit"). In Stephanus, vol. 2, col. 1183d, W. Dindorf calls the instance figurative ("Figur. dicitur de opera interpretandis Graecorum veterum scriptis impendenda"). In *Aelii Aristidis Smyrnaei quae supersunt omnia,* vol. 2 (Berlin 1898), B. Keil speaks of interpretation but not of how one arrives at this meaning; he writes (p. 220): "rhetores cum orationes suas declamarent, Graecis ipsorum aequalibus se dabant, Alexander Graecis antiquis, quorum scripta interpretatur. . . ."

To Heaven

1. G. H. Box, *The Testament of Abraham* (London 1927), pp. 13–14 (my italics). Compare "transmit a speech for me" in M. E. Stone, *The Testament of Abraham* (Missoula, Mont., 1972), p. 21, and "serve me (by delivering) a communication" in E. P. Sanders, "Testament of Abraham," in *The Old Testament Pseudepigrapha,* vol. 1, ed. J. H. Charlesworth (London 1983), p. 886.
2. M. Delcoir, *Le testament d'Abraham* (Leiden 1973), pp. 123, 125 ("servir pour moi la parole").
3. H. W. Fowler and F. G. Fowler, *The Works of Lucian of Samosata,* vol. 3 (Oxford 1905), p. 137 (trans. H. W. F.; my italics).

From Heaven with Hermes and Iris

1. H. W. Fowler and F. G. Fowler, *The Works of Lucian of Samosata,* vol. 1 (Oxford 1905), p. 168 (trans. F. G. F.).
2. See earlier Aristophanes, *Pl.* 1170; also *Etym. Magn.* 268.20, where a variant refers to the office of conveying the dead.
3. "Two Curse Inscriptions," *Hesperia* 6 (1937) 382–95.
4. *The Works of Lucian,* vol. 1, p. 67 (trans. H. W. F.).
5. *The Works of Lucian,* vol. 1, pp. 86–87 (trans. F. G. F.).
6. C. H. Oldfather, trans., *Diodorus Siculus,* Loeb vol. 3 (London/New York 1939).
7. M. Hadas, *An Ethiopian Romance: Heliodorus* (Ann Arbor 1957), p. 123.
8. *Aristophanis Aves* (Halle 1882), pp. 132, 384. Later the poet Nonnus calls Iris διάκτορος *(LSJ).*

Messengers Less than Gods

1. F. H. Colson, *Philo,* Loeb vol. 6 (London/Cambridge, Mass., 1935).
2. In vol. 7 of *Philonis Alexandrini Opera quae supersunt omnia,* ed. L. Cohn and P. Wendland (Berlin 1926).
3. *Lexicon Sophocleum,* vol. 1 (Königsberg 1835), p. 422 ("nuntius").

Constantine

1. A. H. M. Jones, "Notes on the Genuineness of the Constantinian Documents in Eusebius' Life of Constantine," *Journal of Ecclesiastical History* 5 (1954) 196–200.
2. "Britain and the Papyri (*P. Lond.* 878)," in the Festschrift for W. Schubart, *Aus Antike und Orient* (Leipzig 1950), pp. 126–32.
3. "Zu Skeat: Britain and the Papyri," *Festschrift für Friedrich Zucker* (Berlin 1954), pp. 343–48.
4. *A Select Library of Nicene and Post-Nicene Fathers,* ed. H. Wace and P. Schaff, 2d ser., vol. 1 (Oxford/New York 1890).

A Saviour and a Trickster

1. E. M. Forster, *Alexandria: A History and a Guide* (1922; repr., London 1982), p. 112.
2. Achilles Tatius, *Leucippe and Clitophon—Book III* (Classical Association of Rhodesia and Nyasaland 1960), pp. 137–38.
3. The customary reading at 3.18.5 is εἰ διάκονός τις εἶ θεῶν, but the last word ("of gods") is poorly attested and may be interpolated; see E. Vilborg, *Achilles Tatius, Leucippe and Clitophon* (Stockholm 1955); *Achilles Tatius, Leucippe and Clitophon: A Commentary* (Göteborg 1962); A. Henrichs, "Achilleus Tatios, aus Buch III (P. Colon. inv. 901)," *Zeitschrift für Papyrologie und Epigraphik* 2 (1968) 211–26. The damaged third-century papyrus cited in the last study does not include the phrase but the space available suggests that the shorter reading had been followed. Henrichs (p. 200) then gives some weight to a seventeenth-century opinion that τις may be the corrupt remains of an original Ἑκάτης. Even without explicit mention of a god, however, Clitophon's meaning would be unmistakable. Vilborg does not appreciate this and for reasons of clarity adopts "of gods" ("The defining genitive seems necessary; the attempts to interpret διάκονος alone are not convincing"), nor do Carney and Henrichs when they trace the term's religious connotation to usage in the inscriptions, where it designates an official in religious guilds. These guild officers, however, were waiters at table and, for all the dignity attaching to their title, were not in practice an awesome lot.

The Making of Aesop

1. *Vita Aesopi* 4, θεασάμενος τὸ τῆς θεοῦ σχῆμα ἀνθρώπινον περικείμενον ("beholding the human form of the goddess [Isis] enveloping [the herald]"). The phrase is similar to Paul's "being found in human form " (Phil. 2:8). The herald, however, unlike Christ Jesus, has an identity distinct from her "form"; she later (5) distinguishes between "myself" and "thy form (τὸ σὸν σχῆμα)."

The Brother of a Reluctant Prophet

1. F. H. Colson, *Philo,* Loeb vol. 7 (London/Cambridge, Mass., 1937).
2. F. H. Colson, *Philo,* Loeb vol. 6 (London/Cambridge, Mass., 1935).

Jeremiah

1. J. Hudson, *Flavii Josephi quae reperiri potuerunt omnia graece et latine* (Amsterdam 1726), "se pro illis apud Deum intercessurum esse." The Hebrews has *mithpallêl* (Jer. 42:4), the Greek, προσεύξιομαι (Jer. 49:4).

An Unlikely Prophet

1. *Josephus,* Loeb vol. 2–3 (London/Cambridge, Mass., 1927–1928). Similarly "servant" ("Diener") in the German translation of O. Michel and O. Bauernfeind, *Flavius Josephus, De Bello Judaico, Der Jüdische Krieg,* vol. 1 (Bad Homburg 1960).
2. H. St. John Thackeray and R. Marcus, *A Lexicon to Josephus,* pt. 2 (Paris 1934).
3. *Church Order in the New Testament,* Eng. trans. (London 1961), p. 174. In *A Complete Concordance of Flavius Josephus,* vol. 1, ed. K. H. Rengstorf and others (Leiden 1973), B. Justus includes "middleman, go-between, messenger" among meanings of the term in Josephus but does not cite passages where they apply. In *Josephus, The Jewish War,* ed. G. Cornfield (Grand Rapids 1982) we read "minister of the voice of God" with a note (p. 312) that this means "its mouthpiece."
4. *TDNT* 2 (1964), p. 91. See A. Schlatter, *Wie sprach Josephus von Gott?* (Gütersloh 1910), p. 14.
5. *Josephus,* vol. 2, p. 675. n. b.
6. Thackeray notes that Suetonius and Dio Cassius testify to the historicity of Josephus's prediction, *Josephus,* vol. 2, p. 690 n. a. See H. W. Attridge, "Josephus and His Works," in *Jewish Writings of the Second Temple Period,* ed. M. E. Stone (Assen/ Philadelphia 1984), pp. 189–92, 206 n. 34.

"How Sweetly You Do Minister to Love"

1. *Euphues,* ed. E. Arber (London 1868), p. 354.
2. *In C. Verrem* V. 30 (actio secunda). No words of the "minister" kind appear in R. Pichou, *Index verborum amatoriorum* (Hildesheim 1966).
3. A. B. Walkley, cited by Sir Arthus Quiller-Couch in the New Shakespeare edition of the play (Cambridge 1953), p. xxiv.
4. Information on Shakespeare's sources is taken from G. Bullough, *Narrative and Dramatic Sources of Shakespeare,* vol. 2 (London/New York 1958), pp. 59–139. On "ambassade," see there p. 79 concerning Sir Thomas Hoby's translation of B. Castiglione, *Il Cortegiano.* Of the latter and its translation Gough Whitlam has had this to say: "Castiglione's polished and sophisticated classic *Il Cortegiano* (1528) enjoyed an immense success and influence throughout Western Europe and not least in England. It was translated into Spanish, French, German, Flemish and even Russian. Sir Thomas Hoby's translation appeared in 1561 and went through further editions in 1577, 1588 and 1603. . . . He [Castiglione] must also be blamed for Shakespeare's puns. Quite apart from such literary reflections, however, *The Courtier* affected the whole cultural, social and political attitudes and aspirations of the Elizabethans." See *The Italian*

Inspiration in English Literature (Canberra 1980), pp. 3–4. Other interesting detail was presented earlier in the century by S. L. Wolff in *The Greek Romances in Elizabethan Prose Fiction* (New York 1912).
5. M. Hadas, *An Ethiopian Romance: Heliodorus* (Ann Arbor 1957), p. 26.
6. James Joyce, *A Portrait of the Artist as a Young Man* (London 1968), pp. 175–76.
7. Hadas, *Romance*, p. 75.
8. So G. Dalmeyda *Xénophon d'Ephèse,* Budé (Paris 1926): "mes yeux dont Habrocombès est le maître."
9. Dennis Potter, *Hide and Seek* (London 1973), p. 165.
10. *An Aethiopian History Written in Greek by Heliodorous Englished by Thomas Underdowne anno 1587,* with an introduction by Charles Whibley (London 1895), p. 184: "Cibele her chamberleine and Bawd."
11. The date favoured by E. Feuillatre, *Etudes sur les Ethiopiques d'Héliodore* (Paris 1966), pp. 147–48.
12. Hadas, *Romance*, p. 169.
13. The participle is plural by attraction ("one of her chamber-women ministering"); the singular is given in the Budé edition. Notice the Latin, "rerum Venerearum Arsaces ministra," of J. Bourdelotius, *Heliodori Aethiopicorum Libri X* (Paris 1619), p. 317.
14. For example, Hadas, *Romance*, p. 177, "amatory requirements"; R. Smith, *The Greek Romances of Heliodorus, Longus and Achilles Tatius* (London 1889), p. 166, "private gratification"; W. Lamb, *Heliodorus* (London 1961), "accessory"; J. Maillon, *Héliodore*, Budé vol. 2 (Paris 1938), p. 140, "destiné à l'usage."
15. Lamb, *Heliodorus*, p. 181. Lamb would seem to be correct in taking the noun in the sense of sexual intercourse, avoided here by Hadas (*Romance*, p. 181: "make all arrangements for the meeting"), who then renders the same noun in a phrase from a later development of the context as "illicit commerce with another woman" (p. 188; *Hld.* 7.25).
16. T. Underdowne and F. A. Wright, *Heliodorus* (London 1923), pp. 139–40. (This edition is a revision by Wright of Underdowne, "a superb master of English" who sometimes "totally fails to give the meaning of the Greek original," p. 5.)
17. The phrases in translation are from Hadas, *Romance*, pp. 200, 277.
18. J. Hudson, *Flavii Josephi quae reperiri potuerunt omnia graece et latine,* vol. 1 (Amsterdam 1726): "fidelem operam navando."
19. See A. M. Denis, *Introduction aux pseudépigraphes grecs d'ancien testament* (Leiden 1970), p. 57; J. J. Collins, "Testaments," in *Jewish Writings of the Second Temple Period,* ed. M. E. Stone (Assen/Philadelphia 1984), pp. 342–44; H. C. Kee argues for Greek (noting influence of Hellenistic romances on style), "Testaments of the Twelve Patriarchs," in *The Old Testament Pseudepigrapha,* vol. 1, ed. J. H. Charlesworth (London 1983), p. 777.
20. R. H. Charles, *The Testaments of the Twelve Patriarchs* (London 1908). Kee, in "Testaments," translates: "the spirit of promiscuity has wine as its servant for indulgence of the mind."
21. M. Philonenko, *Joseph et Aséneth* (Leiden 1968), p. 109; and see Denis, *Introduction*, pp. 40–48; S. West, *"Josephus and Asenath:* A Neglected Greek Romance," *Classical Quarterly* 24 (1974) 70–81; K. Berger "Jüdisch-Hellenistische Missionsliteratur und Apokryphe Apostelakten," *Kairos* 17 (1975) 232–48; G. W. E. Nickelsburg, "Stories of Biblical and Early Post-Biblical Times," in *Jewish Writings*, pp. 65–71; C. Burchard, "Joseph and Aseneth," in *Pseudepigrapha*, vol. 2 (New York 1985), pp. 177–247.

22. *Introduction*, p. 44; see Burchard, "Joseph and Aseneth," pp. 186–87; his views there and p. 183 on the connection with erotic literature are interesting in the light of the discussion to be developed in the text.
23. K. Kerenyi, *Die griechisch-orientalische Romanliteratur in religionsgeschichtlicher Beleuchtung* (Tübingen 1927), pp. 69–70.
24. "Neglected Greek Romance," p. 77. Burchard is also opposed to an allegorical reading but recognises motifs from broader culture as well as "a deeper meaning," provided the latter is not seen as "coded" in the text; see "Joseph and Aseneth," p. 189.
25. Lysias 1.16; Lucian, *DDeor*. 7.3; *P. Warren* 21.4; Xenophon of Ephesus 1.9.7; Aristaenetus 2.7.15.
26. A. M. Harmon, *Lucian of Samosata*, Loeb vol. 3 (London/New York 1921), p. 27.
27. H. W. Fowler and F. G. Fowler, *The Works of Lucian of Samosata*, vol. 2 (Oxford 1905), p. 18 (trans. F. G. F.)

Errands

1. P. Steinmetz, despite translating by "to do the shopping," sees the sign of the flatterer's indignity in the verb which he takes to mean "to perform the duties of a slave"; see *Theophrast, Charaktere*, vol. 2 (Munich 1962), pp. 43, 51.
2. Archedicus, *K*. 3.277: a cook grumbles about bustling through the market for the paltry bargains he has brought back, concluding, "No one has completed a trip [δεδιακόνηκεν] with such difficulty." Menander, *Dys*. 213: a lad "fetches" (water) from a temple precinct "for a girl" (διακονεῖ κόρῃ); similar interpretations are provided by W. G. Arnott, *Menander's Dyskolos* (London 1960), and W. E. Blake, *Menander's Dyscolus* (New York 1966), but F. Stoessel wrongly argues that the lad is procuring the girl for Sostratus, in *Menander Dyskolos, Kommentar* (Paderborn 1965), p. 77. Achilles Tatius 2.31.4: "Conops, who used to lie in ambush for us, was absent, on an errand for his mistress [τῇ δεσποίνῃ διακονησόμενος]." Achilles Tatius 7.9.11: Leucippe, a free woman in Melitte's protection, is sent into the countryside "on an errand for her" (διακονησομένην αὐτῇ). So too Sophocles, *Ich*. 144: Silenus upbraids his sluggish and fearful satyr sons "unwillingly going to their task" (κανελεύθερα διακονοῦντες) through the forest in search of the lost oxen of Apollo; R. J. Walker's phrase, "doing idle and careless service," misses the connotation of movement (*The Ichneutae of Sophocles* [London 1919], p. 453). Probably to be included here is the papyrus *BGU* 261.26 in which gratitude is expressed to Heras for coming "in order that he might minister for us (διακονέσσι ἱμῖν [sic])"; compare *UPZ* 18.23 below, and the recent *P. Oxy*. 3313.5 (second century CE): Apollonios and Sarapias are apologising in their letter to Dionysia for not attending Sarapion's wedding (they were sending a thousand roses and four thousand narcissi—by arrangement—and wanted no money); E. A. Judge translates: "Greetings. You filled us with joy when you announced the good news of most noble Sarapion's marriage; and we would have come straightaway *to give him our support* [διακονήσοντες αὐτῷ] on a day most prayed for by us, and to join in the rejoicing. But in view of the judicial circuit. . . ."; the phrase I have italicised is rather "on that errand for him" (i.e., with the roses). See G. H. R. Horsley (who drew my attention to this passage and kindly made it available to me in an extract from), *New Documents Illustrating Early Christianity* (The Ancient History Documentary Research Centre, Macquarie University, Sydney 1983), pp. 10–15.
3. In *Elegy and Iambus*, Loeb vol. 2 (London/New York 1931), J. M. Edmonds translates, "I serve Anacreon for that price," namely, the price Anacreon had paid Venus;

since this was "a little song," the transaction is hard to follow, and one could have expected a genitive instead of an accusative in line 14. The verb there, in fact, picks up the idea "he sent me" from line 7. Edmonds dates the ode 50 BCE–50 CE.
4. So R. G. Ussher, *Aristophanes Ecclesiazusae* (Oxford 1973). Ussher, however, notes of the word at 1116, "a comparatively rare word for a servant."
5. In "Index to the Speeches of Isaeus," *Hermathena* 28 (1938) 136, W. A. Goligher gives "public service" here, which is to miss the idea of "errand." In the same note, nonetheless, he gives " 'errands,' 'missions' " for the instance recorded immediately above at Thucydides 1.133.
6. This type of dative (ἀγαθοῖς) is rare, except in Josephus (see examples in the discussion of *AJ* 18.280); see on *Hld.* 5.15 and Alexander Aprh., *Mixt.* 5.16.
7. The adverb qualifies the noun ("of all sorts"). See E. B. England, *The Laws of Plato*, vol. 2 (Manchester 1921), p. 332, who cites παντελῶς παίδων, σφόδρα γυναικῶν (*Lg.* 791c, 639b). He translates in a pejorative sense, however, with "mere servants"; this notion would not be consistent with that of acquiring wealth.

A Letter from Chaerea

1. R. F. Weymouth, *The New Testament in Modern Speech*, 3d ed. (London 1910).
2. A. D. Papanikolau, *Chariton-Studien* (Göttingen 1973), pp. 9–12: with W. Schmid, Kerenyi, and Lesky, Papanikolaou dates the work in the first century BCE; others place it in the second century CE, and Rohde as late as the sixth. G. Molinie in the Budé edition, *Chariton* (Paris 1979), places it 75–125 CE (p. 1).
3. *Charitonis Aphrodisiensis de Chaerea et Callirrhoe amatoriarum narrationum* (Leipzig 1783), p. 191.
4. *Charitonis Aphrodisiensis de Chaerea et Callirhoe amatoriarum narrationum libri octo* (Oxford 1938).
5. Reiske, *Charitonis*, p. 641. Molinie, the most recent editor, omits the comma, adopts Reiske's syntax, but misses the exact sense: "Because of the negligence of the servant who was entrusted with it [serviteur qui en etait chargé], the letter fell right into the hands of Dionysios."

The Twins' Petition

1. *Urkunden der Ptolemäerzeit*, vol. 1 (Berlin/Leipzig 1927). The papyri relevant to the affair are numerous; see *UPZ* 17–58.
2. On the history and for a description of the vast complex, with its shops, hostels, and royal residence, see M. Guilmot, "Le Serapieion de Memphis: Etude topographique," *Chronique d'Egypte* 37 (1962) 359–81.
3. Wilcken, *UPZ*, pp. 46–47.
4. On the twins' attitude, see *UPZ* 20.61–62. On Ptolemy I's promotion of the cult of Sarapis, see H. C. Youtie, "The *Kline* of Sarapis," *Harvard Theology Review* 41 (1948) 9–29.
5. *UPZ*, pp. 179–80.
6. *UPZ*, p. 180; see also his note on *UPZ* 18.22.

In High Places

1. Josephus does not mean "servant." This bears emphasising. He has consistently referred to Achratheus as "the eunuch," even avoiding *"her* eunuch" as at Esth. 4:5,

LXX (see *AJ* 11.223). Nowhere does he use this noun in the sense of "personal attendant at court" but, with the exception of *AJ* 8.354; 9.54, 55 (personal attendant on a prophet) and *AJ* 6.52; 11.188 (waiters at royal banquets), regularly in a sense related to "messenger" (*BJ* 3.354; 4.388, 626; *AJ* 1.298; 7.201, 224; 11.255; 12.187). Comparison of the story of Esther in Josephus and the Septuaginst shows the following points of usage. Josephus uses διακον- (noun and verb) of cupbearing at *AJ* 11.188 where Esth. 1:8 does not ("waiters": οἰκονόμοι). Where Esther uses it indiscriminately of royal attendants (1:10; 6:1R, 3, 5) and royal advisers (2:2), Josephus omits it in the former sense (190) and on other occasions uses more exact designations (γραμματεύς, 248 and 250; θεράποντες, 252; φίλοι, 195). His other uses (228, above; 255, below) are not paralleled in Esther, and may be said to be improvements on the Septuagint in the interest of succinctness, probably also of elegance.
2. R. Marcus, in the Loeb edition, vol. 6, writes "intercede" in both instances, which is more than Josephus is saying (see the observation on *AJ* 10.177). For the first instance Hudson is closer to the meaning with "that she might act for him" ("ut cum fratre pro se ageret"), but his "promised to devote herself to his interests ("se illi . . . operam daturum") goes beyond the Greek; see *Flavii Josephi* . . . (Amsterdam 1726).
3. Arthmius is called "a slave of the king" in Asia (ibid.) because in Athenian estimation kingly rule engendered slavery; hence also the word "despot" in the Greek. Arthmius was not a slave in the ordinary sense, however, and the connotation of servility does not carry over to διακονῶν. In Aristides' later version of the story we read simply τῷ βασιλεῖ διακονῶν (13.254, "on a mission for a king"). See other discussion of the construction under *Asenath* 15.7 (with n. 25 there) and in the preceding section of this chapter.
4. So A. N. W. Saunders, *Demosthenes and Aeschines* (Harmondsworth 1975), p. 73.
5. See notes on the dative at Alexander Aphr., *Mixt.* 5.16, and compare with those on Heliodorus 5.15.
6. P. J. Rhodes, "Thucydides on Pausanias and Themistocles," *Historia* 19 (1970) 387–400.

Communicating

1. *The Ion of Euripides* (London 1894). Contrast, "doing this service in secret for my friend," in D. W. Lucas. *The Ion of Euripides* (London 1949).
2. M. Hadas, trans., *An Ethiopian Romance* (Ann Arbor 1957), p. 95. Compare "making this writing my messenger," T. Underdowne and F. A. Wright, *Heliodorus, An Aethiopian Romance* (London 1923), p. 119.
3. τὸ γὰρ πονοῦν τῆς ψυχῆς ἡ γλῶττα πρός ἱκετηρίαν διακονουμένη τῆς τῶν ἀκουόντων ψυχῆς ἡμεροῖ τὸ θυμούμενον. πρός is taken above periphrastically; alternatively, and if ἱκετηρία equals ἱκεσία, "In conveying the heart's anguish to form a plea, the tongue appeases the anger in the heart of those who listen."
4. Other writers employ διακον- in designating the operation of human faculties; we recall Xenophon of Ephesus 1.9.7–8, but see further on Philo, *Vita Moysis* 199, and such a passage as Themistius 125.9.
5. *Achilles Tatius*, Loeb vol. (London/Cambridge, Mass., 1917).
6. *Achilles Tatius, Leucippe and Clitophon, A Commentary* (Göteborg 1962), p. 72.
7. *Achilles Tatius, Leucippe and Clitophon* (Stockholm 1955).
8. *Achilles Tatius, Leucippe and Clitophon—Book III* (Classical Association of Rhodesia and Nyasaland 1960), p. 91.

9. διακονεῖν πρός at Josephus, AJ 3.128, would not be a parallel.
10. Comentary, p. 72.

CHAPTER 6

A Virgin's Ransom from Hermes

1. T. Underdowne and F. A. Wright, trans., *Heliodorus, An Aethiopian Romance* (London n.d.), p. 142.
2. Ibid., pp. 147–48.
3. Ibid., p. 149.
4. See above on *Hld.* 5.16.
5. "Truly it was through the fire that he sent this stone, through your ministry"; M. Hadas, *An Ethiopian Romance* (Ann Arbor 1957), p. 122.
6. "He has in fact conveyed this treasure trove to you through that fire"; W. Lamb, *Heliodorus: Ethiopian Story* (London 1961), p. 121.
7. "Je suis sûr qu'elle me vient de Mercure, . . . qui me prouve une fois de plus sa générosité. C'est lui qui m'a fait parvenir ce présent que tu as trouvé à travers les flammes." See J. Maillon in R. M. Rattenbury, T. W. Lumb, and J. Maillon, *Héliodore, Les Ethiopiques,* Budé vol. 2 (Paris 1938), p. 55.
8. "Persuasum habens et hoc inventum mihi venire more solito a Mercurio . . . , qui tibi omnino donum per ignem subministravit"; W. A. Hirschig, *Erotici scriptores* (Paris 1856)
9. The sentence at Heliodorus 5.15 reads in the Budé edition cited in n. 7: δέχομαι τὴν θεόπεμπτον ταυτηνὶ λίθον πειθόμενος παρ' Ἑρμοῦ τοῦ καλλίστου καὶ ἀγαθωτάτου τῶν θεῶν ἥκειν μοι συνήθως καὶ τόδε τὸ εὕρημα διὰ τοῦ πυρός σοι τῷ ὄντι τὸ δῶρον διακονήσαντος.

Discrepancies appear in relation not only to less significant details like the force and reference of συνήθως and τὸ ὄντι but also to more significant details like the meaning of εὕρημα and whether this word is the object of πειθόμενος or of διακονήσαντος (so Lamb: impossible if, as editors strongly prefer, δῶρον is also read), and the agreement (as well as the meaning: "conveyed," Lamb; "ministry", Hadas) of διακονήσαντος (with Ἑρμοῦ or πυρός).

10. Calisiris has said to Nausicles (5.13): ταῦτά σοι . . . λύτρα . . . οἱ θεοὶ δι' ἡμῶν προσάγουσι.
11. Notes on this translation: "heaven-sent" is strongly emphasised in the Greek (for which see n. 9); "godsends" were, in fact, popularly associated with Hermes (generating, indeed, the word ἕρμαιον), hence, it would seem, the significance of συνήθως ("appropriately" in the translation); "and" is introduced before mention of the fire to mark a new thought; "gift" is omitted in only one of the eight manuscripts used in the Budé edition, but the sense of the Greek would hold up without it ("which acted at your behest"); "at your behest" is the force of the dative σοι ("for you"). This dative is normal and has been discussed in connection with Lucian, *Cont.* 1, and observed, in order, at Herodotus 4.154.3; Lysias 1.16; Lucian, *Icar.* 20; *Test. Ab.* 9.24; Lucian, *DDeor.* 7.3; Josephus, *AJ* 10.177; Xenophon Eph. 1.9.7; Heliodorus 7.9; *P. Warren* 21.4.8; *SB* 4947.2; Aristaenetus 2.7.2; 2.7.15; *Asenath* 15.7; Menander, *Dys.* 219; Achilles Tatius 2.31.4; 7.9.11; *Anacreontea* 15.14; *UPZ* 18.23; 19.25; Josephus, *AJ* 8.5; Demosthenes 9.43; Aristides 13.254; Priscus, *HGM* 276.7. In the light of usage, the dative cannot be taken in Lamb's sense (n. 6 above), "Hermes has conveyed this to you, Calisiris, through the fire," an idea requiring πρός σε (and this construction

with the preposition also occurs in the following passages from among those just cited: Lucian, *Icar.* 20; *Test. Ab.* 9.24; *P. Warren* 21.8; *SB* 4947.2; Josephus, *AJ* 8.5; 10.177). Indeed, if the participle διακονήσαντος—in whatever sense (e.g., Hirschig, n. 8 above, "subministravit")—is taken as agreeing with "Hermes," the dative is inexplicable.

A Little Philosophy

1. See identical uses of the verb (*Mixt.* 6.11) and the noun (adjectivally; *Mixt.* 6.9,20). Interestingly, to judge by a German report of medieval Arabic translations of this work, the Arabic retained in these passages the sense "transmit" ("weiterzuleiten"); see H. Gätje, "Die arabische Ubersetzung der Shrift des Alexander von Aphrodisiens über die Farbe," *Nachrichten der Akademie der Wissenschaften in Göttingen aus dem Jahre 1967*, Philol.-Hist. Kl. (Göttingen 1967), pp. 341–82. In the light of the preceding note on the dative, the dative at *Mixt.* 5.16 above is to be noted because it is different from those illustrated previously. As in the objective genitive at *Mixt.* 5.14 ("of colours"—compare the subjective genitive "of the translucent object" at *In Sens.* 25.21), what is designated by the dative "colours" is the object to be moved. Similar datives of respect occur at Lucian, *DMort.* 30.3 ("good deeds"); Achilles Tatius 3.10.2 (v. 1. "suffering"); Aristides, the Christian apologist, 15.2 ("justification"). Somewhat similar is the construction διακονεῖν κελεύματί τινος which is illustrated in connection with Josephus, *AJ* 18.280 ("to carry out someone's orders"), but noticeably different is the dative "disagreements" at Priscus, *HGM* 276, because this indirect object does not designate an effect that has been transferred from one area to another but simply an area in respect to which the activity of διακονεῖν is carried out; the same will apply to the dative τραπέζῃ at Heliodorus 5.8 and Acts 6:2, also, it would seem, to σώματι at Alexander Aphr., *in Mete.* 19.6.
2. "If certain materials pass on to the next material objects [ταῦτα τὰ σώματα . . . τοῖς μετ' αὐτὰ διάκονα γίνεται, 18.25]) effects which the materials are themselves immune to, should we be surprised to find that the moon, itself remaining unaffected by the motion of the sun, passes [διαδίδωσι] to bodies naturally susceptible to the sun's motion the effect generated within it by that motion? . . . A material body coming within range of the sun's motion will be affected by the heavenly body; there is nothing unusual, however, should the effect not be that that material body catch fire but that by reason of the sun's influence it becomes a medium for the next material body to become warm and then burst into flame οὐ μὴν οὕτως πάσχειν ὥστε ἐκπυροῦσθαι, ἀλλὰ διακονεῖσθαι διὰ τούτου τοῦ πάθους τῷ ὑπ' αὐ τὸ σύματι, ὡς θερμαίνεσθαι, 19.6]."
3. A. O. Smet, *Alexandre d'Aphrodisias: Commentaire sur les Météores d'Aristote, traduction de Guillaume de Moerbeke* (Louvain/Paris 1968), p. 31.
4. No other instance of this cognate is recorded; *LSJ* is unhelpful in giving its meaning as "pertaining to service." For other instances of the word group in the later philosophers, see Simplicius, Elias, Joannes Philoponios, Olympiodorus (all sixth century), and Michael Ephesius (twelfth) in the index of *Commentaria in Aristotelis libros* (Berlin 1883–).
5. Ears, *Post. Caini* 165: διακόνοις ἀκοαῖς χρώμενοι (people use the ears of the young to pass on idolatrous thoughts); *Vita Moysis* 199: [τὴν ψυχὴν] διακόνοις δὲ τοῖς ὠσὶ χρησαμένην (the soul receives outrageous ideas through the undiscerning ears); hands and teeth are ὑποδιάκονος, *de spec. leg.* 1.204 and 3.301, and there at 1.17 the same

cognate designates the sense faculty as the agent of the mind gathering information from sense objects about the mind behind the universe.
6. "Agent" because no ὡς but after χρῆται as in Philo in the preceding note and at Philostratus, *Im.* 1.3.2, in discussion immediately above; compare the generic use of the verb at 2.23.11 to follow.
7. The term designates public servants; see on 1.9.16 and Plato, *Plt.* 290b.
8. The nuance is missed in the following translation by A. Fairbanks, *Philostratus, Imagines,* Loeb, (London/New York 1931): "Aesop uses him as a slave in developing most of his themes." By contrast O. Schönberger captures the sense through the phrase "durch ihn . . . ausführt" ("Aesop *presents* most of his ideas *through it*"); see *Philostratus, Die Bilder* (Munich 1968).

Double-Crossing Agents and Others

1. Hld 6.2: διάκονον εἰς τὴν ἐπιβουλὴν καθεῖσα τὴν Θίσβην.
2. G. A. Williamson, trans., *Josephus: The Jewish War* (Harmondsworth 1969), p. 254. The Greek reads, οἷς οὐκ ἀπιστήσαντες οἱ ζηλωταὶ διακόνους ἑαυτοὺς ἐπέδοσαν.
3. τοὺς διακονήσαντες τὴν προδοσίαν οἰκέτας.
4. διάκονος αὐτῷ τῆς πρὸς αὐτὸν ἀφίξεως. According to Hudson, seventeenth-century editors emended αὐτῷ to ἀπό, which allowed them to say that Rachel had been Jacob's servant since his arrival at Laban's establishment.
5. ἴσθι διάκονος ὢν χρηστὸς σύμβουλος ἐγένου. R. Marcus translated vol. 6 of Thackeray's Loeb *Josephus* (London 1937).
6. ὁ νόμος διακόνῳ τινὶ χρώμενος Translation after T. C. Skeat and E. P. Wegener, "A Trial before the Prefect of Egypt Appius Sabinus, c. 250 A. D. (P. Lond. Inv. 2565)," *Journal of Egyptian Archaeology* 21 (1935) 224–47.
7. βηχὶ χρησάμενος διακόνωι. See P. Perdrizet, "Le mort qui sentait bon," in *Mélanges Bidet, Annuaire de l'Institut de Philologie et d'histoire orientales,* 2 (Brussels 1934), pp. 719–27.
8. αὐτὸν εἰς ταῦτα παράσχειν διάκονον.
9. Jowett missed the decisive shift here and has Plato repeating himself: "they . . . will not have other slaves and servants for their own use, neither will they use those of the villagers and husbandmen for their private advantage."
10. The ὑπηρέται of the poets are the dancers, rhapsodists, and entrepreneurs; they enact the poet's mind.
11. Oddly, Davies and Vaughan (*The Republic of Plato* [London 1925]), who at 371e render the term by "operatives" here misleadingly give us "domestic servants." Similarly Jowett (*The Dialogues of Plato,* vol. 3 [Oxford 1875]): "And we shall need more servants. Will not tutors be also in request . . . ," who thus puts διακόνων in one category, and the tutors, etc. in another; his "also," however, is not in the Greek.
12. διακονεῖν occurs in this connection at 2.154; διάκονος at 2.152, 154 (cf. also 187, 193, 221, 223); διακονία at 2.152, 153, 154.
13. 2.199 contains one further instance of the verb and one of διάκονος in the context of agency.

Doing Caligula in

1. AJ 19.41: οὐκ ἐπιτάγμασι τοῖς Γαίου διακονούμενοι γνώμῃ δὲ τῇ αὑτῶν.
2. AJ 19.42: διακονούμεθα.
3. Ibid.: τις καὶ καθ' ἡμῶν διακονήσηται τοιῦτα Γαίῳ.

4. H. Feldman, *Josephus*, Loeb vol. 9 (London 1965).
5. διακονία τῆς κατ' ἐκεῖνον ὕβρεως.
6. δεδιακονῆσθαι τοῖς ἀπεσταλκόσι.
7. διακονεῖσθαι ταῖς ἐντολαῖς.
8. διακονεῖσθαι τοῖς ἐκείνῳ προανεψηφισμένοις.
9. διακονήσασθαι τῷ Γαίῳ τὴν ἀνάθεσιν.
10. μανίᾳ τῇ Γαίου διακονούμενος.
11. δεδιακονημένον αὐτοῦ ταῖς ἐντολαῖς. (The text is uncertain in some details but not in a way to obscure this point of syntax.)
12. τὰ πάνθ' ὑπὲρ ὑμῶν . . . διακονεῖν.
13. τὸ δὲ πᾶν . . . δεδιακονήσεταί σοι. The passive is awkward and comparatively rare. For an emperor to be said to undertake a task at the behest of a client king would approach lèse-majesté; the passive construction removes the suggestion of such an impropriety.
14. διακονεῖσθαι τὰ πάντα ἡδονῇ ἐκείνων.

The Piety of Petronius

1. *AJ* 9.25: Ahaziah's third emissary (cf. 4 Kings 1:13–14) begs to be spared the fate of the previous two, explaining that he has come to Elisha not of his own will but "carrying out a royal command" (βασιλέως διακονῶν προστάγματι).
   AJ 14.358: Herod is forced to desist from attempted suicide "by the number of those who would not allow his hand to do what he contemplated [διακονεῖν οἷς ἐντεθύμητο]."
   AJ 17.74: "Antipater arranged a murderous plot and provided a poison by which he might carry it out [ᾧ διακονήσει αὐτῇ (γνώμῃ)]." The main variant here (*o* instead of the dative relative pronoun) would require the less usual impersonal subject, "which would put the plan into effect."
   AJ 17.140: The slave girl Acme informs Antipater that "following his instructions" (διακονουμένη κελεύματι τῷ ἐκείνου) she has written to Herod.
   AJ 18.125: "One of the commanders would die, either the one who gave orders for the war or the one who was hastening to implement the strategy of the other [τὸν γνώμῃ τῇ ἐκείνου ὡρμημένον διακονεῖσθαι]."
2. Callistratus describes a statue of Medea with (13.4) "the hand ready to execute her impulse [διακονεῖν . . . τῷ θυμῷ]" to kill her children. Comparable phrasing occurs with ὑπηρετεῖν: Callistratus 13.2 ("purposes"); Aristides 2.187 ("desires").
3. H. St. John Thackeray, *A Lexicon to Josephus*, pt. I (Paris 1930), preface, p. viii; see also *Josephus: The Man and the Historian* (New York 1929), pp. 106ff. R. J. H. Shutt argues that Thackeray's theory of the Thucydidian assistant "ignores the characteristics common to *Ant.* xvii–xix and the rest of the work" and attributes peculiarities largely to Josephus's liking for a phrase; see *Studies in Josephus* (London 1961), pp. 72 and 74. Inconsistency of style between *AJ* 17–19 and the rest of the work has long been noted, as in W. Schmidt, *De Flavii Josephi Elocutione Observationes criticae*, Jahrbücher für classische Philologie, Supp. Bd. xx (Leipzig 1894), pp. 343–550, in particular p. 366, but is no longer seen as supporting Thackeray's position; see H. Schreckenberg, *Bibliographie zu Flavius Josephus* (Leiden 1968), p. 213.
4. *The Works of Flavius Josephus* (London 1811). The Greek reads: οὐ μὴν δίκαιον ἡγοῦμαι ἀσφάλειάν τε καὶ τιμὴν ἐμαυτοῦ μὴ οὐχ ὑπὲρ τοῦ ὑμετέρου μὴ ἀπολουμένου τοσούτων ὄντων ἀναλοῦν διακονούμενον τῇ ἀρετῇ τοῦ νόμου, ὃν πάτριον ὄντα περιμάχητον ἡγεῖσθε, καὶ τῇ ἐπὶ πᾶσιν ἀξιώσει καὶ δυνάμει τοῦ

θεοῦ τὸν ναὸν οὐκ ἂν περιιδεῖν τολμήσαιμι ὕβρει πεσεῖν τῆς τῶν ἡγεμονευόντων ἐξουσίας.
5. "operam vestram legi adeo praestanti addicitis"; *Flavii Josephi . . . opera omnia* (Amsterdam 1726). The emendations of Cocceius are cited from this edition.
6. *The Works of Flavius Josephus* (London 1906).
7. *AJ* 18.125, 262, 265, 277 (bis); 19.41

The Loving Cup

1. The text is uncertain in some details but not in a way to obscure the point being illustrated.

A Little Flattery

1. πρᾶξις, πρᾶγμα, ἔργον, πράττειν, ὑπουργεῖν, ὑπηρετεῖν, σύνεργος, ὑπουργός, χρεία, ὑπουργία.

At the Behest of a Deity

1. On Epictetus, see chapter 8.
2. C. W. Emmet, *The Fourth Book of Maccabees* (London/New York 1918). The English term "minions" is derogatory of itself, which the Greek word it translates is not. Nor does the Greek word denote royal attendants because armed guards are not that sort of διάκονοι.
3. There are perhaps traces of the idea of an agent of the god Asclepius in the dedicatory inscription of the Athenian Asclepieion (420 BCE), but the text is uncertain. *IG* II, 2d ed., 4960. 6–7: [μεταπεμ]-ψάμενος δ(ρ)ά[κ-οντα ἤγ]αγεν δεῦρε. E. J. Edelstein and L. Edelstein (*Asclepius* 1 [Baltimore 1945], p. 375) translate: "When the god came from Zea at the time of the great mysteries, he took up his residence in the Eleusinium, and having summoned from his home the snake he brought it hither in the chariot of Telemachus." *Michel* 1529A. 6–7 reads (after Dragoumes, 1901): διακ-ονον. A. Koerte, who made the first conjecture ("Die Ausgrabungen am Westabhange der Akropolis," *AthMitt* 21 [1896] 314), cites Pausanias 2.10.3 and 3.23.6, but these are folklore in which the snake is identified with Asclepius and thus hardly illustrate a situation where the god, already arrived in Athens, sends for something (the god is probably but not certainly the subject of the sentence). Michel's reading is certainly more appropriate for an inscription and, taken with a further conjecture, could mean that the god sent for his "agent," namely the snake, which Telemachus, the donor of the shrine, then met in his chariot; this would also be a more direct tribute to Telemachus.
4. F. C. Babbitt, trans., *Plutarch,* Loeb vol. 5 (London/Cambridge, Mass., 1936).
5. T. Taylor, *The Mysteries of the Egyptians, Chaldeans, and Assyrians* (1821; repr., London, 1968), pp. 77–78.

CHAPTER 7

Housebound

1. See also Sophocles, *Ph.* 287: Philoctetes lives alone and has "to provide everything for himself" (τὶ . . . διακονεῖσθαι). (This is a true middle voice, which occurs in

relation to domestic chores also at Plato, *Lg.* 763a, and in relation to table service at Aristophanes, *Ach.* 1017, and Diogenes Laertius 6.31.) Crates, *K* 1.133: in a world where furniture and utensils are mobile and can take orders from their master an old man will be able to attend to his own needs (αὐτῷ . . . διακονήσει). Plato, *R.* 467a: children will accompany parents on military expeditions to learn relevant skills, to look after their parents (θεραπεύειν), and to διακονεῖν καὶ ὑπηρετεῖν πάντα περὶ τὸν πόλεμον: the last two verbs are perhaps to be distinguished, after the manner of Lucian, *DMort.* 30.3, as "fetch" and "do," thus, "to fetch whatever their parents might need on the field and carry out all instructions relating to the exercise"; Plato goes on to compare these children with the potter's boys who "look on διακονοῦντες," i.e., fetching whatever the potter needs, before they are allowed to practise at the wheel (ibid.) *P. Oxy.* 275.11 is a contract of apprenticeship to a weaver where the same distinction perhaps applies: διακονοῦντα καὶ ποιοῦντα πάντα τὰ ἐπιτασσόμενα, "(standing by and) fetching and doing everything he is told in regard to the whole art of weaving."

2. The adjective διακονικός is not so much designating a task as "menial" as "a kind of task normally performed by a διάκονος," i.e., done at the behest of another.

3. Plato, *Lg.* 805e: household activities as distinguished from tilling and herding, all of them as activities normally carried out by slaves (γεωργεῖν τε καὶ βουκολεῖν καὶ διακονεῖν). Plato, *Lg.* 763a: the wardens of the country must dispense with slaves and attend to their needs themselves (διακονοῦντες τε καὶ διακονούμενοι ἑαυτοῖς). Menander, *Dys.* 206: an old family retainer explains that he should join the master in the fields because he has spent enough time on household chores for the mistress (σοι διακονῶν). Epictetus 4.7.37: the worldly man feels the need of slaves and freedmen to do things for him (σοι . . . διακονῶσιν). Lucian, *Philops.* 35: the Egyptian wonderworker Pancrates could turn a broom into a servant for drawing water, buying and providing, and in general "doing everything for us" (πάντα . . . ὑπηρετεῖ κὰι διακονεῖτο ἡμῖν), all this being designated τῆς διακονίας (ibid.). Plutarch, *Mor.* 63e: an old woman tidying up the sick room of Apelles (ἡ διακονοῦσα πρεσβῦτις). (Since Apelles was very poor, the woman may not have been a slave. The anecdote occurs in the context shortly alluded to in the discussion in the text.) In Timaeus we read that the ancient and revered custom was not to be served by bought slaves (*FHG* 1.207: διακονεῖσθαι; on the passive see later) and that the Phocian custom had been for the younger to attend on the older (*FHG* 1.208: διακονεῖν τοὺς νεωτέρους τοῖς πρεσβυτέροις); the reference in both instances seems to be broader than to the table.

4. This usage is not easily distinguishable from that in the preceding note. Lucian, *Philops.* 34: a traveller leaves his household slaves (οἰκέται) behind, being assured that on the way he will not lack personal attendants: τῶν διακονησομένων. Lycurgus 14.55: merchants travel without a mistress but not without "a slave as a personal attendant" (παιδὸς τοῦ διακονοῦντος). Heliodorus 10.7: a queen requests the services of a particular attendant (διακονουμένης μοι). Josephus, *AJ* 13.314: one of the attendants who had to carry away the basins during Aristobolus's vomiting attacks (τῶν διακονονουμένων τις παίδων).

5. See in the preceding note Lycurgus 14.55; Josephus, *AJ* 13.314; and elsewhere Theophrastus, *Char.* 30.16; see also similar expressions at Josephus, *AJ* 2.129; Plutarch, *Mor.* 63e; Aristophanes, *Av.* 73.

6. In the section *Ath.* 262–75 these are Dieuchidas, *FHG* 4.389 (*Ath.* 263a), Timaeus, *FHG* 1.207, 208 (*Ath.* 264c, d), and Crates, *K.* 1.133 (*Ath.* 267e). For the abstract and common nouns there see Posidonius, *FHG* 3.265 and Cleitarchus (*Ath.* 267c).

7. Sophocles, *Ph.* 287; Menander, *Dys.* 206; Crates, *K.* 1.133; Aristophanes, *Av.* 1323,

and see also 838: παραδιακόνει ("to work alongside," in the context of carrying up rubble).
8. Plato, *R*. 467 (twice); *Lg*. 763a, 805e; *Tht*. 175e; *Grg*. 521a; Aristotle 1333a8; Aristides 2.154, 199; Epictetus 4.7.37; Plutarch, *Mor*. 677e.
9. Lycurgus 14.55 (oration); *P. Oxy*. 275.11 (contract).
10. Heliodorus 10.7; Josephus, *AJ* 13.314.

Staff

1. Aristophanes, *Av*. 73 (used as an adjective), 74; Prov. 10:49. These three instances are discussed below.
2. Demosthenes 24.197 (female), and two speeches of doubtful authenticity, Demosthenes 40.14: τοὺς παῖδας τοὺς διακόνους, where the adjective distinguishes the household slaves (who could be expected to give an inventory of the house in an estate under dispute) from those who worked in fields and elsewhere; Demosthenes 59.42: θεραπαίνας δύο καὶ οἰκέτην διάκονον (again adjectival). In "Diáconos helénicos y bíblicos," *Burgense* 4 (1963) 41, n. 109, M. Guerra cites a further adjectival use, Hyper. *[sic]*, fr. 70: τοὺς διακόνους παῖδες *[sic]*, which I have not been able to find.
3. Aristotle, *EN* 1149a27; Teles 41.12; Xenophon, *Oec*. 8.10, 14 (the latter—"the skipper's servant"—is from an anecdote of shipboard life illustrating the domestic attendant's responsibility for neatness and is coloured by this context); *Mem*. 1.5.2: διάκονον δὲ καὶ ἀγοραστήν, the latter a rare word for one who goes to market and perhaps pointing to the literary character of the former and indicating that it too might designate a person engaged in running errands; compare the role of the attendant bird in Aristophanes, *Av*. 74, 75, and the use of the verb for going to market in Lysias 1.16; Theophrastus, *Char*. 2.9.
4. See on *AJ* 11.228.
5. *Alex*. 5, as a youth Alexander of Abonutichus was under the tutelage of a fraudulent magician who used him in the three following nominated roles: ὑπουργῷ καὶ ὑπηρέτῃ καὶ διακόνῳ; of these the first two are not easily differentiated, the second probably indicating that Alexander was an associate in the practice of magic, and the last that he also attended to the personal needs of his patron.
6. Gen. 39:4; 40:4 are minor and, since they refer to Joseph, probably only apparent exceptions.
7. In *Recherches sur le vocabulaire du culte dans la septante* (Paris 1966), p. 98, S. Daniel distinguishes profane uses of λειτουργ- in Kings and Chronicles from strictly cultic uses elsewhere; these profane uses are nonetheless royal and prophetic in their reference.
8. See the remarks on *BJ* 3.354. At *AJ* 9.54, 55, Josephus uses the term for Elisha's attendant, interchanging it with θεράπων. LXX uses λειτουργός and παιδάριον (2 Kings 6:15a and 15b, 17; MT: *mšrt* and *n'r*, although Josephus expands on his biblical source here and it is difficult to trace a clear relationship between the two usages.
9. B. H. Kennedy, trans., *The Birds of Aristophanes* (London 1908), p. 6.
10. See on *Etym. Magn*. 311.11.
11. *LSJ* records this comparative under "servile, menial," which would accord with A. A. Benois's translation, "A struck Phrygian becomes a better slave" (*Les fragments d'Epicharme traduits en français* [Nice, n. d.], p. 68, n. 292). But Suidas had already related the proverb to the Phrygian reputation for idleness and slothfulness (A. Adler, *Suidae Lexicon*, 4 [Leipzig 1935], p. 768, n. 772), and Stephanus, 2.1187a, notes "prompter in serving, more expeditious, and quicker." The comparative would

not be irregular if derived from διάκων, but this form is not attested until centuries after Epicharmus. The common noun is listed with six other terms under οἱ δοῦλοι ("slaves") in the glossary attributed to Cleitarchus (*Ath.* 267c): at least three of the terms are rare or specialised; Pollux 3.76 also records it among terms applicable to slaves (see chapter 8).

In the Abstract

1. Aristotle 1255b27; 1261b37; 1263a20; Xenophon, *Oec.* 7.41; Plutarch, *Mor.* 677e; Aristides 2.152, 153, 154.
2. Josephus, *BJ* 3.70; *AJ* 4.109 (the labours of Balaam's ass); 18.21 (of the Essenes, who provided their own services rather than impose on others the injustice of serving); Posidonius, *FHG* 3.265. Also in the satirical narrative, Lucian, *Philops.* 35.
3. B. Perrin, trans., *Plutarch, Lives,* Loeb vol. 10 (London 1921), n. 357a.
4. F.-M. Abel, *Les livres des Maccabées* (Paris 1949), p. 216. Abel's statement that Stephanus "established by examples the sense 'utensils' " is misleading because Stephanus cites only Moschio (2.1184c) and provides a generic definition, "household gadgets" ("instrumenta ad suppelectilem pertinentia"). In regard to 1 Macc. 11:58 the idea of "golden vessels and furniture for the table" *(RV)* is widely represented (*RSV;* E. J. Goodspeed, *The Apocrypha* [Chicago, 1938]; *NAB;* etc.); hendiadys, as in "a service of gold plate," is preferred by some (*JB; NJB; NEB; GNB;* E. Kautzch, *Die Apokryphen und Pseudepigraphen des Alten Testaments,* vol. 1 [1900; repr. Hildesheim 1962]; etc.). The Sistine-Clementine Vulgate reads "in ministerium" (hence, "to be served in," *AV;* H. Wace, *The Holy Bible, Apocrypha,* vol. 2 [London 1888]; "for his use," Douay, 1609; Knox, 1949), but "in" is a correction for "et," and does not predate the tenth century; cf. R. Weber, *Biblia Sacra iuxta vulgatam versionem* (Stuttgart 1969); D. De Bruyen and B. Sodar, *Les anciennes traductions latines des Machabées* (Maredsous 1932). S. Tedesche (*The First Book of Maccabees* [New York 1950]) translates simply, "and service," and the Anchor Bible (1977) similarly abandons any attempt at precision in providing "gold and other gifts."
5. *The Apocrypha* (London 1892), n. 38, as an alternative to "a service of gold plate."
6. The distinction here is between attendants on the royal person and other (civil service?) staff. The collective sense possibly occurs also at Athenaeus 171b and 439c (the latter with reference to Polybius 30.26.5), although on both occasions the word is perhaps better understood as "ministrations at the banquet." This is the meaning also in the first of two occurrences in Josephus, *AJ* 8.169, a passage telling of the queen of Sheba's admiration of the court of Solomon; the second instance there, in the phrase τὰ τῆς παρασκευῆς αὐτοῦ καὶ διακονίας, could well embrace the Septuagint's reference to the palace staff (3 Kings 10:5).
7. Alternatively, "with the service one gets"; the piece concludes with a recommendation that as circumstances require one should get by with "self-service" (6.34: αὐτοδιακονία). This typically Cynic teaching echoes the phrasing in his discourse on poverty and wealth, "neither extravagant as previously nor in need but satisfied with things as they are, not longing to have one's own servants (4A.130: διακόνους).
8. Cf. *AJ* 8.101; 19.129; Demosthenes 18.206.
9. The later technical use of the word in the Christian church for "alms" or "offerings" (*Apostolic Constitutions* 2.25.8; 3.8.2; 3.13.2; 4.7.1) developed by way of "godly commission" or "duty" (cf. there 4.1.1).

Waiters at Work

1. See the discussion on Athenaeus 192b and other instances in the introduction to part II. In regard to Plutarch's usage note the alternative terms at *Mor.* 620e, 635b, 641b, 643b, 644a, 657d, f, 664b, 678f, 683b, 686a, 703e, 716e; the single designation of a waiter in our terms occurs in an anecdote at 628c (see below), while two instances in the grammatical aside at 677e refer to household chores.
2. Lucian, *Merc. Cond.* 26. In the ode *P.* 906 the noun occurs by virtue of the vocative, and in Philostratus minor, *Im.* 4, the passage is describing a lifeless work of art.
3. Josephus, *AJ* 6.52; 11.188; Lucian, *Sat.* 17; Athenaeus 173b; 420e; Iamblichus, *VP* 146.15.
4. Euripides, *Cyc.* 31 (Odysseus and Cyclops); Polybius 30.26.5 (in a procession); Philo, *Vit. cont.* 75 (standing by during the president's address; the variant reading is participial); Xenophon, *Hier.* 4.2 (tasting for poison; and see the discussion below on Athenaeus 171f); Theopompus, *FHG* 1.315 (introducing female company after a drinking session); Athenaeus 139c (announcing the name of a benefactor); Nicomachus, *K* 3.386 (a cook boasting of his skills). See also συνδιάκονοι, ὑποδιάκονοι in Posidippus, *K* 3.342 (associates of a professional cook).
5. Demetrius in Athenaeus 173f; Xenophon, *HG* 5.4.6; Demosthenes 59.33; Dio Cassius 54.23.4; Diodorus Siculus 5.26.3.
6. Theophrastus, *Char.* 22.4; 30.16 (οἱ διακονοῦντες παῖδες); Plutarch, *Mor.* 440b; 628c; Posidonius, *FHG* 3.260; Hegesander, *FHG* 4.420; Libanius, *Or.* 53.9; Diodorus Siculus 5.40.3 (τῶν διακονούντων οἰκέτων); Lucian, *Merc. Cond.* 16; *Test. Job* 15.3 (τοῖς δούλοις τοῖς διακονοῦσιν); Achilles Tatius 4.15.4 (cited in the following sentences).
7. Other direct objects occur at Lucian, *Asin.* 53 (wine); *Philops.* 35 ("everything," probably as object); Josephus, *AJ* 15.224 (a draught of poison).
8. Athenaeus 659d; Menander, *Dys.* 490; *K* 3.78; Euphron, *K* 3.322; Hegesippus, *K* 3.312; Posidippus, *K* 3.342, 343; probably the fragmentary *P. Heid.* 184 fr. 7.20.
9. See the first note to this section.

A Sense of Occasion

1. The verb occurs in a description of the Persian court, and special note is taken of the white raiment and careful toiletry (Heraclides, *FHG* 2.96: οἱ θεραπεύοντες . . . διακονοῦσιν), and in the courts of tyrants, with mention of libations to the gods (Athenaeus 171f, a citation of Xenophon, *Her.* 4.2). Instances of the verb in Josephus relate to Belshazzar (*AJ* 10.242), Xerxes (*AJ* 11.163, 166), Artaxerxes (*AJ* 11.188), Herod (*AJ* 15.224), Agrippa (*AJ* 18.193), and the Davidic prince Amnon (*AJ* 7.165), and are discussed below in a comparison with usage of the Septuagint; that discussion includes also instances of the common noun in connection with King Saul (*AJ* 6.52) and Artaxerxes (*AJ* 11.188), as well as of the abstract noun in connection with the pharaoh in the story of Joseph (*AJ* 2.65), Solomon (*AJ* 8.169: two instances), Xerxes (*AJ* 11.163), and the captive Agrippa (*AJ* 18.193). Elsewhere the common noun designates waiters at the court of a tyrant (Xenophon, *Hier.* 4.2) and in the residence of the polemarch (Xenophon, *HG* 5.4.6) and at the splendid games of Antiochus IV Epiphanes at Daphne near Antioch (Polybius 30.26.5). The last instance is cited twice in Athenaeus (195e, 439c), and on the latter occasion the reader is informed that "the management of the ministration [τῆς διακονίας]" was in the hands of Antiochus

himself; on the former occasion the phrasing is modified to "management of the activities [πραγμάτων]" because there Athenaeus is referring to a broader range of proceedings, which included a procession, and not just to the banquet. The abstract noun occurs finally as a designation of the duties of waiting in the court of Cotys, king of Thrace (Plutarch, *Mor.* 174d).

2. In Lucian, *DDeor.*, Ganymede attends on Zeus (4.4), as do Hebe and Hephaestus (5.2), while Hermes attends on Hercules and Dionysus (24.2). These passages instance the verb, which occurs also of attendance on Zeus in the comic poet Anaxandrides, *FAC* 2.57, and of Eros in the poet of *Anacreontea* 32.6. In a fragment of Strato we hear of a youth transported to the home of the gods "for ministrations to the blessed ones" (πρὸς μακάρων . . . διακονίας, *APl.* 12.194) of their glorious nectar.

3. Thus, while entertaining the emperor Augustus, Vedius Pollio—notorious for displays both of wealth and of cruelty: he would throw slaves into a pool of lampreys—was about to punish a cupbearer who had broken a crystal goblet when Augustus, to deter his host from brutal retaliation against the διάκονος, called for all Pollio's best goblets and smashed the lot (Dio Cassius 54.23.4). This common noun occurs also in the attack on the profligate excesses of Neaera's party for the games champion (Demosthenes 59.33: probably not authentic) and in a report of the sleeping arrangements that rounded off certain Etruscan drinking sessions (Theopompus, *FHG* 1.315). Lucian's biting satire of the professional aesthete betraying his limitations among the social elite at dinner includes both noun and verb (*Merc. Cond.* 26 and 16). The verb occurs in a picture of the idyllic luxury of the Tyrrhenians (Diodorus Siculus 5.40.3), and the abstract noun in moral discourse of Philodemus, *Ir.* 28.4: "getting upset about the food, the drink, the company, the service you are given [διακονίαν] and such like." In his life of Pythagoras, Iamblichus remarks on avariciousness in respect of fine appointments and attendants (*VP* 146.15).

4. To Anaxandrides, *Anacreontea*, and Strato in the last note but one, add Nicomachus, *K* 3.386, recording the boast of a cook that his fish dish is superior to that of the ordinary waiter; Euripides, *Cyc.* 31: Silenus abhors waiting on Cyclops; in prose the Cynic Diogenes writes θέραψ (rare, poetic) διακονούμενος (*Ep.* 37.4); a drinking song bids the waiter pour to one more noble man (P 906).

"The God Took Joy"

1. *Smaller Dictionary of Greek and Roman Antiquities,* 12th ed. (London 1884), pp. 95–96. One might like to compare with this outline of the ancient dinner the following about twentieth-century custom provided by Theonie Mark in her introduction to *Greek Islands Cooking* (London 1978): "Village eating customs differ considerably from the city, with the largest meal served in the evening, after the men return from the fields and have washed and relaxed. This dinner is served on low, round, wooden tables placed on the raised floor of the kitchen area and . . . surrounded with pillows for everyone to sit on while dining. Two or three dishes are served in beautifully designed earthenware bowls for everybody to choose and eat from, accompanied by thick slices of round wheat bread, olives and pickles, cold water from the clay jug for the children and wine from the home supply for the husband and wife."

2. Robert Fitzgerald, trans., *Homer: The Iliad* (Oxford 1984), p. 13.

3. To attribute to the ancients a faith in the presence of the gods at their religious meals or even to speak of their sense of communion with the gods on these occasions is not to be taking sides in the debate about whether the ancients held a sacramental view of such meals (that is, saw them, in W. L. Willis's description as "a means of acquiring

the deity and/or its special powers and traits''); it is however to assert their belief in and celebration of the gods' willingness to be among them. Indeed, in arguing against sacramentalism, Willis would seem to go too far towards a merely social interpretation of even the public sacrificial meals. See *Idol Meat in Corinth,* SBL Dissertation Series no. 68 (Chico, Calif., 1985), pp. 7–64 (citation from p. 63).

Decorum

1. Philemon, *K* 3.312 (Ath. 170e); Menander, *K* 3.78 (Ath. 245c); Hegesippus, *K* 3.312 (Ath. 290c); Diphilus, *K* 2.553 *bis* (Ath. 292c); Posidippus, *K.* 3.343 (Ath. 377a); Euphron, *K* 3.322 (Ath. 377d). Also Aristophanes, *Ach.* 1017; Menander, *Dys.* 490; *P. Heid.* 184 fr. 7.20.
2. On the history and religious significance of the exclusion of slaves from cult, see F. Bömer, *Untersuchungen über die Religion der Sklaven in Griechenland und Rom,* vol. 4 (Wiesbaden 1963), pp. 81–100.
3. *Leisure and Pleasure in Roman Egypt* (London 1965), p. 13: "The garland played an important role in Greek life and had a sort of divinising value. At a banquet or a festival even the humblest person to some extent partook of the life of the gods if he were garlanded; he shared richly in what we may call the momentary eternity of the rite."

Making Friends

1. Posidonius, *FHG* 3.261 (Ath. 152e): a lavish feast provided by the Celtic chief Lovernius; *FHG* 3.265 (Ath. 153c): Heracleon's disciplined troops group by thousands in the fields and armed guards serve bread, meat, and wine in silence; Timaeus, *FHG* 1.196 (Ath. 153d and 517d): among the Etruscans young girls served naked; Hegesander, *FHG* 4.420 (*Ath.* 419e): when Pyrrho of Elis, the Sceptic who disapproved of indulgence, wished to decline a dinner invitation, he was assured by Plato that the company was more important than the food, which the attendants waste anyway; see also two passive verbs discussed below, Juba, *FHG* 3.472 (Ath. 229c), which refers to ancient Greek custom, and Diodorus Siculus 5.28.4, where customs of the Gauls are seen in the heroic light of Homeric custom.
2. "The Cult of Heroes," in *Essays on Religion and the Ancient World,* vol. 2 (Oxford 1972), p. 588. On these communal meals see the literary parallels and observations presented by W. Bauer in his introduction to his *A Greek-English Lexicon of the New Testament,* trans. W. F. Arndt and F. W. Gingrich (Cambridge/Chicago 1957), pp. xxiii–xxiv. On the sacrificial dimension and divine presences, see D. Gill, *"Trapezomata:* A Neglected Aspect of Greek Sacrifice,'' *Harvard Theological Review* 67 (1974) 117–37. Also W. Burkert, *Greek Religion,* trans. (Oxford 1985), p. 107. And see the note on W. L. Willis' view at the end of the penultimate section.
3. *Hymn to Sarapis* 27, cited in L. Koenen, "Eine Einladung zur Kline des Sarapis (P. Colon. inv. 2555)," *Zeitschrift für Papyrologie und Epigraphik* 1 (1967) 121–26 (p. 122); the invitation reads: "The God invites you to a banquet to be held in the temple of Thoeris tomorrow from 9 o'clock." See also H. C. Youtie, "The *Kline* of Sarapis," *Harvard Theological Review* 41 (1948) 9–29; M. Vandoni, *Feste pubbliche e private nei documenti greci* (Milan/Varese 1964); and Nock's examination of the fellowship in "Cult of Heroes," pp. 585–86.
4. *Leisure and Pleasure in Roman Egypt* (London 1965), p. 49.
5. Macrobius 1.10.22 (citing Philochorus), and compare 1.7.37 (citing Accius); also Ath-

enaeus 639b, who writes of the children taking on these "liturgies." See M. P. Nilsson, *Geschichte der griechischen Religion*, vol. 1 (Munich 1967), pp. 511–13.
6. Cretan slaves feasted alone, Euphorus, *FHG* 1.242 (Ath. 263f); the Sacaea gave slaves mastery over the household, Berosus, *FHG* 2.498 (Ath. 639c); Dio Chrysostom 4.65–70, but cf. Strabo 11.8.4–5; in Arcadia masters feasted with their slaves, Harmodus, *FHG* 4.411 (Ath. 149c); Theopompus, *FHG* 1.319 (Ath. 149d). In addition to the Saturnalia, masters feasted slaves at the Hermaea of Crete, Carystius, *FHG* 4.359 (Ath. 639b: "the masters carried out all the duties relating to administering the banquet [οἱ δεσπόται ὑπηρετοῦσιν πρὸς τὰς διακονίας]"), and at a festival of Poseidon in Troezen, Ath. 639c. For another and fuller treatment of material in this and related sections, see W. L. Willis, *Idol Meat in Corinth* (Chico, Calif., 1985), pp. 7–64, which is part I: "Sacrifice, Cultic Meals, and Associations in Hellenistic Life"; but note the reservation about the tenor of his argument in the last note to the penultimate section, "The God Took Joy."

A Ritual Fit for a King

1. *FHG* 4.349–350 (Ath. 639e–640a). On the aetiology of this cult see M. P. Nilsson, *Griechische Feste von Religiöser Bedeutung* (Leipzig 1906), p. 37; also F. M. Heichelheim, "On Athenaeus XIV, 639e–640a," *Harvard Theological Review* 37 (1944) 351; "Zeus Peloris," *HTR* 40 (1947) 69–70.
2. See comment on ὑπηρετεῖν in relation to Plato, *Plt.* 289c–290b; Epictetus, 3.22.95; and the *Charta Borgiana*, *SB* 5124, n. 6.
3. The first general account of this inscription was published by O. Hamdy-Bey, *Le Tumulus de Nemroud-Dagh* (Constantinople 1883). The inscription was edited by (K. Humann and) O. Puchstein, *Reisen in Kleinasien und Nordsyrien*, Textband (Berlin 1890), pp. 272–78. For commentary see H. Dorrie, *Der Königskult des Antiochos von Kommagene im Lichte neuer Inschriften-Funde* (Göttingen 1964); see also H. Waldmann, *Die kommagenischen Kultreformen unter König Mithradates I* (Leiden 1973), for the complete epigraphical sources (pp. xix–xx). On Antiochus's claims to divinity, see Puchstein, pp. 338–40 (p. 339: "for his realm he is the only divine manifestation; he is also the embodiment of the Commagenian pantheon"). Waldmann (pp. 208–9) speaks of "God-king" ("Gottkönig") as distinct from "Almighty God" ("Hochgott"). Dorrie (pp. 224–30) stresses the originality of his claim in comparison with those of other Hellenistic kings ("of the same nature" [ὁμοουσία] as against "of like nature" [ὁμοιουσία]), and (pp. 138–83) describes the language of the inscription as elevated ("Hochsprache"), noting its solemnity ("gravitas") and the absence of the vernacular (koine).

Temple Liturgies

1. Havercamp in the introduction (p. 10) to Hudson's Amsterdam edition of 1726: "Imprimis notandum est, auctorem nostrum sollicite cavisse ne narrationes de rebus sacris minus placeant Gentilibus."
2. LXX 1 Esd. 1:4,13: θεραπεύετε τὸ ἔθνος . . . ἀπήνεγκαν πᾶσι τοῖς ἐκ τοῦ λαοῦ; 2 Chron. 35:3: λειτουργήσατε . . . τῷ λαῷ. Here as elsewhere it is not merely a matter of Josephus preferring a term with an established place in Greek religious language but also of the authors of the Greek Bible avoiding it.
3. *AJ* 7.363–67 compresses four chapters of 1 Chronicles (23–26) so that for this instance a parallel is difficult to establish. 1 Chronicles frequently uses λειτουργ-, although

δουλεία occurs as a variant at 25:6 in reference to musicians. δουλεύειν appears also in *AJ* 7.367 for the service that priests and levites owe to God day and night (otherwise the word generally refers to political servitude) and thus contrasts with the liturgical sense of διακονεῖν.

4. It is seen again in the phrase that describes the rings attached to the sanctuary's curtain "serving for" (διακονούντων πρός, *AJ* 3.128) its opening and shutting. The English "serving for" misses the connection with sacred ministry that the Greek makes here; the verb is not elsewhere predicated of inanimate objects except in the abstractions of the philosophers (so Themistius, *de An.* 42.6: the body's organs "operate for [πρός] the purpose of" sustaining life). The Bible (Ex. 26 and 36) provides no parallel to Josephus's phrase, and the Latin (F. Blatt, *The Latin Josephus*, vol. 1 [Copenhagen 1958]) does not represent it.

5. That Josephus has temple ministry in mind is further clear from his use of a conventional word for the sexual act (ὁμιλιῶν). See earlier on the agent sense in the same story at *AJ* 18.70, 77.

6. The biblical narrative speaks of the Nazirite vow (LXX 1 Kings 1:11), but it is doubtful how closely Josephus depends on that source, Marcus (Loeb vol. 5) seeing signs of an Aramaic source for the account of Samuel's birth. A vow is a devotional act, and would be of less moment to the Greek than dedication to cult after the manner of a temple initiate.

7. The editors translate an inscription from Doura Europos (*Doura* 875; 181/2CE): "Solaeus . . . and Gornaeus . . . built the peristyle and the wine cellar for the rites of Adonis at their own expense in performance of a vow for the use of themselves [εἰς διακονίαν αὐτοῖς, 875.11] and their descendants in perpetuity." The meaning (of some poor Greek: "descendants" is nominative) is rather "for ministry [of pouring wine] in the rites of Adonis"; the editors themselves suggest that "for themselves" could depend on the verb. *P. Fouad* 25v.i.1 reads δια]κονία θεῶν but lacks context.

8. The few passive instances of the verb are closely related to such ritual usage, occurring in statements about the sacred or special character of utensils or attendants. Thus Belshazzar "was ministered to [διηκονεῖτο, *AJ* 10.242] from God's vessels"; at the feast of "the hundred and fourscore days" presented by Artaxerxes, "they were ministered to [διηκονοῦντο, *AJ* 11.188] from cups of gold"; in praise of the simple life Juba claimed that until the time of Macedonian rule "diners were served [διηκονοῦντο, *FHG* 3.472] in earthenware"; Diodorus Siculus pays attention to the age of attendants among the Gauls, "they are served [διακονοῦνται, 5.28.4] by very young children, boys and girls of a suitable age"; Timaeus similarly was considering the type of attendant, noting "it was not customary for Greeks in the early times to be served [διακονεῖσθαι, *FHG* 1.207] by bought slaves"; the reference here is possibly wider than service at table.

More Lasting than Bronze

1. The inscriptions and the places where they were found are the following (for bibliographical details, see the list of abbreviations); in Troezen, Argolis, southeast of Corinth: *IG* IV, 824a; 774; at various sites in Acarnania, on the west coast: *IG* 2d ed., IX, 247; 248; 250; 251; 252; 451; *CIG* add., 1793b; *CIG* 1800; at Magnesia, Lydia: *IMag.* 109; 217; at Metropolis, Lydia: *CIG* 3037; at Cyzicus, Propontis, on the Sea of Marmara: *Michel* 1226. *MB* 93 and 100 are cited from Guerra, "Diáconos helénicos y bíblicos," *Burgense* 4 (1963) 54–55, after Μυσεῖον καὶ βιβλιοθήκη τῆς εὐαγγελικῆς σχολῆς 2, nos. 2-3 (1876–1878) 93 and 100. In all cases the emended text is cited.

2. *The Organisation of the Early Christian Churches* (London 1881), p. 50.
3. "Zur altchristlichen Verfassungsgeschichte," in *Kleine Schriften*, vol. 1 (Berlin 1958 = *Zeitschrift für wissenschaftliche Theologie* 1914), pp. 148–49. Similarly Lemaire, *Les ministères aux origines de l'église* (Paris 1971), p. 33; Beyer, *TDNT* 2 (1964) 92, although he is influenced by his view of the earliest meaning of the verb: "If the inscriptions teach us anything, it is that the original meaning of διακονεῖν ('to wait at table') persisted."
4. *Geschichte des griechischen Vereinswesen* (Leipzig 1909), pp. 391–92: "The diakonos could have filled roles of quite different kinds; the fact nonetheless that this designation for assistants to priests in public and private cult was widely dispersed suggests the possibility that the Christian title of deacon originated in the pagan title." See also pp. 42 and 534. In a note to p. 392 he is critical of the much narrower view of Hatch, *Organisation*.
5. E. Ziebarth, *Das griechische Vereinswesen* (Leipzig 1896), p. 153; G. Thieme, *Die Inschriften von Magnesia am Mäender und das Neue Testament* (Göttingen 1906), pp. 17–18; and the lexicons, s.v., *LSJ*, Moulton-Milligan, Bauer *AG*.
6. "Diáconos helénicos," pp. 51–56; p. 55: "It becomes very difficult, not to say impossible, to knit together pieces of evidence which might have the same epigraphical character and sometimes present the same words when one attempts to bring down to a common denominator the several diakonoi of these inscriptions which are few in number and, on top of that, are incomplete and damaged by age."
7. The title occurs once before the cook, *IG*, 2d ed., 252; twice after the herald (κῆρυξ), to whom an older tradition allotted the ritual slaying (cf. Athenaeus 660a; Poland, *Geschichte*, p. 395), *IG* 824; *IMag*. 217; otherwise immediately after the cook, *IG* 774; *IG*, 2d ed., 247; 248; 250; 251; 451; *IMag*. 109; *CIG* add. 1793b.
8. Cf. P. Le Bas and W. H. Waddington, *Voyage archéologique*, Deuxième partie, Mégaride et Péloponnèse (Paris 1847), p. 86; P. Foucart, *Des associations religieuses chez les grecs* (Paris 1873), pp. 32–33: "1. In these appointments there was no hierarchy; all the appointments are annual, one is independent of the next, and they are made by the assembly. 2. There is no distinction between civil and religious functions; the same individual was successively treasurer and priest [in a particular case]. . . ."
9. As suggested by E. Legrand, "Inscriptions de Trézème," *Bulletin de Correspondance Hellénique* 17 (1893) 121, and implied by Poland, *Geschichte*, p. 391.
10. An exception could be *CIG* add. 1849c; otherwise (the list is not complete) see *CIG* 1243; 1256; 1271; *IG* XII, 7; *Bulletin de Correspondance Hellénique* 9 (1885) 513, no. 4; *Michel* 990; also *CIG* 2416, where the title stands at the end of a club of gymnasts to indicate its secretary, and *OGI* 352, where his duty is to προγράψαι, in the manner of a γραμματεύς of an association who is commonly instructed to ἀναγράψαι an inscription (the latter meaning more precisely "inscribe," the former "write up"), e.g., Foucart, *Des Associations*, nos. 4.16–17; 5.23–24. The title is used in a similar way of the *hazan* of the Jewish synagogue in Rome, *CIJud*. 172 (see J. B. Frey's comments there, p. xcix); in a later period this synagogue official is called a διάκονος, but probably under Christian influence: Epiphanius, *Haer*. 30.11 (*PG* 41.424): "Among them hazzanim are understood as διακόνων or ὑπηρέτων"; *IGLS* 1321 (Apamea, 391 CE): ἀζζάνα καὶ τοῦ διάκονος ("also called"). The Christian term does not appear among terms for offices in the Jewish synagogue in Egypt; see P. E. Dion, "Synagogues et temples dans l'Egypte hellénistique," *Science et esprit* 29 (1977) 45–75. An almost exclusively clerical denotation will be noted of ὑπηρέτης also in the papyri.
11. For example, *Michel* 1564.
12. See M. San Nicolò, *Ägyptische Vereinswesen zur Zeit der Ptolemäer und Römer* (Munich

1915); see n. 3 in the section above, "Making Friends." The only comparable instance from the papyri is *P. Mil. Vogliano* III, 188.i.15, as discussed in chapter 8.
13. Poland even found titles of some of the higher officials quite childish in their boastfulness ("Äusserungen kindischer Prahlsucht"); see *Geschichte*, p. 411.
14. "Diáconos helénicos," pp. 52–53. F. Dunand also allows that the same group may have been attached to a temple, in *Le culte d'Isis dans le bassin oriental de la Méditerranéen*, vol. 3 (Leiden 1973), p. 162 n. 3.
15. Notably at the Sarapieion of Delos, cf. P. Roussel, *Les cultes égyptiens a Delos du iiie au i^{er} siècle av. J.-C.* (Paris/Nancy 1915–1916), pp. 253–55; note the regularity of their offerings in the year 156/155 BCE (pp. 213–26) and the inscription no. 26 (*IG* IX, 1217) commemorating a banquet in honour of the god which was held by οἱ θεραπεύοντες ἐν τῷ ἱερῷ ("those serving in the temple"); the banquet was served by a temporary epimelete (p. 85).
16. See Pauly-Wissowa, suppl. 4, cols. 914–41, esp. 914–18 (Kornemann); Poland, *Geschichte*, pp. 163–67.
17. *Geschichte*, p. 165 (with other examples); cf. p. 42.
18. See the example in n. 15 above.
19. As suggested by E. W. Hasluck in notes on the municipal status of the "borough," in "Unpublished Inscriptions of the Cyzicus Neighbourhood," *Journal of Hellenic Studies* 24 (1904) 21–23.
20. This spelling of the noun is unique in the inscriptions but common in the papyri. See the second last note in chapter 9.
21. *IGB* 1487 reads --- ης διακονος . . . , the editor suggesting a proper name. Unique in inscriptions, this use will be discussed in regard to the papyri.

CHAPTER 8

"Authorised Representative"

1. *Die Gegner des Paulus im 2. Korintherbrief* (Neukirchen-Vluyn 1964, pp. 31–38; Eng. trans. (with new epilogue and bibliographies), *The Opponents of Paul in Second Corinthians* (Philadelphia 1986), pp. 27–32.
2. *Opponents*, p. 28. In this phrase "authorised" is for the German "vollmächtigen," which in this context I would think of as "plenipotentiary."
3. See W. Schmithals, *Das kirchliche Apostelamt* (Göttingen 1961), pp. 100–3, with other literature there.
4. G. Bornkamm, "Die Vorgeschichte des sogenannten Zweiten Korintherbriefes," Sitzungsberichte der Heidelberger Akademie der Wissenschaften, Philosophisch-historische Klasse (1961), pp. 15–16 (in reference to Georgi's unpublished dissertation, as is the next); G. Friedrich, "Die Gegner des Paulus im 2. Korintherbrief," in Festschrift for O. Michel, *Abraham unser Vater* (Leiden/Köln 1963), pp. 195–96, 205, 210–12; W. Schmithals, *Die Gnosis in Korinth* (Göttingen 1965), p. 196; J. Gnilka, *Der Philipperbrief* (Freiburg im B. 1968), p. 39; idem, "Geistliches Amt und Gemeinde nach Paulus," in *Foi et Salut*, by M. Barth et al. (Rome 1970), pp. 244–45; H. H. Koester, "One Jesus and Four Primitives," *Harvard Theological Review* 61 (1968), p. 233; H. Conzelmann, *Der erste Brief an die Korinther* (Zurich 1969), p. 17. In noticing this and other literature, G. Barth has emphasised the pivotal role which Georgi's (and Bornkamm's) understanding of the opponents has played in subsequent scholarship, "Die Eignung des Verkündigers in 2 Kor 2,14–3,6," in Festschrift for G. Bornkamm, *Kirche* (Tübingen 1980), pp. 257–60.

5. The text has been edited by E. Bethe (1900–1937; repr., Stuttgart 1967). Bethe points out that Pollux's works perished ("not even a scrap of the original onomasticon is known to me," he says in his Latin introduction, p. v), but a digest of this work survived which, with the tenth-century annotations of Arethas, formed the basis of the subsequent textual tradition and was further enriched by later Byzantine scholiasts. Bethe's text shows the glosses (unlike the editions of I. Bekker, 1846, and W. Dindorf, 1824, and the phrases cited in Stephanus), and these include instances of our words at 4.28; 8.115 (cited by Georgi); 8.138 (cited by Stephanus); they are accordingly omitted from the discussion in the text. Outside glosses, instances occur at 3.76 (the common noun in a section on masters and slaves); 7.210 (a citation of Plato, *Plt.* 290d—not *Grg.* 518a, as indicated by Bethe); and 8.137, which is discussed in the text.
6. *De legationibus graecorum publicis* (Leipzig 1885), pp. 9–12. Some discussion of terminology is also provided by D. J. Mosley in *Envoys and Diplomacy in Ancient Greece* (Wiesbaden 1973) and in the same author's contribution to F. Adcock and D. J. Mosley, *Diplomacy in Ancient Greece* (London 1975).
7. Note the force of δέ and especially of καί. The writer of a gloss seems to have taken a lead from the shift of meaning here to extend the consideration of usage even further at this point in instancing designations of ritual officers (ἑτέρας δὲ χρείας κῆρυξ καὶ σπονδοφόρος: "other uses again are 'herald' and 'bearer of libations' ") who functioned as symbolic guarantors of the diplomats' integrity under God; on heralds see Poland, *De legationibus*, pp. 13–14; Mosley, *Envoys*, pp. 84–89; Adcock and Mosley, *Diplomacy*, pp. 152–54.
8. "πρεσβεύω," *TDNT* 6 (1968) 682.
9. The terms are not discussed in Poland, *De legationibus*; Mosley, *Envoys*; or H. J. Mason, *Greek Terms for Roman Institutions* (Toronto 1974).
10. Demosthenes 19.69; Lucian, *Icar.* 20; Philo, *de Ab.* 115, the two latter referring to messages to or from the gods.
11. In Demosthenes 18.311 the juxtaposition of the terms is purely rhetorical, πρεσβεία designating Aeschines' embassy to Philip and διακονία designating civic undertakings that Aeschines had himself called variously ἐπιμέλεια, πραγματεία, διακονία (3.13,16). Plutarch, *Mor.* 794a, designates civic duties within the state διακονικὰς λειτουργίας, these being tax farming and the supervision of harbours and markets, and contrasts them with deputations to foreign parts, πρεσβείας καὶ ἀποδημίας πρὸς ἡγεμόνας. Most significant of any is Priscus, *HGM* 276, where both terms relate to exchanges between states, the one in a diplomatic sense, the other in a nondiplomatic sense. See the earlier discussion of these passages.

Stoic Attitudes

1. *The Opponents of Paul in Second Corinthians*, Eng. trans. (Philadelphia 1986), p. 29.
2. On the revived Stoic interest in Cynicism in the first and second centuries CE, see D. R. Dudley, *A History of Cynicism* (London 1937), pp. 187–201.
3. See A. Bodson, *La morale sociale des derniers Stoiciens* (Paris 1967), pp. 41–44. Citations from Epictetus in the following are from W. A. Oldfather, *Epictetus*, Loeb vol. (London 1926–1928).
4. The originality of Epictetus's teaching here has been emphasised by A. Delatte, "Le sage-témoin dans la philosophie stoic-cynique," *Bull. Acad. Roy. Belg. cl. Lettres* 29 (1953) 166–86; cf. J. Souilhé, *Epictète*, Budé vol. 1 (Paris 1948), p. lvii. Delatte notes that not even ἀπολογεῖσθαι implies oral witness (p. 180).

5. Here as elsewhere in Epictetus κηρύσσειν has an almost derogatory quality (1.16.12; 4.5.24; 4.6.23; especially 4.8.28). Otherwise it is used only of the acclaim of the Olympic champion (1.2.26) and of God's proclamation of peace (3.13.12), which is after the manner of an imperial decree.

Civil Agents and Servants in God's Commonwealth

1. See the earlier discussion of these passages.
2. δουλεία contrasts sharply with ὑπηρεσία in being bondage to the external and in keeping man from following God, e.g., 2.16.41; 4.1.152. δοῦλος is generally derogatory, e.g., 3.22.40; 3.23.71.
3. This distinction was not observed in my article, "Georgi's 'Envoys' in 2 Cor 11:23," *Journal of Biblical Literature* 93 (1974) 88–96, and I consequently misinterpreted 3.22.69; the error is corrected in the exposition which is to follow. (The discussion above of Plato, *Plt.* 289c–290b, has presented another interplay between the two sets of words.) My error in this no doubt facilitated Georgi's judgement (*Opponents of Paul in Second Corinthians,* Eng. trans. [Philadelphia 1986], p. 352) that my "limiting of the meaning of διάκονος/διακονία in Epictetus to 'servant/service' is completely unjustified."
4. See other similar uses of this word group at 3.24.36, 113, 114; 4.7.30.
5. W. A. Oldfather, *Epictetus,* Loeb vol. (London 1926–1928), translates "servants."
6. Oldfather translates "service."
7. Oldfather translates, "He must share with him his sceptre and kingdom, and be a worthy ministrant," but notes that "sceptre" could be "staff," as at 3.22.57.
8. Thus there is no Cynic terminology of διάκονος to contribute to a sketch of Hellenistic religious propagandists. Georgi (*Opponents,* p. 352) has pointed out that material I have presented strengthens the idea of "messenger"—and it clearly does, although, as shown in the text, definitely not in the case of Epictetus—and he suggests he is satisfied with that kind of meaning: "It is not the status but the function that is the major issue for me here." But I am not convinced that this holds true for what he has written. We read in translation frequent phrases about "divine messenger[s]" (p. 156), about their "godlike function" (ibid.), about "the spirit, the models and the standards which the opponents of Paul appealed to and by which they were to be measured" (p. 164), about the envoy "in the sense of responsible, fateful representation and manifestation" (p. 29)—all of these saying a lot about status beyond the mere function of transmitting a message—and the link words for Georgi between Paul and this mysterious world are διάκονος and διακονία (e.g., pp. 245, 299 n. 173). The former word has to be removed from that connection, and the latter, of which Georgi misleadingly writes "διακονία in Hellenistic times frequently meant responsible, fateful representation" (p. 279), is now reviewed in the section to follow.

The Cynic's Mission

1. ἄγγελος (messenger): 3.1.37 (of Hermes: see the earlier note on this passage; it is striking that Epictetus does not use διάκονος even in this connection); 2.23.3 (of the eye); 3.22.23,38 (of the Cynic). κατάσκοπος (scout): 3.22.28 (of the Cynic); 1.24.3–10 (of Diogenes); the cognate verb appears at 3.24.31 in the image of life as a battle; κῆρυξ (herald): 3.21.13 (attached to a shrine); on the cognate verb, see the note on 1.29.64.

2. παρρησιάζεσθαι. This notion is not prominent in Epictetus; the abstract noun occurs only in fr. 36 (of truth), and is used by Arrian of Epictetus himself, *G* 2.
3. Georgi (*Opponents of Paul in Second Corinthians,* Eng. trans. [Philadelphia, 1986], p. 220 n. 471, takes κύριε as a Cynic title here ("Lord")—which must be to stretch a point—rather than as a term of address ("Sir").
4. *P. Lond.* 878; Achilles Tatius 4.15.6. These passages instance delivery and proclamation of a sacred message; similar usage in nonreligious affairs occurs at Isaeus 1.23; Plutarch, *Mor.* 50c; 62d; 64e; Josephus, *AJ* 12.188; Thucydides 1.133; Alciphro 21.2. Instances of task or commission without connotation of movement occur at Josephus, *AJ* 5.349 (in sacred affairs) and, in the political arena, Plato, *R.* 493d; Dio Cassius 54.21.4; Aeschines 3.13; Demosthenes 18.206, 311; Josephus, *AJ* 19.129. The instance above does not designate just a service of or to God, as in A. Bonhöffer, *Epiktet und das Neue Testament* (Giessen 1911), p. 322, although the Cynic's mission is an expression of the service that, as Bonhöffer observes (p. 159), characterises all Stoic life.

CHAPTER 9

1. The index refers to earlier discussion of the verb in prayers: *P. Warren* 21 and *SB* 4947; in a contract: *P. Oxy.* 275; in a petition: *UPZ* 18 and 19; and in two personal letters: *BGU* 261 and *P. Oxy.* 3313. (Also above *P. Heid.* 184, but this is a literary fragment.) The abstract noun occurs in a religious connection: *P. Fouad* 25.
2. *Church Order in the New Testament,* Eng. trans. (London 1961), p. 176.
3. *The Vocabulary of the Greek Testament Illustrated from the Papyri and Other Non-Literary Sources* (London 1930), s.v.: "There is now abundant evidence that the way had been prepared for the Christian usage of this word by its technical application to the holders of various offices."
4. "Diáconos helénicos y bíblicos," *Burgense* 4 (1963) 42 n. 119.
5. *Les ministères aux origines de l'église* (Paris 1971), p. 33 n. 122.
6. "Zur altchristlichen Verfassungsgeschichte," in *Kleine Schriften,* vol. 1 (Berlin 1958), p. 149.
7. For abbreviations of papyrological publications, see the list of abbreviations. References are made in the following manner, which may not always be the style of the edition cited: capital Roman numerals (III) indicate volume; the first Arabic numeral (118) indicates the number of the papyrus, the small Roman numeral (i) indicates the column, and the second Arabic numeral (11) the line; *v* and *r* indicate verso and recto. In citations resolved readings are given unless original abbreviations, lacunae, etc., affect interpretation.
8. M. Vandoni (the editor) in *Acme* 15 (1962) 140–41.
9. See W. Otto, *Priester und Tempel im hellenistischen Ägypten,* vols. 1 and 2 (Leipzig/ Berlin 1905–1908): this work covers literature, inscriptions, and papyri; also, *Beiträge zur Hierodulie im hellenistischen Ägypten* (Munich 1949); P. Bottigelli, "Reportorio topografico dei templi e dei sacerdoti dell' Egitto tolemaico," *Aegyptus* 21 (1941) 3–54; 22 (1942) 177–265; H. Kees, *Das Priestertum im ägyptischen Staat* (Leiden 1953).
10. *P. Hib.* 226 fr. 3.2 (third century BCE) is too tiny a scrap to provide sense; it reads [δ]ιακονον, and is said by the editor to be "perhaps from an ἐρωτικὸς λόγος," which would align the instance with other instances in romantic literature.
11. *Charta papyracea graece scripta musei Borgiani Velitris* (Rome 1788), p. 72: "Vocem διακων e versione omisi, nam quid hic significet, definire nequeo: saltem de diaconis

christianis, ut spero, nemo sanus cogitabit." Among modern dictionaries and onomasticons, *WB* indexes the word under "Ämter, Beamte" (offices, officials); *NB* includes two instances (*BGU* 1046; *SB* 5662) as names, *OAP* adding three others (*P. Bour.* 42; *P. Berl. Leihg.* 8; *SB* 7515). Among editors Lesquier writes lower-case δ instead of the upper-case Δ printed in *BGU* 1046, Kalen also noting the lower case as an alternative for *P. Berl. Leihg.* 8. Upper or lower case is printed here according to edition cited.

A Personal Name

1. See R. Calderini, "Ricerche sul doppio nome personale nell' Egitto greco-romano," *Aegyptus* 21 (1941) 221–60 (a summary of the formulas is given on pp. 221, 224); 22 (1942) 3–45 (the chart on p. 5 shows eleven hundred instances of double names from the second century).
2. See W. Swinner, "Problèmes d'anthroponymie ptolemaïque," *Chronique d'Egypte* 42 (1967) 156–71, esp. 170–71.
3. On ε for α, see F. T. Gignac, "The Language of the Non-Literary Greek Papyri," *Proceedings of the Twelfth International Congress of Papyrology* (Toronto 1970), p. 143.
4. Repeated interlinearly at xiv.332. Other examples are *P. Berl. Leihg.* 8.6 (165 CE): Πᾶσεις Διακ(ονος); *SB* 9370.iii.22 (ca. 150 CE): ὁ καὶ φόμβω(νος) Διάκ(ονος); *SB* 7515.663 (155 CE): καμῆς Διάκ(ονος).
5. *O. Mich.* 1046.1 (third century CE): Πεταῦς διάκων. (Four receipts in this collection, nos. 1037–48, show patronyms.) *O. Mich.* 782.2 (fourth century CE): Πετρόνις (2) διάκον(ος). *P. Mich.* 596.5 (fourth century CE): Ἀμάεις (5) διάκων. (Other names in this and a second receipt issued by the same officer are written with case endings.) *SB* 5662.29: [. . .]ις Διάκων. (Nearby names in this list do not show case endings.)
6. See the solitary possible example *IGB* 1487: ---ης διακονος, the editor suggesting a proper name.
7. My class was taught the Greek verb at school by a teacher named McCowage whom we took to be of the same Irish background as the teacher who had not succeeded at the same task. Twenty years later he explained to me that his father had been a Polish migrant who found he could not compete for work on the Sydney docks during the great depression until he changed his name from Mackiewicz to an equivalent Irish form.
8. See the permutations in Calderini, "Ricerche," pp. 23–40.
9. "Callimachus in the Tax Rolls," *Proceedings of the Twelfth International Congress of Papyrology* (Toronto 1970), pp. 545–51.
10. "Callimachus," pp. 547–48.
11. *Die Personennamen der Kopten* (Leipzig 1929), p. 69, where examples are also given.
12. Ibid., p. 1.
13. Ibid., pp. 28, 33, 63, 69.
14. Ibid., p. 95.
15. The style is recognisable in *P. Strasb.* 333r.12 (second century CE): Μῶρος διάκων Πκι, which is either a triple name (on these see Calderini, "Ricerche," pp. 254–55) or, like some others in the list that have Egyptian characteristics, conceals a patronymic. Similarly *P. Baden* II, 31.25 (Rom.-Byz.): Παῦλος διάκονος, in a list of names, including Egyptian, of residents of a house. *P. Dura* 50r.1 (third century CE) and *P. Lond.* II, 266.267 are too fragmentary or abbreviated to be sure about; in the latter case, at least, avocations are not mentioned in connection with other names. Traces of the name surface in later Coptic remains; see W. E. Crum and H. G. Evelyn White,

The Monastery of Epiphanius at Thebes, pt. II, *Coptic Ostraca and Papyri* (New York 1926), p. 205, n. 192 with the corrigendum of p. xvi, "Pdiacon may, then, be a name." Also W. C. Till, *Datierung und Prosopographie der koptischen Urkunden aus Theben* (Vienna 1962), p. 77 (seventh to eighth centuries CE). Compare the Roman cognomen from an undated inscription, M. TITIUS DIACONUS, in H. Forcellini, *Lexicon totius latinitatis,* vol. 2 (Padua 1940), s.v.

A Coptic Nickname

1. See lines 172, 231, 300, 321, 355.
2. Viereck *(SB)* identifies two, Wilcken *(WO,* p. 340) only one at lines 267, 372.
3. So Wilcken, *WO,* pp. 339–40.
4. *WO,* p. 683.
5. "Verwalter," *WO,* p. 341.
6. On ὑπηρέτης in the bureaucratic language of the papyri, see F. Oertel, *Die Liturgie: Studien zur ptolemäischen und kaiserlichen Verwaltung Ägyptens* (Leipzig 1917), p. 412; N. Hohlwein, *L'Egypte romaine: Recueil des termes techniques relatifs aux institutions politiques et administratives* (Brussels 1912), p. 412; F. Preisigke, *Fachwörter des öffentlichen Verwaltungsdienstes Ägyptens in den griechischen Papyrusurkunden der ptolemäisch-römischen Zeit* (Göttingen 1915), s.v.; P. M. Meyer, *Juristische Papyri* (Berlin 1920), s.v.; H. Kupiszewski and J. Modrzejewski, "ΥΠΗΡΕΤΑΙ. Etude sur les fonctions et le rôle des hyperètes dans l'administration civile et judiciaire de l'Egypte greco-romaine," *Journal of Juristic Papyrology* 11–12 (1957–1958) 141–66; A. I. Pavlovskaja, "Die Sklaverei im hellenistischen Ägypten," in *Die Sklaverei im hellenistischen Staaten im 3.–1. Jh. v. Chr.,* ed. T. V. Blavatskaja et al., Ger. trans. (Wiesbaden 1972), pp. 231, 236–37; W. Peremans and E. Van 't Dack, *Prosopographia Ptolemaica,* vols. 1–6 (Louvain 1950–1968), nos. 390, 1441, 1748, 1749 (all civil administration and finance; *P. Leit.* 5.38 may be added); 2432–52 (military). The latter work, for its period at least, puts the technical use of the word into perspective, for in nos. 1–1824, for example, in reference to subordinate officials, there is an overwhelming predominance of the formula ὁ παρὰ τοῦ δεῖνα; while having its technical applications, the word would not have been such standard terminology as Kupisjewski and Modrejewski perhaps imply. The same inference is probably to be drawn from the local rarity of the word at Oxyrhynchus; see P. Mertens, *Les Services de l'Etat Civil et le Controle de la Population à Oxyrhynchus* (Brussels 1958). Preisigke's definition in *Fachwörter,* s.v., would seem to be adequate: "a civil servant, but often enjoying personal responsibility in public matters and so better called 'assistant officer' who might be engaged in the broadest range of duties." We can compare the English expression, "local government officer." The word thus smacks of bureaucracy while διάκονος retains a brush of the abstract and a touch of poetry.

Odd Jobs

1. βάληι: it is difficult to know what this word means here. In *Papyrusbriefe aus der Frühsten Römerzeit* (Uppsala 1925), p. 137, B. Olsson translated "drive" ("treibt"), which is rare: only in poets, according to *LSJ,* and not attested elsewhere in the papyri. The sense "pay for" would require ὑπέρ.
2. No cattle appear in the documents collected by S. Avogadro, "Le ΑΠΟΓΡΑΦΑΙ di proprietà nell'Egitto greco-romano," *Aegyptus* 15 (1935) 131–206 (see esp. 193–99);

see further H. G. Gundel, "Der Gaustratege Hierax," *Chiron* 1 (1971) 319–24 ("calf" occurring only *BGU* 356; *P. Oxy.* 245).
3. Olsson, *Papyrusbriefe,* p. 137: "sprich mit dem Diener Petheus."
4. The normal words are νομεύς and ποιμήν.
5. "Le ΑΠΟΓΡΑΦΑΙ," p. 194.
6. See D. Comparetti in *P. Flor.* II, pp. 41–42, 59.
7. See for example N. Hohlwein, *L'Egypte romaine* (Brussels 1912), p. 133.
8. *P. Giss.* 101 (third century CE) presents a similar association of terms with three individuals listed as managers (φροντιστής; προνοητής) and one as (line 10) 'Ερμάμμωνι διάκονι. The role of this last person, however, is not clear; it is not necessarily to be related to that of the managers because individuals with other avocations are also listed (a cook, two vineyard workers; the editor also identifies a slave) and this could be just one more among them. The papyrus is p. 33 of a housekeeping book, the rest of which has not survived, and lists outgoings of a small landowner or estate manager for the first eight days of a month; some of the payments went to the persons listed. The missing words at line 10 are perhaps crucial for determining the avocation of the person there; without them one might be inclined to see the person as a member of some estate's household staff.
9. In the translation by T. C. Skeat in *P. Lond.* VII.
10. *A Large Estate in the Third Century B. C.* (Madison 1922), pp. 16–21. Skeat was not convinced; see *P. Lond.* VII, p. 200.
11. The name "Pindaros" does not otherwise occur in W. Peremans and E. Van 't Dack, *Prosopographia Ptolemaica,* vol. 5 (Louvain 1963).
12. As described—albeit ironically—by Plutarch, *Mor.* 50e and so designated 63b; Herodotus 4.71.4; Josephus, *AJ* 8.354.

Mistaken Identity

1. *P. Berl. Leihg.,* p. 358.
2. *P. Col.* V, pp. 98–99.
3. The published photograph of the papyrus (pl. IV) suggests the reading might not be certain; ακω is clear of holes but the pointed letters hardly show at all.
4. From *NB* and *OAP* Κῶς (Tebtynis, second century BCE), Κῶνος (Karanis, second and third centuries CE), 'Ακῶς (Thebes, first century CE) are available. Although there is a problem of space after the ν of the published text, the editors more than once remark on the irregularity of the spacing.

Panopolis

1. "Relevé topographique des immeubles d'une métropole (P. Gen. Inv. 108)," *Recherches de Papyrologie* 2 (1962) 37–73.
2. *Une description topographique des immeubles à Panopolis,* Fr. trans. (Warsaw 1975). The papyrus is cited from the reconstructed columns and lines of this edition; upper case A refers to sections not placed in sequence with the rest.
3. *Une description,* p. 43. The tables there exemplify the difficulty of reaching firm figures.
4. Ibid., p. 13.
5. The property of a man recorded in a document of this year is listed in the survey as belonging to his sons and heirs; Skeat's argument is reported and supported by J. D. Thomas, "Chronological Notes on Documentary Papyri," *Zeitschrift für Papyrologie*

und Epigraphik 6 (1970) 177–81; H. C. Youtie, "P. Gen. Inv. 108 = SB VIII 9902," *Zeitschrift für Papyrologie und Epigraphik* 7 (1971) 170–71.

6. J. van Haelst brackets this with three other documents illustrative of Diocletian's persecution (*SB* 7315; *P. Oxy.* 2601; *P. Lond.* 1920), "Les sources papyrologiques concernant l'Eglise en Egypte à l'époque de Constantin," in *Proceedings of the 12th International Congress of Papyrology* (Toronto 1970), pp. 497–503 (see his nos. 15–18).
7. *The History of Christianity,* vol. 2 (London 1867), p. 222.
8. "P. Gen. Inv. 108 = SB VIII 9902," p. 171 n. 7: "The specimen of P. Gen. Inv. 108 provided in facsimile by V. Martin . . . suggests a date much closer to 299 than to 355."
9. For a review of eastern discipline, see I. Doens, "Ältere Zeugnisse über den Diakon aus den östlichen Kirchen," in *Diaconia in Christo,* ed. K. Rahner and H. Vorgrimler (Freiburg/Basle/Vienna 1962), pp. 31–56. The earliest canons of a synod are from Ancyra (314 CE); older material appears in *Apostolic Constitutions* and similar compilations.
10. *P. Oxy.* 2376 in the translation by Jack Lindsay, *Leisure and Pleasure in Roman Egypt* (London 1965), pp. 243–44.
11. "Relevé," p. 61.
12. The poem is celebrated and analysed within its cultural setting in the concluding chapters of Jack Lindsay's *Leisure and Pleasure.* Kuros is another poet son of Panopolis who, Lindsay writes, "through his poetic talents, rose to high office at Byzantion, but, in the retirement of his patroness Eudokia in 445 CE, became a bishop. . . . and when he became a bishop, he is said to have shown gross ignorance of theology" (pp. 361–62), thus perhaps embellishing the picture of an un-Christian city. See also G. M. Browne, "A Panegyrist from Panopolis," in *Proceedings of the XIV International Congress of Papyrology* (London 1975), pp. 29–33, who notes of the period around 340 CE that "Panopolis was especially famous as a centre of Hellenic culture" (p. 31).
13. *Leisure and Pleasure,* p. 362.
14. *Alexandria: A History and a Guide* (Alexandria 1922; repr., London 1982), p. 80.
15. Canon 15 restricts the number of deacons for even large cities to seven; in practice the prescription may have been waived, see Doens, "Ältere Zeugnisse," p. 39.
16. "Les sources," pp. 500–501.
17. *Une description,* p. 13.
18. H. I. Bell, *Cults and Creeds in Graeco-Roman Egypt* (Liverpool 1953), p. 87; Thomas, "Chronological Notes," p. 179; van Haelst, "Les sources," p. 498; O. Montevecchi, *La Papirologia* (Turin 1973), p. 291.
19. "Chronological Notes," p. 179.
20. "Relevé," p. 66.
21. *Une description,* pp. 72–73.
22. Reviewing Martin in *Chronique d'Egypte* 38 (1963) 165. She accepted Martin's early date.
23. *Pauly-Wissowa,* s.v. The editors of *P. Oxy.* 43v themselves expressed some reservation in printing "churches (?)."
24. *P. Stud. Pal.* XX, 103 (331 CE); *P. Oxy.* 2344 (336 CE); *SB* 9622 (343 CE); *P. Osl.* 113 (346 CE). The presence of the work $\kappa\alpha\theta o\lambda\iota\kappa\acute{\eta}$ ("catholic") in each of these may mean that the word "church" designates a community rather than a building; the possibility is raised below in connection with *P. Abi.* 55. In ecclesiastical writers of the third century the sense "church building" is rare; see G. W. H. Lampe, *A Patristic Greek Lexicon* (Oxford 1961) s.v.

25. *Manuscript, Society and Belief in Early Christian Egypt* (London 1979), p. 24. See also H. I. Bell, "Evidences of Christianity in Egypt during the Roman Period," *Harvard Theological Review* 37 (1944) 201–2; K. Treu, "Christliche Papyri 1940–1967," *Archiv fur Papyrusforschung* 19 (1969) 169–206; Montevecchi, *La Papirologia*, pp. 296–321; B. F. Harris, "Biblical Echoes and Reminiscences in Christian Papyri," in *Proceedings of the XIV International Congress of Papyrology* (London 1975), pp. 155–60. See as well the so-called Christian letters in collections by G. Ghedini, *Lettere Cristiane dai papiri greci del III e IV secolo* (Milan 1923); his "Paganismo e cristianesimo nelle lettere papiracee greche dei primi secoli d. Cr.," in *Atti del IV Congresso Internazionale di Papirologia* (Milan 1936), pp. 333–50; and M. T. Cavassini, "Lettere cristiane nei papiri greci d'Egitto," *Aegyptus* 34 (1954) 266–82; M. Naldini, *Il Cristianesimo in Egitto* (Florence 1968). Bell underlines the difficulties in using either the letters ("Evidences," pp. 192–99) or the "libelli" of 250 CE (*Cults*, p. 85), and van Haelst ("Les sources," p. 497) is sceptical in regard to both. On churches and monasteries, see L. Antonini, "Le chiese cristiane nell' Egitto dal IV al IX secolo secondo i documenti dei papiri greci," *Aegyptus* 20 (1940) 129–208; P. Barison, "Ricerche sui monasteri dell' Egitto bizantino ed arabo secondo i documenti dei papiri greci," *Aegyptus* 18 (1938) 29–148. The existence of Pachomius's cenobitic foundations in the environs of Panopolis has no direct bearing on the vigor of Christian life in that city; in any case, Pachomius was not baptised until 313 CE, according to D. J. Chitty (*The Letters of St. Antony the Great*, Eng. trans. [Oxford 1980], pp. vii–viii) and his first foundation dates from 325 CE. For later developments, see B. R. Rees, "Popular Religion in Graeco-Roman Egypt. II. The Transition to Christianity," *Harvard Theological Review* 36 (1950) 86–100. And see the several studies in *The Roots of Egyptian Christianity*, ed. B. A. Pearson and J. B. Goehring (Philadelphia 1986).
26. See H. Masurillo, "Early Christian Economy," *Chronique d'Egypte* 61 (1956) 124–34.
27. "Les sources," p. 498; of those listed by him, nos. 2–12, the following are pertinent here: *P. Alex.* 29; *P. Got.* 11; *P. Oxy.* 2603; and two of the letters of Sotas, *PSI* 208 and 1041 (cf. *P. Oxy.* 1492); see as well *P. Gren.* II. 73.
28. τοῖς κατὰ τόπον συνλιτουργοῖς πρεσβυτέροις καὶ διακώνοις.
29. Στεφάνῳ δι[ακόνῳ ἀγ]απήτῳ υἱῷ; the letter concludes with a greeting ἐν κ(υρί)ῳ.
30. *P. Stud. Pal.* XX, 103; *P. Osl.* 113; *SB* 9622; *P. Abi.* 55; *P. Würz.* 16; cf. also *CPR* 227 (undated fourth century). The full phrase is: ὁ δεῖνα διάκονος τῆς καθολικῆς ἐκκλεσίας ἀπὸ κωμῆς. *P. Würz.* omits the words "catholic church," the praescript referring to the deacon as "residing in the village (as) a deacon." The praescript of *P. Abi.* 55 refers to the writer as "deacon from the village of . . . ," and only later (line 14) does the writer say, "for I am a deacon of the catholic church." This is probably an indication of how "catholic" is to be taken in these phrases; the deacon, not the church, is identified with the village, and this could be taken to imply that καθολική means something like "orthodox" rather than, as editors translate, "principal." As "orthodox," the deacons perhaps feel more advantageously placed to win favours from the authorities.
31. Later documents regularly refer to deacons in phrases like "deacon and oilseller," "deacon and breadseller," "deacon and notary."
32. *Une description*, p. 69.
33. "Relevé," pp. 61–63.
34. It is interesting that the form διάκονος occurs in this document from Upper Egypt rather than διάκων of the papyri from the Fayum and elsewhere in Lower Egypt; the

324   Notes

same distribution of forms is to be observed in the text of the Sahidic (Upper Egypt) and Bohairic (Lower Egypt) New Testament.

35. "Relevé," p. 64: "fonctions subalternes pour la plupart sans doute liturgiques"; he aligns the word with terms designating surveyors, inspectors of corn, taxation officers.

PART III

1. *Church Order in the New Testament,* Eng. trans. (London 1961), pp. 173–78.

CHAPTER 10

The Apostle as Spokesman

1. The ὑπηρέται of 4:1 is a distinct image. Again, "servants" *(RSV)* is not adequate, no more is "ministers" *(RV)*; "subordinates" *(NEB)* is closer to the administrative personnel in mind, as required by the word's link with οἰκονόμοι, "stewards." Suitable is "officials."
2. The advantage Paul seeks to obtain by the use of the title here and at 3:6 may be compared particularly with that taken by Josephus at *BJ* 3.354; 4.626.
3. Some of the more recent literature is cited previously in the section "Authorised Representative," n. 4, of chapter 7. In addition see C. K. Barrett, "Christianity at Corinth," *Bulletin of the John Rylands Library* 46 (1963–1964) 269–97; "Paul's Opponents in 2 Corinthians," 17 (1970–1971) 233–54, with Barrett's comment on p. 233: "It is not too much to say that a full understanding both of New Testament history and of New Testament theology waits on the right answering of this question." Other literature to 1973 is cited in J. J. Gunther, *St. Paul's Opponents and Their Background* (Leiden 1973). See further, K. Berger, "Die impliziten Gegner," in Festschrift G. Bornkamm, *Kirche* (Tübingen 1980), pp. 273–400; M. E. Thrall, "Super-Apostles, Servants of Christ, and Servants of Satan," *Journal for the Study of the New Testament* 6 (1980) 42–57; V. P. Furnish, *II Corinthians* (Garden City, N.Y., 1984), pp. 48–52, with current bibliography.
4. More or less after the manner advocated by E. A. Judge ("St. Paul and Classical Society," *Jahrbuch für Antike und Christentum* 15 [1972] 19–36): "The solid ground in the problem of 2 Cor. 10–13 is not to be found between the lines. It lies in the actual reaction of Paul as we possess it, especially in its intensity and elaboration" (pp. 34–35).
5. Barrett underlines the dangers ("Christianity in Corinth," p. 270): "a certain and unambiguous interpretation of a particular verse would give one a clear insight into part at least of the Corinthian history; yet only if one has a clear picture of the history is it possible to interpret the verse with confidence. There is a trap here. . . . How easy to make a hurried inference from a text of one of the epistles to historical circumstances, and then to use the supposedly known historical circumstances to confirm the interpretation of the text!"
6. *Apostolat-Verkündigung-Kirche* (Gütersloh 1965), p. 122. In many passages Roloff speaks of "lowly service" as integral to apostleship, e.g., p. 233. Similarly, Furnish, *Corinthians,* p. 533.
7. Cf. Rom. 14:18: "he who serves Christ is acceptable to God and approved by men"; when Apelles is said to be "approved in Christ" (Rom. 16:10), the meaning is that his worth as a Christian is known to the brethren.

8. In addition to the discussion that follows, see Paul's remarks at 10:7; 11:6b; also the "signs" at 12:12, and the largely ironical "Examine yourselves," etc., at 13:5–7.
9. 10:13–18 argues that apostolic authority is not open-ended; God is not so prodigal of evangelisers that they should be commissioned to work over one another's ground.
10. Roloff, *Apostolat*, p. 234; S. H. Travis, "Paul's Boasting in 2 Corinthians 10–12," *Studia Evangelica* 6 (1973) 527–32 (by bringing out Paul's emphasis on the Damascus incident, Travis also supports the analysis of the whole section that has been presented here); Furnish, *Corinthians*, pp. 532–41.
11. So for example D. Georgi, *The Opponents of Paul in Second Corinthians*, Eng. trans. (Philadelphia 1986), pp. 279–80, although the focus for him is less the suffering than the boasting; R. Bultmann, *Der zweite Brief an die Korinther* (Göttingen 1976), p. 219, and "Exegetische Probleme des zweiten Korintherbriefes," *Exegetica* (Tübingen 1967), p. 317.
12. The parallel with "angel of light" also indicates that the genitive "of righteousness" is not objective (as that at 2 Cor. 3:6, "ministers of a new covenant," will be seen to be) but descriptive ("spokesmen from the realm of righteousness").
13. Compare Hecate's emissary in Achilles Tatius 3.18.5; also Charon as συνδιάκτορος of Hermes in Lucian, *Char.* 1.

The Spokesman as Medium

1. See Chariton 8.8.5; Josephus, *AJ* 6.298; *Test. Ab.* 9.24; Lucian, *Icar.* 20; *Cont.* 1; Theophrastus, *Char.* 2.9; *Anacreontea* 15.14. Commentators occasionally adduce Josephus, *AJ* 6.298, after W. Baird, "Letters of Recommendation: A Study of 2 Cor. 3.1–3," *Journal of Biblical Literalism* 80 (1961) 169. The idea of the amanuensis, however, is still common, for example, M. Rissi, *Studien zum Zweiten Korintherbrief* (Zurich 1969), p. 20, and *JB*, "drawn up by us"; see V. P. Furnish, *II Corinthians* (Garden City, N.Y., 1984), p. 182 (who is himself undecided and translates "cared for by us"; cf. *NJB*, "entrusted to our care").
2. The letter in 3:3 is certainly to the Corinthians, and so also surely that in 3:2. The fact that the latter is probably to be understood as written on "our" hearts is entirely consistent with the doctrine underlying the image, for Paul, by reason of his position, has first to receive the word of faith, which, when passed on, is then recognised as his commendation. The Greek perfect participle "written" supports this for Paul never loses the word but holds it, and stands in a permanent relationship to those who have believed it. If "your" hearts is to be read, the doctrine is the same although the metaphor is less consistently applied.
3. The genitive "of the new covenant" is objective; see illustrations in Alexander, *Mixt.* 5.14; Philostratus, *Im.* 1.3.2; Josephus, *AJ* 1.298; 11.255; *BJ* 4.626; Philo, *Jos.* 242; Tatian, *Or.* 19.2.
4. Several versions show a curious shift in this direction. The Vulgate's "ministratio" (but "ministerium iustitiae" for the last of the four phrases) and the *AV, RV* "ministration" is improved to the *RSV* "dispensation"; *NEB* echoes this in "The law . . . dispensed death" but is led to say "must not even greater splendour rest upon the divine dispensation of the Spirit?" as if God now replaces the law as the dispensing agent. A similar interpretation underlies *NJB*'s revision of *JB* from "the administering of the Spirit," which means effecting the Spirit's presence, to "the ministry of the Spirit," which is a ministry effected by the Spirit.
5. See the objective genitive after the abstract noun in Josephus, *AJ* 19.129 (possibly also at 8.101).

6. A partial exception is 2 Cor. 11:8, discussed below. The journeys of 11:16 are not, of course, showing that Paul is precisely an itinerant διάκονος but are among a number of evidences establishing him as the authentic διάκονος among the Corinthians.
7. Plato, *Plt.* 290c; 299d; Philo, *de vita Moysis* 84; Josephus, *BJ* 3.354; 4.626; *AJ* 10.177; Achilles Tatius 3.18.5; 4.15.6.
8. Philo, *Gigant.* 12; Plutarch, *Mor.* 416f; Iamblichus, *de Myst.* 1.20.
9. Alexander Aphr., *Mixt.* 5.14, 16; *in Mete.* 18.25; 19.6; *de An.* 59.14; Ammonius, *in Cat.* 15.9; Themistius, *in de An.* 125.9; Philostratus maior, *Im.* 1.3.2; Heliodorus 5.15.
10. The transition from 11:4 to 11:5 is not so awkward that the "superlative apostles" of verse 5 could not have preached the "different gospel" of verse 4 and were not in consequence the same as the "false apostles" of verse 13. The thought supposed by γάρ (11:5) may be supplied without the necessity of linking the statement syntactically with 11:1; thus, in paraphrase: "(verse 1) Bear with me, (2) for it was I who betrothed you to Christ in the first place, (3) although I know that you are wayward like Eve (4) and that if someone comes along with another gospel you will accept it; (5) [this is foolish of you] for I tell you I am inferior to none."
11. E.-B. Allo, *Seconde épître aux Corinthiens* (Paris 1956), p. 259.

Paul's Sacred Mission to the Gentiles

1. The sense of mission from God to proclaim or preach is expressed by the abstract noun at Epictetus 3.22.69 and *P. Lond.* 878; and with a single message at issue at Achilles Tatius 4.15.6. The idea of task or commission is also relevant here whether under God, as at Josephus, *AJ* 5.349; in civic affairs, as at Demosthenes 18.311, Aeschines 3.13, Dio Cassius 54.21.4, Plato, *R.* 493d; or by private undertaking, as at Plato, *Lg.* 919; a range of uses in reference to profane errands is also relevant; Isaeus 1.23; Thucydides 1.133; Plutarch, *Mor.* 50c; 62d; 64e; Josephus, *AJ* 12.188; Alciphro 21.2.
2. The genitive after the abstract noun usually indicates the person or agent carrying out a task: Josephus, *AJ* 2.65; 18.193; Alexander Aphr., *in Sens.* 25.21; Rom. 11:13; 15:31; Acts 21:19; 2 Tim. 4:5; Rev. 2:19; Hermas, Sim. 9.26.2; 9.27.2. It can also be descriptive, indicating the person in whose name the task is carried out: Epictetus 3.22.69; Josephus, *AJ* 5.344; Ignatius, Mag. 6.1; Smyr. 12.1; extending to nonpersonal uses: Josephus, *BJ* 3.70; Julian 305d. Or it can be objective: Josephus, *AJ* 8.101 (probably); 19.129; 2 Cor. 3:7–9; 5:18; 9:12; Acts 6:4. Thus it is not used to indicate that a person is the receiver of an action called διακονία. Strato's "of the blessed" (*APl.*12.194) refers to ministrations at table and is possibly descriptive (like Josephus, *AJ* 5.344).

Paul and the Twelve in Acts

1. See references in preceding section to usage at Rom. 11:13.
2. See the index of other Greek terms. C. Spicq provides a succinct treatment in *Notes de lexicographie néo-testamentaire,* vol. 1 (Fribourg/Göttingen 1978), pp. 901–6, but could go further in applications to Chrsitian usage; R. J. Dillon does not make enough of the word in *From Eye-witnesses to Ministers of the Word* (Rome 1978).

The Later Christian Mission

1. For parallels in which the dative signifies the object delivered, see Lucian, *DMort.* 30.3 and Alexander Aphr., *Mixt.* 5.16.

CHAPTER 11

The Collection for Jerusalem

1. The words occur at Rom. 15:25, 31; 2 Cor. 8:4, 19:20, 9:1, 12, 13.
2. *TDNT* 2 (1964) 87.
3. V. P. Furnish, *II Corinthians* (Garden City, N.Y., 1984), p. 401, with reference to K. Berger, "Almosen für Israel: Zum historischen Kontext der paulinischen Kollekte," *New Testament Studies* 23 (1977) 180–204.
4. *Die Geschichte der Kollekte des Paulus für Jerusalem* (Hamburg-Bergstedt 1965), p. 60.
5. For relevant instances of the abstract noun, see the note to Rom. 11:13 in the preceding chapter.
6. See uses of the abstract noun in the sense of executing a task at Demosthenes 18.206; Josephus, *AJ* 8.101; 19.129 (objective genitive). At Plutarch, *Artist.* 21.4, connotations possibly include carrying. See further on the idea of commission and errand as in the note to Rom. 11:13 in the preceding chapter.
7. See Lucian, *Cont.* 1; Josephus, *AJ* 6.298; Demosthenes 9.43; Achilles Tatius 7.9.11.
8. The variant δωροφορία ("bringing of presents") harmonises with the idea of delegation with gifts.
9. The following note summarises and illustrates the point of grammar. Various combinations of verb and dependent words occur; here τί represents a direct object; τινί an indirect object or dative; πρός or εἰς τινά to or towards a person or place. The datives in all instances (1) to (4) indicate the person who despatches the courier.
   1. τι τινι πρός τινα: *Test. Ab.* 9.24; Lucian, *Icar.* 20.
   2. τινι πρός/εἰς: Josephus, *AJ* 8.5; 10.177; *P. Warren* 21.8; *SB* 4947.2.
   3. τι τινι: Herodotus 4.154.3; Lucian, *Cont.* 1; Heliodorus 7.9.4; Aristaenetus 2.7.2; Anacreontea 15.14.
   4. τινι: Lysias 1.16; Demosthenes 9.43; 19.69; Menander, *Dys.* 219; Lucian, *DDeor.* 7.3; *P. Warren* 21.4; *UPZ* 18.23; 19.25; Achilles Tatius 2.31.4; 7.9.11; Xenophon Eph. 1.9.7; Aristaenetus 2.7.15; Priscus, *HGM* 276.7 (in this instance the personal datives only).

   An exception to (3) would seem to be Plato, *Plt.* 290a and 1 Pet. 1:12. Much other material illustrates the role of the dative as outlined above, in particular usage in the area of agency; see some in the section discussing Josephus, *AJ* 19.41, and others in the section discussing *AJ* 18.280. One of these reads, "Lupus carried out the task for those who sent him": τοῖς ἀπεσταλκόσι (*AJ* 19.194); the senders in the case of the collection are "God's people" in Macedonia and Achaia.

The Delegation from Antioch to Jerusalem

1. The variant "to Jerusalem" (12:25) does not affect the following discussion.
2. Acts 2:23; 10:42; 17:31; Luke 22:22. Compare Rom. 1:4.

Emissaries of the Apostle

1. See comparable uses at Lysias 1.16; *UPZ* 18.23.
2. We read of them at Rom. 16:3 (co-workers) and 1 Cor. 16:19 (house church); also Acts 18:1–3, 18, 26. Other co-workers are mentioned at Rom. 16:9, 21; 1 Cor. 3:9; 2 Cor. 1:24; 8:23; Phil. 2:25; 4:3; Col. 4:11; Philem. 1 and 24.
3. As in Thucydides 1.133.

4. In addition to the previous note, see couriers and messengers of diverse rank and role in Aristophanes, *Ec.* 1116; Sophocles, *Ph.* 497; Plato, *R.* 370e; Lucian, *DMort,* 30.2; Josephus, *AJ* 7:201; 7.224; 11:228; 12.187.
5. See the discussion in B. M. Metzger, *The Text of the New Testament* (Oxford 1964), pp. 240–42.
6. The point is exemplified in the usage at Demosthenes 19.69.
7. See comment on the other occurrences of this expression at 1 Tim. 1:12; Heb. 1:14; Acts 11:29; 1 Cor. 16:15.

Emissaries of the Community

1. See the preceding note.
2. The abstract noun (in a sense other than "mission") occurs with this dative at Plato, *Lg.* 919d ("holding a commission from private individuals") and at *R.* 493d ("a task undertaken in the name of the state"). Plato also uses the expression "to set themselves" a task (again not in the sense "mission").
3. The common noun has no feminine ending, and there is no significance for church order or affairs in its application to women, of whom it is freely predicated in various senses as Aristophanes, *Av.* 1253; *Ec.* 1116; Xenophon, *Oec.* 8, 10; Demosthenes 24.197; Josephus, *AJ* 1.298; Epictetus 2.23.7, 8. For usage illustrating Rom. 16:1, see n. 4 to the preceding section.
4. Usage taken from these letters is not affected by questions of authenticity; the letters remain early Christian documents.
5. See references above to Rom. 11:13. Note the translation of M. Staniforth, "suitable for such a mission," in *Early Christian Writings* (Harmondsworth 1968), p. 114.
6. See references to usage in n. 4 of the preceding section.

CHAPTER 12

Christ and Rome

1. For various uses of the common noun in the sense of agent: Plato, *R.* 371e; 373c; *Lg.* 763a; Philostratus maior, *Im.* 1.3.2; Plutarch, *Mor.* 63b; Aristides 2.199; Lucian, *JConf.* 11; *Dem.Enc.* 31; Epictetus 3.22.63; 3.24.65; 3.26.68; 4.7.20; Philo, *Jos.* 242; *de spec. leg.* 1.116; *de Gigant.* 12; *de decalogo* 178; Josephus, *BJ* 4.388; *AJ* 1.298; 11.255; *SB* 7696.31; 7871.15; 4 Macc. 9:17.

Local Office

1. Usage is to be compared with usage in civic affairs in passages like Plato, *Lg.* 955c (verb), and Dio Cassius 54.21.4 (abstract noun), and with wide-ranging uses in Josephus, *AJ* 18 and 19.
2. The normal personal dative has been illustrated in the n. 9 to the discussion of Rom. 15:25. The dative in Heb. 6:10 is probably best compared with the singular datives at Plato, *Plt.* 290a and 1 Pet. 1:12, which designate not the term or beneficiaries of the action but the arena where the ministering takes place (datives of respect).

Extending Ministry

1. Attendance: Philodemus, *Ir.* 28.4; Plutarch, *Mor.* 174d; arrangements: Athenaeus 439c; Josephus, *AJ* 8.169; approaching ritual: Athenaeus 639b; *Anacreontea* 32.6; *Test. Job.* 11.2; 15.1.
2. Thus *Test. Job.* 12.2: "to attend on the poor at table."
3. See Isocrates, *Trapeziticus* 17.2. (Plato uses the verb of retailers at *R.* 371d, because it is his generic term for the process of interchange.)
4. A biblical word, LXX Deut. 6:18; 12:25, 28; 13:18; Exod. 15:26.
5. Luke's verb with dative of table occurs in Heliodorus 5.8; similarly to Luke the writer is designating the particular kind of royal service that is performed at the table.
6. Historical problems remain: Who carried out the ministry prior to the dispute if not the Twelve? Why do we only hear of the Seven as evangelisers? The concept in Acts 6:1 of "sacred charge" underlies the naming of early monastic establishments in Egypt and subsequently the "diaconiae" of Rome; compare "mission" for a gospel hall. Relevant papyri are *VBP* 94 (fifth century CE); *P. Cairo Masp.* 67003; 67096; 67111; 67138 (sixth century). In addition, the inscriptions *SB* 6009; 6010 (fifth/sixth century).

Gifts

1. See comparable uses in Lucian, *Merc.Cond.* 27; Achilles Tatius 3.10.2.

CHAPTER 13

Philippi

1. See in chapter 7 the sections "More Lasting Than Bronze" and those beginning with "And the God Took Joy"; in chapter 5 the sections beginning "To Heaven."
2. *The Opponents of Paul in Second Corinthians*, Eng. trans. (Philadelphia 1986), p. 30.
3. *Les ministères aux origines de l'église* (Paris 1971), pp. 99. W. R. Schoedel is sympathetic to this view; see on IgEph. 2.1 in *Ignatius of Antioch* (Philadelphia 1985), p. 46.
4. H. von Campenhausen, *Ecclesiastical Authority and Spiritual Power*, Eng. trans. (London 1969), p. 69.

The Church of Timothy

1. Note that the verbs here (1 Tim. 3:10, 13) mean "perform this office," "operate as deacon"; they do not express the idea of service to other people, even to the overseers. See on Hermas, Sim. 9.26.2; 1 Pet. 4:11.

Clement of Rome

1. See a comparison of usages in the discussion of Josephus, *AJ* 10.72, and related passages in the section "Temple Liturgies" of chapter 7.
2. 1 Clem. 42.5: "I will establish their bishops in righteousness and their deacons in faith." The first part of this corresponds to Isa. 60:17 but the second part is unbiblical, it being impossible that anyone could have translated either of the parallel terms in the Hebrew or Greek (*RSV* "overseers," "taskmasters," LXX "rulers") by διάκονος. The phrase "in faith" is also without parallel (cf. "peace," "righteousness") and

might play on the classic ideal of the "faithful" deacon. The reference to "deacons" shows signs of being a Christian gloss, whether Clement's or traditional, illustrating the intimate tie between "overseer" and "deacon" in early Christian circles.
3. Lemaire (*Les ministères aux origines de l'église* [Paris 1971] p. 150, with other literature in n. 17 supporting the inclusive sense) takes the two titles as a general expression for "ministers of the community," all of whom are presbyters.
4. The exclusively liturgical character of the title "deacon" belongs also to "office of bishop" and "bishops." It is not entirely clear, however, that the latter title is co-extensive with that of "presbyter." It may be that the presbyters, as well as being the natural custodians of tradition, and possessing disciplinary powers (57.1), were an electoral college (44.3, "eminent men?") advancing some of their members to liturgical office. While the congregation had the right to rescind such appointments, they have, in Clement's view and at the instigation of a few, used it unjustly against some of the appointees, and Clement would be upholding the presbyters' right to a reasonable discretion. Whatever the case (and it cannot be investigated here), his usage would not illustrate a simple "fusion of two titles" (H. von Campenhausen, *Ecclesiastical Authority and Spiritual Power*, Eng. trans. [London 1969] p. 84) but a studied differentiation; as a liturgical role, the "office of bishop" is in itself irrelevant to questions about authority, consequently also, as has often been argued on other grounds (von Campenhausen, ibid., pp. 88–92) to questions about apostolic succession.

Ignatius and Polycarp

1. IgEph. 5.3; Mag. 4; Tr. 2.1; Phld. 7.2; Smyr. 8.1–9.1.
2. Phld. Int.; 7.2; Smyr. 8.1–9.1.
3. IgEph. 2.2; 20.2; Tr. 2.2; 13.2; Smyr. 8.1.
4. See the section "Emissaries of the Community" in chapter 11.
5. See his summary in *Les ministères aux origines de l'église* (Paris 1971), pp. 197–99.
6. Philo, "the deacon from Cilicia" (Phld. 11.1), unlike Burrhus, who also travels but is the deacon of the bishop of Ephesus (IgEph. 2.1) and has been commissioned to join Ignatius, seems to be an unattached deacon, there being no sign that in following Ignatius he is deputising for his church and is so designated on that ground. Ignatius says of him that "he now officiates [$\dot{\upsilon}\pi\eta\rho\epsilon\tau\epsilon\hat{\iota}$] for me in the word of God," meaning that he performs secretarial, liaising, or preaching work on behalf of the captive bishop. Because he came from distant Cilicia, there may have been no point informing the people of Ephesus of the particular bishop and church he belonged to.
7. See Schweizer on "the remarkable fact" of the parallel between deacons and Jesus, *Church Order in the New Testament*, Eng. trans. (London 1961), p. 154 n. 563, and the comments on Tr. 3.1 in W. R. Schoedel, *Ignatius of Antioch* (Philadelphia 1985), p. 141. More significantly, but erroneously, J. Rius-Camps presents the parallel as evidence against the authenticity of the letters as they stand; see *The Four Authentic Letters of Ignatius, the Martyr* (Rome 1979), pp. 220–40, and the discussion in C. Munier, "A propos d'Ignace d'Antioche," *Revue des Sciences Religieuses* 54 (1980) 55–73, in particular 64–65.
8. See the discussion on the perception of the deacons' role in *Apostolic Constitutions* and in Hippolytus in chapter 3 above, concluding section "Deacons of Old."
9. Kirsopp Lake, trans., *The Apostolic Fathers*, Loeb vol. 1 (London/Cambridge, Mass., 1959).
10. Of the variants of the gospel saying to which the phrase may also allude, only Mark

9:35 has "of all διάκονος" (in that order; cf. 10:44: "slave of all"); otherwise "of you" (Mark 10:43; Matt. 20:26; 23:11).

Hermas and Justin

1. *The Concept of the Church in the Shepherd of Hermas,* Eng. trans. (Lund 1966), p. 145.
2. The phrase καί φιλόξενοι ("and hospitable") is added to their title at 9.27.2 and would seem to be a case of hendiadys prompted by the wholly laudatory view of this group; being "hospitable" is among attributes required of "overseers" at 1 Tim. 3:2.
3. In O. De Gebhardt and A. Harnack, *Hermae Pastor* (Leipzig 1877), pp. 40–41.

Origins of the Title

1. See above on Rom. 13:4; 15:8; Gal. 2:17. The notion of agency is expressed more frequently by the verb (e.g., Heb. 6:10; 1 Pet. 4:11) and is related to what Paul says in respect of his role in the collection for Jerusalem.
2. These have been described in the section of chapter 7, "More Lasting Than Bronze."
3. In such a case there might be a connection with the recurrence of liturgical imagery in the letter (2.17, 25, 30; 3.3, 10; 4.6, 18; cf. 2.5–11).

CHAPTER 14

1. See on Matt. 25:44 in the section "Works of Mercy" of chapter 3.
2. This is the abstract noun; see comparable uses at Carystius, *FHG* 4.359; Athenaeus 439c; Philodemus, *Ir.* 28.4; Strato, *APl.* 12.104; Josephus, *AJ* 2.65; 8.169; 11.163; 18.193; Plutarch, *Mor.* 174d; *Test. Job* 11.2; 15.1.

Jesus as Waiter

1. See the report and further reference in relation to Lucian, *Sat.* 18, and Bato of Sinope, *FHG* 4.349–50.
2. Thus Schürmann and Roloff, above chapter 2 in the section "Jesus' Service at the Supper," and Jeremias, in the section "The Isaian Servant" there.
3. See the discussion in chapter 7 in the section "Waiters at Work." Of itself the participle does not, as Roloff (see preceding note) urges, point to the historical character of the tradition.
4. Because the parallel points to a waiting role, reference to other roles of "young ones" would not seem to be relevant, but see the material in M. Guerra, "Diáconos helénicos y bíblicos," *Burgense* 4 (1963) 82–89, and earlier F. Poland, *Geschichte des griechischen Vereinswesen* (Leipzig 1909), pp. 93, 96, 370–71. Philo had aimed to show that the Therapeutae honoured the ideal of this ancient rite better than the Greeks (see above on *Vit. Cont.*), while the Greeks themselves were conscious of the custom having gone by default with the growth of slavery. See the discussion around Athenaeus 192b; Timaeus, *FHG* 1.207–8 remains relevant even though the attendance is broader than service at table.

Servant of All

1. See Herodotus 4.72.2; 9.82.2; Esth. 1:10; 6:1*R*, 3, 5; but the sense "household attendant" is not common (see the section "Staff" of chapter 7).
2. Instances of the sense occur mainly in poetry, oratory, and philosophical discourse (again see section "Staff" of chapter 7).
3. Compare the prophet's attendant in Josephus, *AJ* 8.354; 9.54, 55; see also Lucian, *Alex.* 5.

The Son of Man

1. Other passives are constructed from (1) the transitive verb in senses of conveying a message or of achieving an effect (2 Cor. 3:3; 8:19, 20; Demosthenes 50.2; 51.7; Josephus, *AJ* 18.293); (2) from the intransitive verb in the sense of "to attend at table," in these cases the verb being predicated of persons only for the purpose of introducing into the context a note of sacred ritual or ethnic custom (*OGI* 383.159; Philo, *de vita cont.* 70; Josephus, *AJ* 10.242; 11.188; Juba, *FHG* 3.472); or (3) from the intransitive verb in the sense "to attend on a person," in this case however not being predicated of a person, as at Mark 10:45, but in an impersonal construction (Timaeus, *FHG* 1.207).
2. This has been a recurrent theme of chapters 4–9, but see in particular on Achilles Tatius 3.10.2; Chariton 8.8.5; Josephus, *AJ* 18.280; 11.228; 1.298; 4 Macc. 9:17; 1 Thess. 3:2.
3. See the report of these views in chapter 2.
4. See on Wellhausen, Schürmann, J. M. Robinson, Tödt, Popkes in chapter 2.
5. On the native Greek tradition, see the discussion in the section "The God Took Joy" of chapter 7.
6. See expressions of this view in the following sections of chapter 2: "The New Ethic," "Rule in the Community," and "Functions in the Community."
7. See the section in chapter 2, "In Fealty to God."
8. So Barrett and Gerhardsson, above chapter 2, in the section "In Fealty to God."
9. See authors cited in the section "The Isaian Servant" of chapter 2; some of them combine the idea of serving God with that of serving brethren.
10. εὖ δουλεύοντα πολλοῖς; note λατρεύοντα in Symmachus (without any object), cf. K. F. Euler, *Die Verkündigung vom leidenden Gottesknecht aus Jes 53 in der griechischen Bibel* (Stuttgart 1934), p. 41.
11. Tasks under God: Acts 6:2; 2 Cor. 3:3; 1 Pet. 1:12; 4:10; Mand. 2.6; filling the office of diaconate: 1 Tim. 3:10, 13; Vis. 3.5.1; Sim. 9.26.2; under the charge of an apostle: Philem. 13; Acts 19:22; by authority within a community: Rom. 15:25; 2 Cor. 8:19, 20; 2 Tim. 1:18; Heb. 6:10; 1 Pet. 4:11.
12. See the contrary influential view from 1913 of Wilhelm Bousset in the introduction to chapter 2.
13. We have seen these in our discussion of Matt. 25:44. The only other instance of this use of the verb in early Christian writings is at Justin, *Dial.* 79.2, where angels are reported as ministering on the Son of man; the reference is to Dan. 7:14 where the peoples pay courtly attendance (LXX: λατρεύουσα; Theodotion: δουλεύσουσιν).

AFTERWORD

1. R. P. C. Hanson, "Office and the Concept of Office in the Early Church," in *Studies in Christian Antiquity* (Edinburgh 1985), p. 120.

2. *Hope in the Desert: The Churches' United Response to Human Need, 1944–1984*, ed. K. Slack (Geneva 1986), p. 134.
3. *Called to be Neighbours: Diakonia 2000*, Official Report, WCC World Consultation, Larnaca, 1986, ed. K. Poser (Geneva 1987).
4. See HMand 12.3.3 and other passages discussed with it.
5. *The Eldership in the Reformed Church* (Edinburgh 1984), pp. 8, 14. Terminology and practice in respect of elders and deacons—not so much deaconesses—in different churches is confusing. My comment in the afterword is directed at the kind of diaconate which is instituted by ordination (or its equivalent) as an official, public and, to date, permanent ministry of a church. For a helpful review of current terminology and practice see *Deacons in the Ministry of the Church,* A Report to the House of Bishops of the General Synod of the Church of England (London 1988), pp. 23–38 (with other information on the Anglican communion throughout the book). This report devotes a paragraph to Robert W. Henderson's *Profile of the Eldership,* prepared for the World Alliance of Reformed Churches (1974), in which the author appears to anticipate ideas propounded in Torrance's paper, conceiving of the elder as "a deacon who specializes in spiritual matters" (p. 35), and recommending that "the traditional eldership be remodelled under the rubrics for an amplified diaconate, with an ordination parallel to that of presbyters" (ibid.). Another account of the complex usage among churches in the United States is provided by James M. Barnett in *The Diaconate* (Minneapolis 1981), pp. 153–55.
6. Independent Evangelical Lutheran Church (Federal Republic of Germany and West Berlin), *Churches Respond to BEM,* vol. 6, ed. M. Thurian, Faith and Order Paper 144 (Geneva 1988), p. 56.
7. Ibid., p. 25.
8. Ibid., p. 36.
9. "The Significance and Status of BEM in the Ecumenical Movement," in *Orthodox Perspectives on Baptism, Eucharist, and Ministry,* ed. G. Limouris and N. M. Vaporis, Faith and Order Paper 128 (Brookline, Mass., 1985), p. 89.
10. "The BEM Document in Romanian Orthodox Theology: The Present Stage of Discussions," in *Orthodox Perspectives,* pp. 98, 99. Ten years previously, in a bilingual statement, His Eminence Athenagoras Kokkinakis, archbishop of Thyateira and Great Britain, illustrated the clarity of usage whose lack is lamented above by Metropolitan Antonie; in English he wrote of the "mission" and "ministry" of the Twelve Apostles, and in Greek of διακονία. See *The Thyateira Confession* (Leighton Buzzard, Beds. 1975), pp. 56 and 198.
11. *All Are Called: Towards a Theology of the Laity* (London 1985), p. 3.
12. Ibid., p. 4.
13. C. Duquoc, *Provisional Churches,* Eng. trans. (London 1986), pp. 48–49.
14. "The Recognition of Ministry: What is the Priority?" *One in Christ* 23 (1987) 34–35.
15. "The Point of Departure in Theology for Determining the Nature of the Priestly Office," in *Theological Investigations,* Eng. trans., vol. 12 (London 1974), p. 36.
16. *Evangelii Nuntiandi,* Eng. trans. (Homebush, NSW, 1979), para. 18.
17. " 'Waiting at Table': A Critical Feminist Theological Reflection on Diakonia," *Concilium* 198 (Edinburgh 1988) 86, 87.
18. Ibid., p. 91.
19. G. D. Henderson, *Church and Ministry* (London 1951), p. 198.
20. T. M. Lindsay, *The Church and The Ministry in The Early Centuries* (London 1903), p. 62.
21. This book does not attempt a bibliography; the index of authors is alone forbidding. In declining to provide a bibliography for his *Ministry: A Case for Change,* Eng. trans.

(London 1981), Edward Schillebeeckx noted that in just the previous ten years the bibliography extended to more than four thousand titles. Things were no better for students in earlier periods; James Ainslie, author of *The Doctrines of Ministerial Order in the Reformed Churches of the 16th and 17th Centuries* (Edinburgh 1940), complained on introducing his bibliography that "The ordinary mortal cannot hope to be acquainted with all, or even most of it" (p. 261).

22. V. T. Stokes in *The Tablet,* London (15 July) 1989.

Appendix I
Meanings of the Greek Words for Ministry

1.0 The words occur in contexts of three kinds:
   i. message;
   ii. agency;
   iii. attendance upon a person or in a household.
2.0 The underlying notion in these three areas is of activity of an in-between kind; thus
   i. in the area of message:
   [common noun] go-between; spokesperson; courier;
   [verb] to be a go-between; to perform an errand; to deliver;
   [abstract noun] errand;
   ii. in the area of agency:
   agent; instrument; medium;
   to effect; to officiate; to mediate;
   commission; execution of task; mediation;
   iii. in the area of attendance:
   attendant;
   to attend; to fetch; to go away to do something;
   act of attendance; performance of a task; task; staff (collectively).
2.1 These meanings are approximate. Modern languages do not seem to possess single equivalent terms. In English "minister," "to administer," "ministry," "ministration," etc., are more satifactory than "servant," "to serve," etc. In almost all instances a special nuance is present by reason of context (see 3 below).
2.2 In the long history of the usage there is no evidence of any change in meanings except in the case of the Christian designation "deacon" (see 5 below).
2.3 The meaning "to wait at table" is not basic (the German "Grundbedeutung") but is merely one expression of the general notion of "go-between"—that is, the table attendant goes between diner and kitchen.
2.4 The words speak of a mode of activity rather than of the status of the person performing the activity. Thus they are not expressing notions of lowliness or servitude, nor in Christian usage did the idea of doing a benevolent action accrue to the idea of ministering.

Appendix I

2.5 The words are thus equally applicable to positions of authority and dignity and to those of lowly esteem: a Roman procurator "effects" the will of the emperor, a general "prosecutes" a war, etc.

3.0 The context in which the words are used is as significant for an understanding of their uses as are the preceding descriptions of their field of meaning.

3.1 By reason of their connotations the words also contribute to the context.

3.2 The words are comparatively rare, occurring in the more formal types of literature—poetry, oratory, speculative and moral philosophy, religious discourse and prayer—or may be used to grace less formal types like historical narrative and high romance.

3.3 In accord with this the words also occur in commemorative inscriptions.

3.4 It thus appears that the words were not part of the vernacular or everyday language and did not have an ordinary or unadorned meaning.

3.5 An exhaustive study of usage in the papyri supports this conclusion.

3.6 By virtue of their formal quality the words occur in passages of a profoundly religious nature.

4.0 Christian usage is indistinguishable from non-Christian except in the instance of the designation "deacon" (see 5 below).

4.1 Christians used the words because of their currency in religious, ethical, and philosophical discourse.

4.2 In Christian sources the words refer mainly to:
  i. message from heaven;
  ii. message between churches;
  iii. commissions within a church.

4.3 Examination of usage in Paul and Acts establishes that "ministry of the word" is a prerogative of the apostle and of those whom the apostle commissions.

4.4 Paul's conviction that "ministering the word" is to expose the hearer to the immediacy of God's revelatory and reconciling activity is at times explicit and is basic to his exposition of apostleship by means of these terms.

4.5 The distinction between an apostolic or evangelistic commission and a commission from the community is clearly discernible.

4.6 Whether the words apply to message or to another type of commission, they necessarily convey the idea of mandated authority from God, apostle, or church.

4.7 Thus the main reference in Christian literature is to "ministry under God," and the notion of "service to fellow human beings" as a benevolent activity does not enter. The latter is true in particular of:
  i. the Son of man at Mark 10:45, whose "ministry" is a commission under God "to give his life as a ransom"; and
  ii. the terms as applied by Paul to the collection for the Christians in Jerusalem (e.g., Rom. 15:25); this was a "commission" in the name of the Asian churches involving an "errand" from Asia, and because the churches were assemblies of God, the commission was a "sacred task," for this reason being designated by these terms from religious language.

4.8 In the gospels the words occur in:
  i. ethical maxims;
  ii. parables; and
  iii. gospel narrative.

In all instances the usage conforms to non-Christian literary convention, and the incidence of the words here cannot be taken as a sign of an intention to inculcate a distinctively Christian style of social behaviour. (See comment on, for example, Matt. 25:44 and Luke 22:27.)

Appendix I

5.0 The designation "deacon" does not derive from attendance at table but from attendance on a person.
5.1 This person is not the needy person or the congregation or community but the episkopos (the later "bishop"), whose "agent" the "deacon" is.
5.2 The word was chosen as a title of this Christian officer because the word had currency in religious language.
5.3 The title is not derived directly from non-Christian religious guilds, in which this common noun designated ceremonial "waiters," but is an original Christian designation for an "agent in sacred affairs."
5.4 The title probably originated in cult.

Appendix II
Uses of the διακον- Words in the New Testament

I. Heaven's Spokesmen
   i. Mediating the word
      1 Thess. 3:2
      1 Cor. 3:5
      2 Cor. 3:3, 6, 7, 8, 9; 5:18, 6:3, 4; 11:15, 23
      Eph. 3:7; 4:12
      Col. 1:7, 23, 25
      1 Tim. 4:6
      1 Pet. 1:12
      Acts 6:4; 21:19
   ii. Mandated for the word
      Rom. 11:13; 12:7
      1 Tim. 1:12
      2 Tim. 4:5
      Acts 1:17, 25; 20:24
II. On the Mission for the Word
   i. Paul
      2 Cor. 11:8
   ii. Collaborators
      Philem. 13
      Eph. 6:21
      Col. 4:7
      2 Tim. 4:11
      Acts 19:22
III. Emissaries of Heaven
      Heb. 1:14
IV. Emissaries of the Church
   i. Paul
      2 Cor. 8:4, 19, 20; 9:1, 12, 13
      Rom. 15:25, 31
      Acts 11:29; 12:25

  ii. Stephanas and Phoebe
    1 Cor. 16:15
    Rom. 16:1
V. Agents of the Other World
  i. The Roman Empire
    Rom. 13:4
  ii. Christ
    Gal. 2:17
    Rom. 15:8
VI. Commissions in the Church
  i. As charisma
    1 Cor. 12:5
    1 Pet. 4:10
  ii. Of churchmen
    Col. 4:17
    2 Tim. 1:18
    Heb. 6:10
    1 Pet. 4:11
    Rev. 2:19
    Acts 6:1, 2
  iii. Deacons
    Phil. 1:1
    1 Tim. 3:8, 10, 12, 13
VII. The Gospels
  i. Gospel narrative
    Mark 1:13 par; 1:31 par; 15:41 par
    Luke 10:40
    John 2:5, 9; 12:2
  ii. Parables
    Matt. 22:13; 25:44
    Luke 12:37; 17:8
  iii. Maxims
    Mark 9:35; 10:43 par
    Matt. 23:11
    John 12:26
  iv. Luke's supper
    Luke 22:27
  v. The mission of the Son of man
    Mark 10:45 par

Appendix III
Key to the Greek Alphabet

English and Greek Letters

A	a	α	A	
B	b	β	B	
C	c			
D	d	δ	Δ	
E	e	ε	E	[short e]
		η	H	[long e]
F	f			
G	g	γ	Γ	
H	h			
I	i	ι	I	
J	j			
K	k	κ	K	
L	l	λ	Λ	
M	m	μ	M	
N	n	ν	N	
O	o	ο	O	[short o]
		ω	Ω	[long o]
P	p	π	Π	
Q	q			
R	r	ρ	P	
S	s	σ	Σ	
		s	ς	[final s]
T	t	τ	T	
U	u	υ	Y	
V	v			
W	w			
X	x	ξ	Ξ	
Y	y			
Z	z	ζ	Z	

th [as in *think*]	θ	Θ
ph [as in *phone*]	φ	Φ
kh [as in *chorus*]	χ	X
ps [as in *pseudo*]	ψ	Ψ

Greek Alphabet

alpha	α	A
beta	β	B
gamma	γ	Γ
delta	δ	Δ
epsilon	ε	E
zeta	ζ	Z
eta	η	H
theta	θ	Θ
iota	ι	I
kappa	κ	K
lambda	λ	Λ
mu	μ	M
nu	ν	N
xi	ξ	Ξ
omicron	ο	O
pi	π	Π
rho	ρ	P
sigma	σ	Σ
tau	τ	T
upsilon	υ	Y
phi	φ	Φ
chi	χ	X
psi	ψ	Ψ
omega	ω	Ω

Abbreviations

Abbreviations have been used sparingly; the following are largely of collections of papyri and inscriptions with some others of dictionaries and of English translations of the Bible.

Anchor Bible	A modern series of translations with commentary (Garden City, NY 1964–)
Ath.	Athenaeus
AthMitt.	Mitteilungen des deutschen archäologischen Instituts, Athenische Abteilung
AV	Authorised or King James Version of the Bible (1611)
Bauer *AG*	W. Bauer, *A Greek-English Lexicon of the New Testament and Other Early Christian Literature,* trans. and adptd. W. F. Ardnt and F. W. Gingrich (Cambridge/Chicago 1957)
BGU	*Aegyptische Urkunden aus den Koeniglichen Museen zu Berlin, Griechische Urkunden,* vol. 2 (Berlin 1898)
CIG	A. Boeckhius, *Corpus Inscriptionum Graecarum* (Berlin 1828–)
CIJud.	J.B. Frey, *Corpus Inscriptionum Iudaicarum,* vol. 1 (Vatican City 1936)
CPR	*Corpus Papyrorum Raineri,* vol. 1 (Vienna 1895)
Douay	The English translation of the Bible from the Latin Vulgate at Douay (Old Testament 1609) and Rheims (New Testament (1582)
Doura	M. I. Rostovtzeff, F. E. Brown, and C. B. Welles, *The Excavations at Doura-Europos* (New Haven 1939)
FAC	J. M. Edmonds, *The Fragments of Attic Comedy,* vols. 1–3 (Leiden 1959)
FHG	C. Muller and T. Muller, *Fragmenta Historicorum Graecorum,* vols. 1–5 (Paris 1878–1885)
GN	*Good News for Modern Man, the New Testament in Today's English Version* (1966)
GNB	*Good News Bible* (1976)
IG	*Inscriptiones Graecae* (for volumes and parts see *LSJ*, p. xxxix)
IG, 2d ed., IX	G. Klaffenbach, *Inscriptiones Graecae,* vol. 9, Pars 1, editio altera, fasc. II (Berlin 1957)
IGB	G. Mihailov, *Inscriptiones graecae in Bulgaria repertae,* vol. 3 (Serdica 1961)
IGLS	L. Jalabert and R. Moutarde, *Inscriptions grecques et latines de la Syrie,* vols 1 and 4 (Paris 1929, 1955)

IMag.	O. Kern, *Die Inschriften von Magnesia am Maeander* (Berlin 1900)
JB	The Jerusalem Bible (1966)
Knox	The English translation of the Bible from the Latin Vulgate by Ronald Knox (1945–1949)
LSJ	H. G. Liddell, R. Scott, and H. S. Jones, *A Greek-English Lexicon*, 9th ed. (Oxford 1973)
LXX	*The Septuagint*, or the Jewish Greek version of the Hebrew scriptures made in the third century BCE
MB	Μυσεῖον καὶ βιβλιοθήκη τῆς εὐαγγελικῆς σχολῆς 2, nos. 2–3 (1876–1878) 93 and 100, cited from M. Guerra, "Diáconos helénicos y bíblicos," *Burgense* 4 (1963) 54–55
Michel	C. Michel, *Recueil d'inscriptions grecques* (Brussels 1900) with *Recueil Supplement* (Brussels 1912)
Moffatt	The English translation of the New Testament by James Moffatt (revised 1934)
Moulton-Milligan	J. H. Moulton and G. Milligan, *The Vocabulary of the Greek Testament Illustrated from the Papyri and Other Non-Literary Sources* (London 1930)
NAB	New American Bible (1970)
NB	F. Preisigke, *Namenbuch* (Heidelberg 1922)
NEB	*The New English Bible* (19)
NIV	*The Holy Bible, New International Version, The New Testament* (1974)
NJB	*The New Jerusalem Bible* (1985)
OAP	D. Foraboschi, *Onomasticon alterum papyrologicum* (Milano/Varese 1967)
OGI	W. Dittenberger, *Orientis Graeci Inscriptiones Selectae*, vol. 1 (Leipzig 1903)
O. Mich. 782	H. C. Youtie and O. M. Pearl, *Papyri and Ostraca from Karanis* (Ann Arbor 1944)
O. Mich. 1046	H. C. Youtie, "Diplomatic Notes on Michigan Ostraca," *Classical Philology* 39 (1944) 28–39
P. Abi.	H. I. Bell and others, *The Abinnaeus Archive* (Oxford 1962)
P. Alex.	A. Swiderek and M. Vandoni, *Papyrus grecs du Musée gréco-romain d'Alexandrie* (Warsaw 1964)
P. Amh.	H. Musurillo, "Early Christian Economy, A Reconstruction of P. Amherst 3(a) (= Wilcken, Chrest. 126)," *Chronique d'Egypte* 61 (1956) 124–34
P. Baden	F. Bilabel, *Veröffentlichungen aus den badischen Papyrus-Sammlungen*, pt. ii, *Griechische Papyri* (Heidelberg 1923)
P. Berl. Leihg.	T. Kalen, *Berliner Leihgabe Greichischer Papyri* (Uppsala 1932)
P. Bour.	P. Collart, *Les papyrus Bouriant* (Paris 1926)
P. Cairo Masp.	J. Maspero, *Catalogue général des antiquités égyptiennes du Musée du Caire; Papyrus grecs d'époque byzantine*, vols. 1–3 (Cairo 1911–1916)
P. Cairo Zen.	C. C. Edgar, *Catalogue général des antiquités égyptiennes du Musée du Caire; Zenon Papyri*, vols. 1–4 (London 1925–1931)
P. Col.	J. Day and C. W. Keyes, *Tax Documents from Theadelphia, Papyri of the Second Century A.D.* (New York 1956)
P. Dura.	C. B. Welles, R. O. Fink, and J. F. Gilliam, *The Parchment and Papyri, being the Excavations at Dura-Europos, Final Report*, vol. 5, pt. 1 (New Haven 1959)
P. Flor.	D. Comparetti and G. Vitelli, *Papiri Greco-Egizii*, vol. 2 (Milan 1911)
P. Fouad	A Bataille and others, *Les Papyrus Fouad*, vol. 1 (Cairo 1939)
P. Gen. inv.	V. Martin, "Relevé topographique des immeubles d'une métropole (P. Gen. Inv. 108)," *Recherches de Papyrologie* 2 (1962) 37–73
P. Giss.	O. Eger and others, *Griechische Papyri im Museum des oberhessischen Geschichtsvereins zu Giessen*, vol. 3 (Berlin 1912)

P. Got.	H. Frisk, *Papyrus grecs de la Bibliothèque municipale de Gothembourg* (Göteborg 1929)
P. Gren.	B. P. Grenfell and A. S. Hunt, *New Classical Fragments and other Greek and Latin Papyri* (Oxford 1897)
P. Heid.	E. Siegmann, *Literarische griechische Texte der Heidelberger Papyrussammlung* (Heidelberg 1956)
P. Hib.	B. P. Grenfell and A. S. Hunt, *The Hibeh Papyri* (London 1906)
Phillips	The English translation of the New Testament by J. B. Phillips (1958)
P. Jews	H. I. Bell, *Jews and Christians in Egypt* (London 1924)
P. Leit.	N. Lewis, *Leitourgia papyri* (Philadelphia 1963)
P. Lond. 266	F. G. Kenyon, *Greek Papyri in the British Museum*, vol. 2 (London 1898)
P. Lond. 2052	T. C. Skeat, *Greek Papyri in the British Museum*, vol. 7, *The Zenon Archive* (London 1974)
P. Mich. 223, 224	H. C. Youtie, *Tax Rolls from Karanis*, pt. 1 (Ann Arbor 1936)
P. Mich. 473	H. C. Youtie and J. G. Winter, *Papyri and Ostraca from Karanis*, 2d series (Ann Arbor 1951)
P. Mich. 596	G. M. Browne, *Documentary Papyri from the Michigan Collection* (Toronto 1970)
P. Mil. Vogliano	I. Cazzaniga and others, *Papiri della Università degli Studi di Milano (P. MIL. VOGLIANO)*, vol. 3 (Milan 1965)
P. Osl.	S. Eitrem and L. Amundsen, *Papyri Osloenses*, vol. 3 (Oslo 1936)
P. Oxy.	B. P. Grenfell and A. S. Hunt, *The Oxyrhynchus Papyri* (London 1898–)
PSI	G. Vitelli and others, *Papiri greci e latini* (Florence 1912–)
P. Strasb.	B. Wagner in *Papyrus Grecs de la biblothèque nationale et universitaire de Strasbourg, No. 301 à 500* (Strasbourg 1973)
P. Stud. Pal.	C. Wessely, *Catalogus Papyrorum Raineri, Series Graeca, Pars I, Textus Graeci* (Leipzig 1921)
P. Warren	M. David. *Papyrologica Lugduno-Batava*, vol. 1, *The Warren Papyri (P. Warren)* (Leiden 1941)
P. Würz.	U. Wilcken, *Mitteilungen aus der Würzburger Papyrussammlung* (Berlin 1934)
Pauly-Wissowa	*Real-Encyclopädie der klassischen Altertumswissenschaft* (1892–)
RSV	Revised Standard Version of the Bible (1946; 2d ed. 1971)
RV	Revised Version of the Bible (1885)
SB	F. Preisigke, F. Bilabel, and E. Kiessling, *Sammelbuch griechischen Urkunden aus Ägypten*, vols. 1–8 (Strassburg, later Berlin/Leipzig, Heidelberg 1915–)
Stephanus	H. Estienne, *Thesaurus graecae linguae*, ed. C. B. Hase and others, vol. 2 (Paris 1833)
TDNT	*Theological Dictionary of the New Testament* (Grand Rapids 1964–1976), being the translation of the following
TWNT	*Theologisches Wörterbuch zum Neuen Testament*, ed. G. Kittel (Stuttgart 1932–1979)
UPZ	U. Wilcken, *Urkunden der Ptolemäerzeit (ältere Funde)*, vol. 1 (Berlin/Leipzig) 1927)
VBP	F. Bilabel, *Griechische Papyri* (Heidelberg 1924)
v.l.	a variant reading in the manuscripts
WB	F. Preisigke, *Wörterbuch der griechischen Papyrusurkunden*, vols. 1–3 (Berlin 1925–1931), vol. 4 in progress, ed. E. Kiessling (Berlin/Marburg 1944–1971), *Supplementum* I, ed. E. Kiessling and W. Rübsam (Amsterdam 1940–1971)
WO	U. Wilcken, *Griechische Ostraka aus Ägypten*, vols. 1–2 (Leipzig/Berlin 1899)

Index of Sources of διακον- Words

Non-Christian and Post-Biblical Writings

The following list is based on the list of authors and and works in the Liddell and Scott lexicon (LSJ). For each author it records known instances of the writer's uses of [a] the verb διακονεῖν, [b] the common noun διακονία (this is occasionally used adjectivally), [c] the abstract noun διάκονος, and [d] other cognate words. (The latter do not occur in the early Christian literature under review.) If the edition used has been other than the standard Budé, Loeb, Oxford, or Teubner, it is indicated either by the name of the editor, as in the first entry for Achilles Tatius (Vilborg), or by the abbreviation for the collection, as for Alexander *(CAG)*. The bibliography for these follows this note.

In regard to the style of the references, when the reference is to a page number, rather than to the usual numbered sections like chapter and paragraphs, "page" is printed after the published source; a number following upon a page number is then the number of the line on the page. When the extant source is Athenaeus, the reference to his *Deipnosophistae* is added to the modern published source as (Ath.) because the latter, as in the Loeb edition, is likely to be within closer reach of the reader. Partly for the sake of consistency, but mainly for my own convenience, all references are in Arabic numerals with the addition in some cases of the letters supplied by editors to subdivide larger sections, as in *R*. 371d. Full stops separate sections of a work; in the text and notes semicolons separate one occurrence from another, except that where the words occur more than once within the same section a comma is used. Thus Epictetus 2.23.7,8 signifies two occurrences. All sources are listed; a few are not discussed in the text.

Bibliography of Sources (other than Budé, Loeb, Oxford, and Teubner editions)

Adler, A., *Suidae Lexicon*, vol. 4 (Leipzig 1935)
Beekby, H., *Anthologia Graeca* (Munich 1958)
Bethe, E., *Pollucis Onomasticon*, vols 1–3 (Stuttgart 1967)
Borgogno, A., *Menandri Aspis* (Milan 1972)
CAG: Commentaria in Aristotelem Graeca (Berlin 1883–)
Charles, R. H., *The Greek Versions of the Testaments of the Twelve Patriarchs* (Oxford 1908)
Dindorf, L., *Historici Graeci Minores*, vol. 1 (Leipzig 1870)
Droysen, H., *Eutropi Breviarium ab urbe condita cum versionibus graecis* (Berlin 1879)
Dübner, F., *Anthologia Palatina*, vol. 1 (Paris 1864)

Edmonds, J. M., *Elegy and Iambus*, Loeb vol. 2 (London 1931)
Etym. Magn.: T. Gaisford, *Etymologicon Magnum* (1848; repr. Amsterdam 1967)
FAC: J. M. Edmonds, *The Fragments of Attic Comedy*, vols. 1–3 (Leiden 1959)
FHG: C. Müller and T. Müller, *Fragmenta Historicorum Graecorum*, vols 1–5 (Paris 1878–1885)
Guerra, M., "Diáconos helénicos y bíblicos," *Burgense* 4 (1963) 41n.109
Hense, O., *Teletis Reliquiae* (Tübingen 1919)
Hercher, R., *Epistolographi Graeci* (Paris 1871)
James, M. R., *The Testament of Abraham* (Cambridge 1892)
Jebb, S., *Aelii Aristidis Adrianensis Opera Omnia Graece et Latine*, vols 1–2 (Oxford 1722–1730)
K: T. Kock, *Comicorum Atticorum Fragmenta*, vols 1–3 (Leipzig 1880–1888)
Kraft, R. A. et al., *The Testament of Job according to the SV Text* (Missoula 1974)
Latte, K., *Hesychii Alexandrini Lexicon*, vol. 1 (Copenhagen 1953)
Nauck, A., *Iamblichi de vita Pythagorica Liber* (Amsterdam 1965)
O'Neil, E. N., *Teles (The Cynic Teacher)* (Missoula 1977)
Oliver, J. H., *The Civilizing Power: Panathenaic Discourse, Aelian Aristides* (Philadelphia 1968)
P: D. L. Page, *Poetae Melici Graeci* (Oxford 1962)
Parthey, G., *Iamblichi de mysteriis liber* (Berlin 1857)
Perry, B. E., *Aesopica* (Urbana 1952)
Philonenko, M., *Joseph et Aséneth* (Leiden 1968)
Reiske, J. J., *Charitonis Aphrodisiensis de Chaerea et Callirrhoe amatoriarum narrationum* (Leipzig 1783)
SA: Supplementum Aristotelicum, vols 1–2, ed. I. Bruns (Berlin 1887–1892)
Steinmetz, P., *Theophrast, Charaktere*, vols 1–2 (Munich 1960–1962)
Vilborg, E., *Achilles Tatius, Leucippe and Clitophon* (Stockholm 1955)

Achilles Tatius, novelist, 4th c.(?) CE [Vilborg]
[a] 2.31.1
    2.31.4   298n.2, 301n.11, 327n.9
    3.10.2   131–32, 302n.1, 329n.1, 332n.2
    4.15.4   155, 309n.6
    7.9.11   298n.2, 301n.11, 327n.9
[b] 3.18.5   108–9, 325n.13, 326n.7
[c] 4.15.6   107–8, 318n.4, 326n.7
Aeschines, orator, 4th c. BCE
[a] 3.15   146
[b] 3.13   145, 316n.11, 318n.4, 326n.1
Aeschylus, tragic playwright, 6th–5th c. BCE
[b] *Prometheus Vinctus*
    942   91, 101, 171
Alcaeus, comic playwright, 5th–4th c. BCE
[a] *K* 1.759
Alciphro, letter-writer, 4th c. CE
[c] 21.2   125, 318n.4, 326n.1
Alexander of Aphrodisias, philosopher, 3rd c. CE [*CAG, SA,* page]
[a] *In Aristotelis Meteorologicorum libros commentaria*
    19.6   137, 302n.1,n.2, 326n.9
    De Mixtione
    5.16   137, 207, 299n.6, 300n.5, 326n.1,n.9
    6.11   302n.1

Non-Christian and Post-Biblical Writings

[b] *in Mete.*
    18.25   137, 326n.9
    Mixt.
    5.14   137, 325n.3, 326n.9
    6.9   302n.1
    6.20   302n.1
[c] *In librum de sensu commentarium*
    25.21   302n.1, 326n.2
    In Aristotelis Topicorum libros octo commentaria
    262.2
[d] *De Anima liber*
    59.14   137, 326n.9
Ammonius, philosopher, 5th c. CE [*CAG* page]
[a] *In Aristotelis Categorias commentarius*
    15.9   137, 207, 326n.9
Anacreontea [Edmonds, *Elegy*]
[a]   15.14   124, 301n.11, 325n.1, 327n.9
    32.6   155, 310n.2,n.4, 329n.1
Anaxandrides, comic playwright, 4th c. BCE *[FAC]*
[a]   2.57 (Ath. 39a)   310n.2,n.4
Anonymous poets
[b]   5th c. BCE *[P]*
    P. 906   309n.2, 310n.4
[c]   Ath. 171b   308n.6
Antipater Sidonius, writer of epigrams, 2nd c. BCE [Dübner]
[d]   7.161   105, 292n.4
Antipho, orator, 5th c. BCE
[a]   1.16   144–45
    1.17   143, 144–45
    1.20   145
Archedicus, comic playwright, 4th–3rd c. BCE [*K*, volume/page]
[a]   3.277 (Ath. 294c)   298n.2
Aristaenetus, teacher of logic
[a]   2.7.2   120, 301n.11, 327n.9
    2.7.15   120, 298n.25, 301n.11, 327n.9
Aristides, teacher of logic, 2nd c. CE [Jebb volume/page]
[a]   1.122   146
    2.154   303n.12
    2.198   140
    2.199 bis   140
    13.254 [Oliver]   300n.3, 301n.11
[b]   2.152   303n.12
    2.154   151, 303n.12, 307n.8
    2.187   146, 303n.12
    2.193   303n.12
    2.199   147, 151, 303n.13, 307n.8, 328n.1
    2.221   303n.12
    2.223   303n.12
[c]   1.82   97–98, 131
    2.152   303n.12, 308n.1
    2.153   303n.12, 308n.1
    2.154   303n.12, 308n.1
Aristophanes, comic playwright, 5th–4th c. BCE
[a] *Acharnenses*
    1017   305n.1, 311n.1
    Aves
    1323   150, 151, 152, 306n.7

[b] *Aves*
    73   152, 306n.5, 307n.1,n.3
    74   152, 307n.1,n.3
    1253   104, 126, 328n.3
    Ecclesiazusae
    1116   125, 328n.3
[d] *Plutus*
    1170   91, 100, 294n.2
    Aves
    838   306n.7
Aristotle, philosopher, 4th c. BCE
[a] *Politica*
    1333a8   151, 307n.8
[b] *Ethica Nicomachea*
    1149a27   152, 307n.3
[c] *Politica*
    1255b27   308n.1
    1261b37   308n.1
    1263a20   308n.1
[d] 1255b25
    1259b23
    1277a36
    1333a7   151
Asenath (anon.) [Philonenko]
[a] 2.11   122, 300n.3
    13.12   122
    15.7   122–24, 166, 300n.3, 301n.11
Athenaeus, linguist, 2nd–3rd c. CE
[a] 171f   155, 309n.1,n.4
    173a   76, 156, 159
    192b   75, 155, 158, 161, 309n.1, 331n.4
    192f   75, 155, 158
    659d   76, 159, 309n.8
[b] 139c   76, 159, 309n.4
    173b   76, 159, 309n.3
    420e   75, 159, 309n.3
[c] 171b   308n.6
    439c   308n.6, 309n.1, 329n.1, 331n.2
    639b   312n.6, 329n.1
[d] 274b   164

Bato of Sinope, historian, 2nd c. BC [*FHG*, volume/page]
[a] 4.349 (Ath. 639f)   156, 161–62, 165, 331n.1
    4.350 (Ath. 640a)   156, 162, 165

Callimachus, epic poet, 3rd c. BCE [*Etym. Magn.*, page]
[d] 268   92, 105
Callistratus, sophist philosopher, 4th c. CE
[a] 13.4   143
Carystius, historian, 2nd c. BCE [*FHG*, volume/page]
[c] 4.359 (Ath. 639b)   312n.6, 331n.2
Chariton, novelist, 1st c.(?) BCE [Reiske]
[a] 8.8.5   126–27, 204, 332n.2, 325n.1
Cleitarchus
[b] Ath. 267c   306n.6, 307n.11
Crates, comic playwright, 5th c. BCE [*K*, volume/page]
[a] 1.133 (Ath. 267e)   305n.1, 306n.7

Demetrius
[b] Ath. 173f   159, 167, 309n.5
Demosthenes, orator, 4th c. BCE
[a] *Epistulae*
    2.11   146
    Orationes
    9.43   129, 301n.11, 327n.9
    19.69   129–30, 316n.10, 327n.9, 328n.6
    50.2   93, 145, 332n.1
    51.7   145, 332n.1
[b] 24.197   307n.2, 328n.3
    40.14   307n.2
    47.52   124
    59.33   309n.5, 310n.3
    59.42   307n.2
[c] 18.206   145, 308n.8, 318n.4, 327n.6
    18.311   145, 316n.11, 318n.4, 326n.1
Dieuchidas, historian, 4th c. BCE [*FHG*, volume/page]
[a] 4.389 (Ath. 263a)   155, 159, 163, 306n.6
Dio Cassius, historian, 2nd–3rd c. CE [Loeb numeration]
[a] 17.10   155
[b] 54.23.4   309n.5, 310n.3
[c] 54.21.4   146, 318n.4, 326n.1, 328n.1
[d] 78.13.1
    65.10.2   164
Dio Chrysostomus, sophist philosopher, 1st–2nd ç. CE
[a] 7.65   156
Diodorus Siculus, historian, 1st c. BCE
[a] 5.28.4   156, 311n.1, 313n.8
    5.40.3   156, 309n.6, 310n.3
[b] 5.26.3   309n.5
Diogenes the Cynic, philosopher, 4th c. BCE [Hercher]
[a] 37.4   310n.4
[b] 37.3   164
Diogenes Laertius, 3rd c.(?) CE
[a] 6.31   305n.1
Diphilus, comic playwright, 4th–3rd c. BCE [*K*, volume/page]
[a] 2.553 bis (Ath. 292c)   311n.1

Epicharmus(?), comic playwright, 5th c. BCE [Adler, page]
[d] 768   153
Epictetus, philosopher, 1st–2nd c. CE
[a] 2.23.11   138, 173
    2.23.16 bis   138
    4.7.37   306n.3, 307n.8
[b] 2.23.7   138, 172, 173, 175, 328n.3
    2.23.8   138, 175, 328n.3
    3.7.28   138, 175
    3.22.63   171, 174–75, 328n.1
    3.24.65   171, 174–75, 328n.1
    3.26.28   171, 172, 174, 175, 328n.1
    4.7.20   147, 171, 174, 175, 328n.1
[c] 3.22.69   107, 171, 175–76, 326n.1
Euphron, comic playwright, 3rd c. BCE [*K*, volume/page]
[a] 3.322 (Ath. 377d)   289n.5, 309n.8, 311n.1
Euripides, tragic playwright, 5th c. BCE

[a]	*Cyclops*	
	406	156
	Ion	
	396	131
[b]	*Cyclops*	
	31	309n.4, 310n.4
	fr.	
	375	

Hegesander, historian [*FHG*, volume/page]
[a]   4.420 (Ath. 419e)   309n.6, 311n.1
Hegesippus, comic playwright, 3rd c. BCE [*K*, volume/page]
[a]   3.312 (Ath. 290c)   309n.8, 311n.11
Heliodorus, novelist, 3rd c. CE
[a]   5.8   130, 302n.1, 329n.5
      5.15   133–36, 141, 299n.6, 300n.5, 326n.9
      7.9.4   118–19, 310n.11, 327n.9
      7.20.2   119
      10.7.5   306n.4, 307n.10
[b]   4.8   131
      6.2   138–39, 303n.1
[d]   7.16.1   119
Hephaestio, astrologist, 4th c. CE
[c]   3.45.9
Heraclides Cumaeus, historian, 4th c. BCE [*FHG*, volume/page]
[a]   2.96 (Ath. 145b)   309n.1
Herodotus, historian, 5th c. BCE
[a]   4.154.3   73, 75, 89, 92, 141, 301n.11, 327n.9
[b]   4.71.4   74, 321n.12
      4.72.2   74, 152, 332n.1
      9.82.2   74, 332n.1
Hesychius, lexicographer, 5th c.(?) CE [Latte, page]
[a]   433 bis   89
[b]   22
Hyperides (?) [Guerra]
[b]   *fr.* 70   307n.2

Iamblichus, philosopher, 4th c. CE
[a]   *Protrepticus*
      6   137–38
[b]   *De vita Pythagorica* [Nauck, page]
      146   309n.3, 310n.3
[d]   *De Mysteriis* [Parthey]
      1.20   148–49, 326n.8
Isaeus, orator, 5th c. BCE
[c]   1.23   108, 125, 318n.4, 326n.1

Josephus, historian, 1st c. CE
[a]   *BJ*
      4.252   139, 143
      AJ
      2.129   146, 306n.5
      3.128   301n.9, 313n.4
      3.155   164
      6.298   129, 130, 325n.1, 327n.7
      7.165   309n.1
      7.365   164
      8.5   129, 301n.11, 327n.9
      8.6   129

9.25	304n.1
10.72	164, 329n.1
10.177	111, 300n.2, 301n.11, 326n.7, 327n.9
10.242	309n.1, 313n.8, 332n.1
11.163	309n.1
11.166	309n.1
11.188	309n.1, 313n.8, 332n.1
13.314	306n.4,n.5, 307n.10
14.358	304n.1
15.224	309n.1,n.7
17.74	304n.1
17.140	304n.1
18.74	164
18.77	313n.5
18.125	304n.1, 305n.7
18.193	309n.1
18.194	
18.262	142, 305n.7
18.265	142, 305n.7
18.269	141, 142
18.277 bis	142, 305n.7
18.280	142–44, 299n.6, 302n.1, 327n.9, 332n.2
18.283	142
18.293	142, 332n.1
18.304	142
19.34	140–41, 146
19.41	141–42, 303n.1, 305n.7, 327n.9
19.42 bis	141, 303n.2,n.3
19.194	141, 327n.9
[b] BJ	
3.354	111–15, 121, 299n.1, 307n.8, 324n.2, 326n.7
4.388	139, 328n.1, 299n.1
4.626	111–15, 121, 299n.1, 324n.2, 325n.3, 326n.7
AJ	
1.298	139, 299n.1, 325n.3, 328n.1,n.3, 332n.2
6.52	299n.1, 309n.3
7.201	129, 130, 299n.1, 328n.4
7.224	129, 299n.1, 328n.4
8.354	152, 299n.1, 321n.12, 332n.3
9.54	299n.1, 307n.8, 332n.3
9.55	299n.1, 307n.8, 332n.3
11.188	299n.1, 309n.1,n.3
11.228	129, 299n.1, 307n.4, 328n.4, 332n.2
11.255	139, 299n.1, 325n.3, 328n.1
12.187	120–21, 299n.1, 328n.4
[c] BJ	
3.70	308n.2, 326n.2
AJ	
2.65	309n.1, 326n.2, 331n.2
4.109	308n.2
5.344	164, 326n.2
5.349	148, 318n.4, 326.1
7.378	154
8.101	164, 308n.8, 325n.5, 326n.2, 327n.6
8.169 bis	308n.6, 309n.1, 329n.1, 331n.2
10.57	164
11.163	309n.1, 331n.2
12.188	120–21, 318n.4, 326n.1

(*Josephus [c] VJ, continued*)

18.21	308n.2
18.193	326n.2, 331n.2
19.129	141, 308n.8, 318n.4, 325n.5, 326n.2, 327n.6

[d] *AJ*

18.70	121, 313n.5

Juba, historian, 1st c. BCE–1st c. CE [*FHG*, volume/page]

[a] 3.472 (Ath. 229c) 311n.1, 313n.8, 332n.1

Julianus, emperor, 4th c. CE

[c] 305d 165, 326n.2

[d] *Oratio*

2.68c	164

Libanius, sophist philosopher, 4th c. CE

[a] *Oratio*

53.9	160, 309n.6

Lucian, sophist philosopher, 2nd c. CE

[a] *Dialogi Deorum*

4.4	310n.2
5.2	310n.2
7.3	101, 298n.25, 301n.11, 327n.9
24.2	102–3, 117, 310n.2

Dialogi Mortuorum

30.3	125, 130, 302n.1, 305n.1, 326n.1

Contemplantes seu Charon

1	91, 100, 102, 118–19, 301n.11, 327n.7,n.9

De Mercede Conductis

16	156, 309n.6, 310n.3
27	124, 127, 131, 329n.1

Tyrannicida

22

Asinus seu Lucius

53	309n.7

Icaromenippus

20	99–100, 104, 301n.11, 316n.10, 325n.1, 327n.9

Philopseudes

34	151, 306n.4, 309n.7
35	151, 306n.3

Saturnalia seu Cronosolon

18	161, 331n.1

[b] *Dialogi Mortuorum*

30.2	125, 328n.4

De Mercede Conductis

26	309n.2, 310n.3

Alexander

5	307n.5, 332n.3

Jupiter Confutatus

11	148, 328n.1

Saturnalia seu Cronosolon

17	147–48, 161, 309n.3

Demosthenis Encomium

31	139, 328n.1

[c] *Philopseudes*

35	306n.3, 308n.2

[d] *Alexander*

33	292n.4

Contemplantes seu Charon

1	91, 325n.1

Lycurgus, orator, 4th c. BCE
[a]   14.55   306n.4,n.5, 307n.9
Lysias, orator, 5th c. BCE
[a]   1.16   96–97, 124, 298n.25, 301n.11, 307n.3,
                                                   327n.1,n.9

Menander, comic playwright, 4th–3rd c. BCE
[a]   *Dyscolus*
      206   306n.3,n.7
      213   298n.2
      219   301n.11, 327n.9
      490   309n.8, 311n.1
      *K*3.78 (Ath. 245c)   309n.8, 311n.1
[b]   *Aspis* [Borgogno]
      12
[d]   *K* 3.34 (Ath. 172c)
Moschio, writer of paradoxes
[c]   Ath. 208a   153

Nicomachus, comic playwright [*K*, volume/page]
[b]   3.386 (Ath. 291a)   309n.4, 310n.4
Nicostratus, comic playwright, 4th c. BCE [*K*, volume/page]
[a]   2.229
Nonnus, epic poet, 5th c. CE [LSJ]
[d]   294n.8

Paenius [Droysen]
[a]   6.18.5   146
Pherecrates, comic playwright, 5th c. BCE *[FAC]*
[a]   1.89
Philemon, comic playwright [*K*, volume/page]
[a]   3.312 (Ath. 170e)   311n.1
Philo, philosopher [Loeb numeration] 1st c. CE
[a]   *De vita contemplativa*
      70   163, 331n.4, 332n.1
      71   163
[b]   75   163, 309n.4
      De Josepho
      242   104, 105, 148, 325n.2, 328n.1
      De gigantibus
      12   104, 148, 326n.8, 328n.1
      De posteritate Caini
      165   302n.5
      De vita Moysis
      199   300n.4, 302n.5
[d]   *De vita contemplativa*
      50   163
      De Abrahamo
      115   104, 110, 316n.10
      De specialibus legibus
      1.17   302n.5
      1.66   104, 148
      1.116   148, 328n.1
      1.204   302n.5
      3.301   302n.5
      De decalogo
      178   104, 148, 328n.1

De vita Moysis
1.84   104, 105, 110, 213, 326n.7
Philodemus, philosopher, 1st c. BCE
[c] De ira
28.4   310n.3, 329n.1, 331n.2
Philostratus Maior, sophist philosopher, 2nd–3rd c. CE
[v] Imagines
1.3.2   138, 303n.6, 325n.3, 326n.9, 328n.1
Philostratus Minor, sophist philosopher, 3rd c. CE
[b] Imagines 4   160, 309n.2
Plato, philosopher, 5th–4th c. BCE
[a] Republic
    371d   79, 80, 86, 125, 329n.3
    476a bis   305n.1, 307n.8
    Laws
    763a bis   139–40, 306n.3, 307n.8
    805e   306n.3, 307n.8
    955c bis   145, 328n.1
    955d   145
    Theaetetus
    175e   151, 152, 307n.8
    Politicus
    290a   81, 84, 327n.9, 328n.2
    Gorgias
    521a   88, 140, 307n.8
[b] Republic
    370e   78–80, 88, 89, 98, 103, 105, 125, 140, 328n.4
    371a   79, 140
    371e   80, 125, 140, 303n.11, 328n.1
    373c   140, 328n.1
    Laws
    763a   305n.1, 328n.1
    782b   105
    831e   125
    Politicus
    290c   81, 85, 92, 98, 100, 103, 105, 111, 216, 326n.7
    290d   81, 85, 92, 316n.5
    Gorgias
    517b   87–88, 89, 92, 138, 150, 152, 153
    518c   87, 150, 152
[c] Republic
    371c   79, 80, 125
    493d   146, 318n.4, 326n.1, 328n.2
    Laws
    919d   125, 290n.4, 326n.1, 328n.2
[d] 633c
    Theaetetus
    175c
    Politicus
    299d   81, 85, 86, 92, 100, 326n.7
    Gorgias
    517b   87–88
    517d   87–88
    518a   88, 316n.5
    fr.   See Plutarch, Mor. 416f

Plutarch, biographer and philosopher, 1st–2nd c. CE
- [a] *Moralia*

63e	151, 306n.3,n.5
301e	156, 160
440b	160, 309n.6
628c	309n.6
677e	151, 308n.1, 309n.1
[b] 63b	147, 321n.12, 328n.1

- [c] *Aristides*

21.4	125, 164, 327n.6

Philopoemen

2.3	153

Moralia

50c	147, 318n.4, 326n.1
62d	147, 318n.4, 326n.1
64e	147, 318n.4, 326n.1
174d	309n.1, 329n.1, 331n.2
677e	152, 307n.8, 308n.1
[d] 416f	149, 207, 326n.8
777b	292n.3
794a	316n.11

Pollux, grammarian, 2nd c. CE [Bethe]

[a]	4.28	316n.5
	8.138	169, 316n.5
[b]	3.76	307n.11, 316n.5
	8.137	105, 169, 316n.5
[c]	8.115	316n.5
[d]	7.210	316n.5

Polybius, historian, 2nd c. BCE

[a]	12.6.7	*See* Timaeus 1.207
[b]	30.26.5	308n.6, 309n.1,n.4
[c]	15.25.21	154

Posidippus, comic playwright, 3rd c. BCE [*K*, volume/page]

[a]	3.336 (Ath. 659c)	156
	3.343 (Ath. 377a)	289n.5, 311n.1
[d]	3.342 (Ath. 376e)	289n.5, 309n.4,n.8
	3.343 (Ath. 376f)	289n.5, 309n.8

Posidonius, historian, 2nd–1st c. BCE [*FHG*, volume/page]

[a]	3.260 bis (Ath. 152b,d)	155–56, 160, 309n.6
	3.261 (Ath. 152e)	311n.1
	3.265 (Ath. 153c)	156, 311n.1
[c]	3.265 (Ath. 266f)	153, 306n.6

Priscus, historian, 5th c. CE [Dindorf, page]

[a]	276	130, 231, 301n.11, 302n.1, 316n.11, 327n.9

Pyrgio, historian [*FHG*, volume/page]

[a]	4.486 (Ath. 143e)	159–60

Sophocles, tragic playwright, 5th c. BCE
- [a] *Ichneutai*

144	298n.2

Philoctetes

	287	305n.1, 306n.7
[b]	497	105, 124–25, 328n.4

fr. [Pearson]

137	105

Soranus, medical writer, 2nd c. CE
[d]   1.80
Strato, writer of epigrams, 2nd c. CE [Beekby]
[c]   12.194   310n.2,n.4, 326n.2, 331n.2
Synesios, poet, 4th–5th c. CE
[d]   8.42   292n.4

Teleclides, comic playwright, 5th c. BCE [Etym, Magn., page]
[b]   755
Teles, philosopher, 3rd c. BCE [Hense or O'Neil]
[b]   41.12 = 4A.130 [O'Neil]   307n.3, 308n.7
[c]   52.7 = 6.10 [O'Neil]   154
[d]   6.34 [O'Neil]   308n.7
Testament of Abraham [James]
[a]   9.24   98–100, 104, 301n.11, 325n.1, 327n.9
Testament of Job [Kraft]
[a]   12.2   165, 329n.2
      15.3   165, 309n.6
      15.8   165
[c]   11.1–2   126, 165, 329n.1, 331n.2
      11.3–4   126, 165
      15.1   165, 329n.1, 331n.2
Testament of Juda [Charles]
[b]   14.2   121, 124, 166
Themistius, sophist philosopher, 4th c. CE [*CAG*, page]
[a]   125.9   137, 207, 300n.4, 326n.9
      42.6   138, 313n.4
Theophrastus, philosopher, 4th–3rd c. BCE [Steinmetz]
[a]   *Characteres*
      2.9   97, 124, 307n.3, 325n.1
      22.4   160, 309n.6
      30.16   160, 306n.5, 309n.6
Theopompus, historian, 4th c. BCE [*FHG*, volume/page]
[b]   1.315 (Ath. 517f)   309n.4, 310n.3
Thucydides, historian, 5th c. BCE
[b]   1.133   130–31, 169–70, 327n.3
[c]   1.133   125, 130–31, 169–70, 299n.5, 318n.4, 326n.1
Timaeus, historian, 4th–3rd c. BCE [*FHG*, volume/page]
[a]   1.196 (Ath. 153d)   311n.1
      1.197 (Ath. 517d = 153d)
      1.207 (Ath. 264c)   306n.3,n.6, 313n.8, 331n.4, 332n.1
      1.208 (Ath. 264d)   306n.3,n.6

Vita Aesopi, 1st c. CE [Perry]
[b]   Vita G 7   109

Xenophon, historian, 5th–4th c. BCE
[a]   *Anabasis*
      4.5.33   156, 160
[b]   *Historia graeca*
      5.4.6   309n.1,n.5
      Hiero
      4.2   155, 309n.1,n.4
      Memorabilia
      1.5.2   307n.3
      Oeconomicus
      8.10   307n.3, 328n.3
      8.14   307n.3

[c] 7.41   153, 308n.1
[d] 7.41
Xenophon of Ephesus, novelist, 2nd c.(?) CE
[a] 1.9.7   117–18, 298n.25, 300n.4, 301n.11, 327n.9
     1.9.8   117–18, 300n.4

Papyri

This list gives only the number of a document. Reference to the line is provided in the text. For the bibliography see the abbreviations.

BGU 261   298n.2, 318n.1
    597   177, 178, 182–83
    1046   179, 318n.11

Charta Borgiana   *See SB* 5124

O. Mich. 782   319n.5
           1046   177, 319n.5

P. Baden II, 31   319n.15
P. Berl. Leihg. 4v.   184
              8   318n.11, 319n.4
P. Bour. 42   179, 190, 318n.11
P. Cairo Masp. 67003, 67096, 67111, 67138,   288n.9, 329n.6
    67139
P. Col. V, 2   184
P. Dura. 50r.   319n.15
P. Flor. 121   177, 178, 183, 190
P. Fouad 25   313n.7, 318n.1
P. Gen. inv. 108   185–90
P. Giss. 101   321n.8
P. Heid. 184   309n.8, 311n.1, 318n.1
P. Hib. 226   318
P. Lond. 266   319n.15
          878   *See SB* 9218
          2052   183–84, 190
P. Mich. 224   180–81
          473   177, 178, 185
          596   319n.5
P. Mil. Vogliano 188   178
P. Oxy. 275   305n.1, 307n.9, 318n.1
         3313   298n.2, 318n.1
P. Strasb. 333r.   319n.15
P. Warren 21   117, 177, 298n.25, 301n.11, 318n.1, 327n.9

SB 4947   117, 301n.11, 318n.1, 327n.9
   5124   177, 178, 181–82, 190
   5662   318n.11, 319n.5
   7515   318n.11, 319n.4
   7621   179, 190
   7696   139, 178, 328n.1
   7871   139, 178, 328n.1
   9218   105–7, 318n.4
   9370   319n.4

Inscriptions

This list gives only the main number used in the text for each inscription. For the line references see the text. For the bibliography see abbreviations, but for *Hesperia* see G. W. Elderkin on p. 101n.3.

CIG 1800	167, 313n.1
3037	166, 167, 313n.1
CIG add. 1793b	313n.1, 314n.7
Doura 875	313n.1
Hesperia 3	101
IG IV, 774	313n.1, 314n.7
824a	167, 313n.1, 314n.7
IG, 2d ed., IX, 247, 248, 250, 251, 252, 451	313n.1, 314n.7
IGB 1487	315n.21, 319n.6
IGLS 1321	314n.10
IMag. 109	313n.1, 314n.7
217	167–68, 313n.1, 314n.7
MB 93	166, 168, 313n.1
100	166, 168, 313n.1
Michel 1226	168, 313n.1
1529A	305n.3
OGI 383	162, 332n.1
SB 6009, 6010	288n.12, 329n.6
7871	139, 178

Biblical and Early Christian Writings

Septuagint Greek Bible, 3rd c. BCE [Rahlfs]

Note: Some observations on the Septuagint's usage in relation to the words under study are made on the following pages: 13, 36, 39, 52, 55, 63, 68, 74, 105, 111, 129, 152, 164, 165, 238, 312n.2.

Prov.	10:4a	152
Esth.	1:10	299n.1, 332n.1
	2:2	299n.1
	6:1	299n.1, 332n.1
	6:3	68, 153, 299n.1, 332n.1
	6:5	68, 153, 299n.1, 332n.1
1 Macc.	11:58	153–54, 293n.1
4 Macc.	9:17	148, 328n.1, 332n.2

New Testament
Matt.	4:11	245
	8:15	245

	20:26	330n.10
	20:28	*See* Mark 10:45
	22:13	245, 283n.14
	23:11	247, 248, 330n.10
	25:44	64–65, 245, 293n.6, 332n.10
	27:55	245
Mark	1:13	245, 283n.14
	1:31	245, 283n.14
	9:35	40, 58, 59, 247, 248, 287n.12, 330n.10
	10:43	247, 248, 330n.10
	10:45	4, 12, 15, 18, 29, 39, 42, 46–61, 65, 71, 248–52
	15:41	38, 59, 245, 283n.14
Luke	4:39	245
	8:3	245
	10:40	245, 246, 283n.14
	12:37	245, 246
	17:8	245, 283n.14
	22:26	247
	22:27	36, 39, 246–47, 248, 250
John	2:5	245
	2:9	245
	12:2	245
	12:26	12, 48, 248
Acts	1:17	53, 213, 230
	1:25	213, 230
	6:1	13, 230–31
	6:2	130, 230–31, 302n.1, 332n.11
	6:4	96, 213, 230–31, 326n.2
	11:29	64, 221, 230, 328n.7
	12:25	64, 221, 230
	19:22	223–24, 229, 230, 332n.11
	20:24	212–13, 228, 230
	21:19	212–13, 215, 230, 326n.2
Rom.	11:13	13, 37, 211, 212, 213, 215, 228, 230, 233, 326n.1,n.2, 327n.5,n.6, 328n.5
	12:7	233, 287n.8
	13:4	13, 228
	15:8	227–28
	15:25	64, 94, 220–21, 222, 224, 229, 327n.1, 328n.2, 332n.11
	15:31	13, 220–21, 230, 326n.2, 327n.1
	16:1	224–25, 226, 287n.8, 328n.3
1 Cor.	3:5	195–97, 202, 205, 207
	12:5	232, 251, 258, 259
	16:15	12, 224, 287n.8, 328n.7
2 Cor.	3:3	126–27, 203–4, 219, 228, 293n.6, 332n.1,n.11
	3:6	195, 197–98, 202, 205, 228, 241
	3:7	204
	3:8	204
	3:9	204, 259
	4:1	205, 230
	5:18	230, 233, 259, 326n.2
	6:3	205, 230
	6:4	64, 169, 195, 197–98, 202, 205
	8:4	218, 230, 293n.6, 327n.1
	8:19	218–19, 327n.1, 332n.1,n.11

(2 Cor., continued)

	8:20	218–19, 327n.1, 332n.1,n.11
	9:1	217, 230, 327n.1
	9:12	219, 326n.2, 327n.1
	9:13	219, 230, 327n.1
	11:8	211, 326n.6
	11:15	202, 228, 241
	11:23	170, 195, 198–203, 205, 215
Gal.	2:17	13, 228
Eph.	3:7	233
	4:12	13, 35, 71, 233–34
	6:21	223
Phil.	1:1	44–45, 235–36
Col.	1:7	209–10, 223, 233
	1:23	209–10, 233
	1:25	210, 233
	4:7	222–23, 226
	4:17	228
1 Thess.	3:2	104, 223, 224, 332n.2
1 Tim.	1:12	215, 221, 216, 237, 328n.7
	3:8	237–38
	3:10	242, 329n.1, 332n.11
	3:12	
	3:13	41, 242, 329n.1, 332n.11
	4:6	215, 237
2 Tim.	1:18	229, 237, 332n.11
	4:5	228, 237, 326n.2
	4:11	224, 237
Philem.	13	222, 223, 224, 230, 332n.11
Heb.	1:14	216, 221, 328n.7
	6:10	229, 232, 287n.8, 328n.2, 332n.11
1 Pet.	1:12	215–16, 328n.2, 332n.11
	4:10	232, 332n.11
	4:11	232, 287n.8, 329n.1, 332n.11
Rev.	2:19	230, 326n.2

Apostolic Fathers

Did.	15.1	238
1 Clem.	40.5	238, 249
	42.4	238–39
	42.5	238–39
IgEph.	2.1	226, 240, 329n.3, 330n.6
IgMag.	2	240
	6.1	240, 326n.2
	13.1	239
IgTr.	2.3	240–41
	3.1	239–40
	7.2	239, 241
IgPhld.	Int.	239
	1.1	228
	4	240, 241
	7.1	239
	10.1	225, 226, 236, 239
	10.2	225, 239
	11.1	226, 330n.6

IgSmyr.	8.1	239, 240	
	10.1	239	
	12.1	226, 239, 326n.2	
	12.2	240	
IgPol.	6.1	239	
PolPhil.	5.2	241, 287n.12	
	5.3	241	
HVis.	3.5.1	242, 332n.11	
HMand.	2.6	231, 332n.11	
	12.3.3	231, 333n.4	
HSim.	1.9	232	
	2.7	232	
	8.4.1	249, 251	
	8.4.2	249, 251	
	9.15.4	215	
	9.26.2	229, 242, 329n.1, 332n.11	
	9.27.2	228, 242	

Apologists

Justin	*Ap.*	1.65.5	242–43
		1.67.5	242–43
	Dial.	79.2	332n.13
Tatian	*Or.*	13.3	sic
		19.2	216
Aristides		15.2	302n.1

Papyri Relating to Early Deacons

CPR 227	323n.30
P. Abi. 55	322n.24, 323n.30
P. Alex. 29	323n.27
P. Giss. 103	189
P. Got. 11	323n.27
P. Gren. 73	323n.27
P. Jews 1913, 1914	189
P. Osl. 113	322n.24, 323n.30
P. Oxy. 1492, 2603	323n.27
P. Stud. Pal. XX, 103	322n.24, 323n.30
P. Würz. 16	323n.30
PSI 208, 1041	323n.27
SB 9622	322n.24, 323n.30

Index of Other Greek Terms

These terms are selected for their interest at various points of the book.

ἄγγελος, 91, 105, 113, 125, 171, 176, 292n.3
ἀκόλουθος, 59, 152, 174, 293n.1
δόκιμος, 200
δουλ-, 52, 55, 59, 61, 82, 92–93, 139–40, 153, 155, 158, 241, 307n.11, 312n.3, 317n.2, 332n.10
δωροφορία, 327n.8
ἐκκλησία, 186, 188–89
θεραπ-, 69, 87, 93, 106–7, 153, 167
κόποι, 200
λατρ-, 93, 241, 332n.10

λειτουργ-, 70, 152, 216, 219, 238, 243, 312n.2, 316n.11
λογεία, 218
παῖς, 124, 151, 293n.1
πρεσβεία, 129–30, 170–71, 225, 316n.11
συνείδησις, 200, 204
ὑπερλίαν, 208–9
ὑπηρετ-, 69–70, 81–84, 92, 94, 106, 123, 125, 138, 147, 153, 162, 165, 166–67, 173–74, 182, 214, 236, 240, 244, 292n.3, 293n.1, 312n.6, 314n.10, 320n.6, 324n.1, 330n.6

Index of Authors

Abel, F.-M., 308n.4
Abrecht, P., 271n.33
Adcock, F., 316n.6,n.7
Adler, A., 307n.11
Ainslie, J. L., 278n.69, 333n.21
Allmen, D. von, 275n.44
Allmen, J. J. von, 277n.68
Amberg, E. H., 271n.3
Amiot, F., 268n.5
Amman, A., 290n.1
Anderson, H., 57, 282n.2
d'Andilly, A., 144
Antonini, L., 323n.25
Arndt, W. F., 293n.6, 311n.2, 314n.5
Arnott, W. G., 298n.2
Audet, J. P., 40, 278n.6
Avis, P. D. L., 274n.38
Avogadro, S., 320n.2

Babitt, F. C., 305n.4
Ball, C. J., 153
Bandello, M., 116
Barison, P., 323n.25
Barnett, J. M., 281n.17, 333n.5
Baronius, C. Soranus, 66
Barr, J., 94
Barrett, C. K., 55, 60, 64, 324n.3,n.4
Barrois, G., 288n.4
Barth, G., 315n.4
Barth, K., 22, 272n.2
Barth, M., 274n.36, 315n.4
Bauer, G., 271n.34
Bauer, W., 293n.6, 311n.2, 314n.5
Bauernfeind, O., 296n.1
Bausch, W. J., 32
Bea, A., 16, 19
Bekker, I., 316n.5
Bell, H. I., 322n.18, 323n.25
Bennett, J. A. W., 291n.5
Benois, A. A., 307n.11
Benoit, P., 55

Berger, K., 297n.21, 324n.3
Best, E., 57, 283n.3
Bethe, E., 316n.5
Beyer, W. H., 6–7, 11, 24, 37, 39, 40, 45, 48, 50–52, 53, 61, 94, 111, 217, 249, 258, 291n.1, 314n.3
Blake, W. E., 127, 298n.2
Blank, J., 278n.6
Blatt, F., 313n.4
Blavatskaja, T. V., 320n.6
Blaydes, F. H. M., 104, 292n.7
Blenkinsopp, J., 278n.6
Bliss, K., 267n.17
Bodson, A., 316n.3
Boeckh, A., 168
Boegner, M., 273n.16
Boff, L., 34
Boisacq, E., 292n.2
Bömer, F., 311n.2
Bonhöffer, A., 318n.4
Borkowski, Z., 185–90
Bornkamm, G., 57, 171, 315n.4, 324n.3
Borret, M., 288n.16
Borsch, F. W., 55
Boswell, J., 3
Botte, B., 70
Bottigelli, P., 318n.9
Boulton, P. H., 51, 56, 249
Bourdelotius, J., 297n.13
Bousset, W., 47
Bowman, J., 282n.3
Box, G. H., 294n.1
Brandt, W., 6–7, 11, 39, 41, 45, 48–50, 53, 61, 249, 254, 258
Branscomb, B. H., 57
Braun, H., 281n.20
Bridel, C., 30, 45, 267n.17, 279n.24
Brown, R. E., 277n.62
Browne, G. M., 322n.12
Bullough, G., 296n.4
Bulteau, M. G., 268n.4

Bultmann, R., 58, 325n.11
Burchard, C., 297n.21, n.22
Burghardt, W. J., 289n.2
Burkert, W., 311n.2
Burrows, W. R., 34
Buttmann, P. C., 12, 90, 94, 292n.10

Calderini, R., 319n.1,n.8,n.15
Calvin, J., 31
Cameli, L. J., 19, 57-58
Campbell, L., 85, 86, 290n.1,n.3,n.9
Campenhausen, H. von, 270n.13, 279n.12,n.25, 329n.4, 330n.4
Carney, T. F., 108, 132
Carpenter, L. L., 55
Carr, A., 38
Carrington, P., 55
Castiglione, B., 296n.4
Cavassini, M. T., 323n.25
Cavert, S. M., 270n.7
Ceretti, G., 272n.2
Chambry, E., 86
Chantraine, P., 291n.2, 292n.2
Charles, R. H., 297n.20
Charlesworth, J. H., 297n.19
Charley, J. W., 276n.54
Chitty, D. J., 323n.25
Cocceius, 143
Coenen, L., 13
Cohn, L., 295n.2
Collange, J. F., 64
Collins, J. J., 297n.19
Collins, J. N., 317n.3
Colpe, C., 57
Colson, F. H., 295n.1, 296n.1,n.2
Colson, J., 7, 20, 44, 289n.4
Combrink, H. J. B., 59
Comparetti, D., 321n.6
Congar, Y., 31, 267n.17, 268n.2
Conzelmann, H., 278n.10, 315n.4
Cooke, B. J., 26, 274n.38, 276n.60, 279n.23
Cooke, L. E., 21, 270n.2
Coppens, J., 278n.3
Corbett, S., 288n.6
Corecco, E., 272n.2
Cornfield, G., 296n.3
Cornford, F. M., 291n.10
Coronia, G., 289n.15
Cranfield, C. E. B., 55, 59, 64
Cremer, H., 94
Cross, R. C., 78
Crowfoot, J. W., 289n.14
Crum, W. E., 68, 319n.15

Dagens, C., 272n.2
Dalman, G., 52
Dalmeyda, G., 297n.8
Daniel, S., 307n.7
Davies, J. L., 291n.10, 303n.11

Davies, W. D., 59, 278n.3
Day, J., 184
De Bruyen, D., 308n.4
De Gebhardt, O., 331n.3
Degen, J., 267n.18
Delatte, A., 316n.4
Delcoir, M., 294n.2
Delhaye, P., 269n.13
Delorme, J., 275n.46, 278n.7, 279n.32
Denis, A. M., 122, 297n.19
Descamps, A., 278n.3
Dillon, R. J., 326n.2
Dindorf, W., 294n.3, 316n.5
Dion, P. E., 314n.10
Dix, G., 70
Dodd, C. H., 55
Doens, I., 322n.9,n.15
Dörrie, H., 312n.3
Dragoumes, 305n.3
Duchaine, M. C., 30
Duchesne, L., 66
Dudley, D. R., 316n.2
Dulles, A., 274n.32
Dunand, F., 315n.14
Duquoc, C., 34

Edelstein, E. J., 305n.3
Edelstein, L., 305n.3
Edmonds, J. M., 298n.3
Ehrenstrom, N., 30
Elderkin, G. W., 101, 292n.8
Ellendt, F., 105
Elmund, G., 267n.17
Emmet, C. W., 305n.2
England, E. B., 299n.7
Epagneul, M. D., 42
Estienne, H. See Stephanus
Euler, K. F., 332n.10

Fairbanks, A., 303n.8
Farrer, A. M., 27
Federici, T., 269n.13
Feiner, J., 269n.37
Feldman, H., 121, 141, 144
Feuillatre, E., 297n.11
Fischer, J. E., 272n.16
Fitzgerald, R., 310n.2
Fliedner, F., 9
Fliedner, T., 9-11
Forcellini, H., 319n.15
Forster, E. M., 188, 295n.1
Foucart, P., 314n.8,n.10
Fowler, F. G., 124, 294n.1,n.5
Fowler, H. N., 290n.1, 294n.3,n.4
Frankl, W., 288n.6
Fransen, P., 276n.60
Frey, J. B., 314n.10
Friedländer, P., 84, 86, 290n.3
Friedrich, G., 72, 315n.4

Frieling, R., 273n.14
Fries, H., 279n.32
Frisk, H., 291n.2, 292n.2
Frutaz, A., 288n.4
Fuller, R. H., 55
Furnish, V., 324n.3,n.6, 325n.10

Gaisford, T. (Etym. Magn.), 89, 92, 294n.2, 307n.10
Ganoczy, A., 275n.39
Gaselee, S., 131-32
Gassmann, G., 29, 30, 273n.22
Gätje, H., 302n.1
Georgi, D., 7-8, 45, 64, 72, 170-76, 206, 236, 325n.11
Gerhardsson, B., 60
Gerke, F., 28
Gewiess, J., 268n.5, 279n.32
Ghedini, G., 323n.25
Gignac, F. T., 319n.3
Gill, D., 311n.2
Gill, D. M., 31
Gingrich, F. W., 293n.6, 311n.2, 314n.5
Gnilka, J., 315n.4
Goebel, A., 90, 292n.2
Goehring, J. B., 323n.25
Goertz, H. J., 273n.22
Goligher, W. A., 299n.5
Goodspeed, E. J., 308n.4
Goppelt, L., 279n.32
Gould, E. P., 285n.20, 293n.4
Gow, A. S. F., 292n.4
Grabner, J., 276n.52
Grelot, P., 268n.5, 278n.3
Gross, H., 59, 268n.5
Grundmann, W., 51
Gschnitzer, F., 92
Guerra, M., 7-8, 44, 59, 61, 166, 167, 177, 290n.3, 291n.1, 307n.2, 331n.4
Guillot, L. B., 27
Guilmot, M., 299n.2
Guitard, A., 268n.4
Gundel, H. G., 320n.2
Gunther, J. J., 324n.3
Gutierrez Garcia, J. L., 268n.5
Gwatkin, H. M., 293n.4

Hadas, M., 294n.7, 297n.5,n.7,n.14,n.15, 300n.2, 301n.5
Haelst, J. van, 188, 189, 322n.6,n.18, 323n.25
Haenchen, E., 58
Haendler, G., 274n.38
Hahn, F., 57
Hamdy-Bey, O., 312n.3
Hanson, A. T., 27, 278n.7
Hanson, R. P. C., 332n.1
Harmon, A. M., 124
Harnack, A., 65, 331n.3
Harris, B. F., 323n.25

Harris, J. R., 215
Härter, F. H., 9
Hasluck, E. W., 315n.19
Hatch, E., 66, 166, 262, 314n.4
Havercamp, S., 312n.1
Hebblethwaite, P., 16, 17
Heichelheim, F. M., 312n.1
Henderson, G. D., 333n.19
Henderson, R. W., 333n.5
Hendriksen, W., 285n.22
Hengel, M., 287n.9
Henrichs, A., 295n.3
Herzog, F., 30
Hess, K., 268n.5
Heuser, G., 181
Higgins, A. J. B., 55
Hirschig, W. A., 301n.8
Hoby, T., 296n.4
Hoffmann, J., 279n.32
Hohlwein, N., 320n.6, 321n.7
Hooker, M. D., 55, 57
Hornef, J., 42
Horsley, G. H. R., 298n.2
Hort, F. J. A., 223
Hotchkin, J. F., 268n.6
Houtepen, A., 34
Hudson, J., 121, 143, 154, 296n.1, 300n.2, 303n.4
Humann, K., 312n.3
Hume, B., 275n.48

Jannasch, W., 13
Jebb, S., 97
Jeremias, J., 56, 331n.2
Jessen, O., 292n.2
John XXIII, Pope, 43
John Paul II, Pope, 34
Johnson, S., 3
Johnson, S. E., 55
Jones, A., 55
Jones, A. H. M., 289n.14, 295n.1
Jowett, B., 291n.9, 303n.9,n.11
Joyce, J., 297n.6
Judge, E. A., 298n.2, 324n.4

Kalen, T., 184, 318n.11
Kalsbach, A., 288n.1,n.4
Käsemann, E., 37, 279n.32
Kasper, W., 39, 276n.60, 279n.32
Kautzch, E., 308n.4
Kee, H. C., 297n.19
Kees, H., 318n.9
Keil, B., 294n.2
Kennedy, B. H., 307n.9
Kerenyi, K., 122, 299n.2
Kertelge, K., 51, 279n.32
Kessler, H., 54
Keyes, C. W., 184
Kirill, Archbishop, 257

Kittel, G., 5, 6, 72, 94, 217
Klauser, T., 268n.3
Klostermann, E., 58
Klostermann, F., 18
Knox, R., 308n.4
Koenen, L., 311n.3
Koerte, A., 305n.3
Koester, H. H., 315n.4
Kokkinakis, A., 333n.10
Kramer, H., 42–43
Krautheimer, R., 67
Krimm, H., 66, 266n.1
Krusche, W., 271n.1
Kuhn, H. W., 58
Küng, H., 40, 275n.46
Kupiszewski, H., 290n.5, 320n.6

Lacombrade, C., 292n.4
Lagrange, M. J., 57
Lamb, W., 297n.14,n.15, 301n.6,n.9
Lamb, W. R. M., 294n.1
Lampe, G. W. H., 57, 65, 322n.24
Lane, W. H., 60
Lawrence, D. H., 32
Le Bas, P., 314n.8
LeBlanc, P. J., 268n.7
Leclerq, H., 288n.4
Lecuyer, J., 269n.34, 282n.22
Lee, H. D. P., 291n.10
Legrand, E., 314n.9
Legrand, H. M., 274n.32
Leisegang, J., 104
Lejeune, M., 291n.2
Lemaire, A., 7, 37, 45, 55, 61, 63, 64, 177, 236, 239, 243, 291n.1, 314n.3, 330n.3
Lesky, A., 299n.2
Lesquier, J., 318n.11
Leudesdorff, G., 267n.15
Liddell, H. G., 93, 101, 177, 292n.5, 302n.4, 307n.11, 314n.5, 320n.1
Lienhard, M., 271n.7
Lietzmann, H., 6, 45, 166, 177, 291n.6
Lindbeck, G. A., 14
Lindsay, J., 161, 188, 311n.3,n.4, 322n.12
Lindsay, T. M., 333n.20
Löffler, P., 271n.34
Löhe, W., 10
Lohmeyer, E., 286n.4
Löhrer, M., 265n.2, 269n.24
Lohse, E., 56
Lucas, D. W., 300n.1
Luther, M., 9, 31
Lyly, J., 116
Lyonnet, S., 55

McAdoo, H. R., 274n.4
McBrien, R., 275n.48
McDonnell, K., 39
McGowan, J. K., 277n.65
McCulloch, J., 269n.15

M'Neile, A. H., 285n.20
Manson, T. W., 55, 278n.3
Manson, W., 56, 286n.1
Marc, G., 277n.67
Marcus, E., 277n.62
Marcus, R., 121, 154, 296n.2, 300n.2, 303n.5, 313n.6
Margoliouth, D. S., 143
Mark, T., 310n.1
Marrou, H. I., 67
Martin, V., 185–90
Mason, H. J., 316n.9
Maspero, J., 67–68
Massaux, E., 287n.12
Masurillo, H., 323n.26
Maynard, G. H., 144
Melzer, F., 268n.5
Menoud, P. H., 277n.68
Merk, A., 223
Merklein, H., 274n.36
Mertens, P., 320n.6
Metzger, B. M., 328n.5
Meyer, P. M., 320n.6
Michel, O., 296n.1, 315n.4
Milligan, G., 45, 177, 291n.6, 314n.5
Milman, J., 186
Minnear, P. S., 59
Mitchell, N., 276n.60
Mitchell, T., 97
Modrzejewski, J., 290n.5, 320n.6
Moede, G., 39, 279n.32
Moeller, C., 18, 20
Mohrmann, C., 291n.6
Moigt, J., 34
Molinie, G., 299n.2,n.5
Molland, E., 279n.32
Motevecchi, O., 322,n.18, 323n.25
Morsdorf, K., 269n.20
Mosley, D. J., 316n.6,n.7,n.9
Moulder, W. J., 285n.28
Moulton, J. H., 45, 177, 291n.6, 314n.5
Müller-Bardorff, J., 268n.5
Munier, C., 330n.7
Murray, G., 272n.12
Mussner, M., 268n.5

Nagy, G., 32, 271n.3
Naldini, M., 323n.25
Nardoni, E., 279n.32
Nazari, O., 291n.2
Neukamm, K. H., 267n.18
Nickelsburg, G. W. E., 297n.21
Niese, B., 154
Nightingale, F., 9
Nilsson, M. P., 311n.5, 312n.1
Nock, A. D., 65, 161, 311n.3
Nordhues, P., 267n.20

O'Meara, T. F., 31, 34
Oldfather, C. H., 294n.6

Oldfather, W. A., 292n.6, 316n.3, 317n.5,n.6,n.7
Olsson, B., 320n.1
Opocensky, M., 271n.3
Osborn, E., 287n.14
Otto, G., 267n.15
Otto, W., 318n.9

Page, D. L., 292n.4
Pannenburg, W., 279n.32
Panvinio, O., 288n.1
Paoli, General, 3
Papaderos, A., 24, 26
Papanikolau, A. D., 299n.2
Paton, D. M., 273n.20
Paul VI, Pope, 41, 260
Paul, R. S., 34–35, 273n.10,n.16
Pauly-Wissowa, 315n.16, 322n.23
Pavlovskaja, A. I., 320n.6
Pearson, B. A., 323n.25
Perdrizet, P., 303n.7
Peremans, W., 320n.6, 321n.11
Pernveden, L., 242
Perrin, B., 308n.3
Perrin, N., 57, 61
Pesch, R., 37, 279n.32
Philippi, P., 267n.15, 279n.24,n.32
Philips, G., 274n.35
Philonenko, M., 297n.21
Pichou, R., 296n.2
Pies, O., 42
Pius XII, Pope, 42
Plămadeălă, A., 24, 257
Pokusa, J., 281n.14
Poland, F., 166, 167, 171, 314n.7,n.9, 315n.3,n.16, 331n.4
Poletti, U., 280n.4
Popkes, W., 54, 332n.4
Poser, K., 267n.18, 333n.2
Potter, D., 297n.9
Pottmeyer, H. J., 275n.46
Power, D. N., 278n.3
Préaux, C., 188
Preisigke, F., 320n.6
Prokes, M. T., 38
Prümm, K., 283n.22
Puchstein, O., 312n.3

Quiller-Couch, A., 296n.3

Rahner, K., 43, 260, 269n.32, 322n.9
Recke-Volmarstein, A. von der, 9
Rees, B. R., 323n.25
Reicke, B., 54–55, 249, 293n.10
Reiske, J. J., 97, 127
Rengstorf, K. H., 296n.3
Reploh, K. G., 59
Rhodes, P. J., 300n.5
Richards, I. A., 291n.9

Richards, M., 277n.61
Rius-Camps, J., 330n.7
Robert, P., 275n.46
Roberts, C. H., 189
Robinson, J. A., 12
Robinson, J. M., 54, 332n.4
Rohde, F., 299n.2
Roloff, J., 7, 53–54, 61, 199, 249, 341n.2,n.3, 325n.10
Roussel, P., 315n.15
Roy, L., 275n.48
Ruether, R., 38
Rufinus, 144
Ruffini, E., 276n.60
Russell, E. A., 279n.32

Sabourin, L., 56
San Nicolò, M., 314n.12
Sanders, E. P., 294n.1
Sartori, L., 272n.2
Saunders, A. N. W., 300n.4
Schäfer, T., 267n.17
Schaff, P., 295n.4
Schamoni, W., 42
Schelkle, K. H., 20, 40, 268n.5, 279n.32
Schillebeeckx, E., 268n.9, 276n.60, 279n.32, 333n.21
Schlatter, A., 60, 111
Schmid, J., 285n.17
Schmid, W., 299n.2
Schmidt, J. H. H., 90, 92–93
Schmidt, W., 304n.3
Schmithals, W., 315n.3,n.4
Schmitz, H., 272n.2
Schniewind, J., 55
Schober, T., 271n.6
Schoedel, W. R., 329n.3, 330n.7
Scholfield, A. F., 292n.3
Schönberger, O., 303n.8
Schow, N., 178
Schreckenberg, H., 304n.3
Schreiber, J., 57
Schubart, W., 105
Schulz, A., 285n.19
Schürmann, H., 52, 55, 278n.11, 279n.32, 331n.2, 332n.4
Schüssler Fiorenza, E., 38, 261, 279n.21
Schutz, R., 17
Schütz, W., 268n.5
Schweitzer, A., 57
Schweitzer, W., 270n.15
Schweizer, E., 7, 35–37, 39, 40, 52, 61, 111, 177, 193–94, 249, 330n.7
Shakespeare, W., 115–17, 123
Shepherd, M. H., 268n.5
Shorey, P., 291n.11
Shutt, R. J. H., 304n.3
Simonson, C., 270n.3
Simpson, M. N., 279n.15

Skeat, T. C., 105, 186–87, 303n.6, 321n.9
Skemp, J. B., 83, 84, 289n.1, 290n.1
Slack, K., 333n.2
Smet, A. O., 302n.3
Smith, R., 297n.14
Smith, W., 157
Smithers, G. V. 291n.5
Smyth, T., 41
Sodar, B., 308n.4
Soden, H. von, 223
Souilhé, J., 316n.4
Spiazzi, R., 268n.5
Spicq, C., 326n.2
Stallbaum, G., 290n.3
Staniforth, M., 328n.5
Steinmetz, P., 298n.1
Stephanus, 93, 169, 292n.2,n.4,n.7, 294n.2, 307n.11, 308n.4, 316n.5
Sticker, A., 266n.4,n.5, 267n.17
Stoessel, F., 298n.2
Stokes, V. T., 334n.22
Stone, M. E., 294n.1, 297n.19
Strong, J., 12, 293n.4
Sturz, F. C., 92
Suidas, 307n.11
Swete, H. B., 59, 251
Swinner, W., 319n.2

Takenaka, M., 21
Tatlow, T., 273n.14
Tavard, G., 34, 259
Taylor, A. E., 86, 290n.1,n.3
Taylor, J. V., 65
Taylor, T., 305n.5
Taylor, V., 56
Tedesche, S., 308n.4
Tetlow, E., 39
Thackeray, H. St John, 111, 121, 143, 296n.2
Thayer, J. H., 293n.6
Thieme, G., 314n.5
Thomas, J. D., 188, 321n.5, 322n.18
Thrall, M. E., 324n.3
Thurian, M., 35, 333n.6
Till, W. C., 319n.15
Tillard, J. M. R., 277n.62
Tischendorf, C., 223
Tödt, H. E., 54, 57, 58, 332n.4
Torrance, T. F., 255
Travis, S. H., 325n.10
Trench, R. C., 90, 93–94
Treu, K., 323n.25
Turner, E. G., 189
Turner, N., 63

Uhlhorn, G., 10
Underdowne, T., 118, 297n.16, 300n.2, 301n.1
Ussher, R. G., 299n.4

Valentini, G., 289n.15
Van 't Dack, E., 320n.6, 321n.11
Van den Born, A., 268n.5
Vandoni, M., 311n.3, 318n.8
Vaughan, D. J., 291n.10, 303n.11
Viereck, P., 320n.2
Vilborg, E., 132, 295n.3
Vinay, V., 267n.17
Vischer, L., 20
Vogels, H. J., 223
Von Hase, H. C., 268n.5
Vorgrimler, H., 43, 269n.32, 322n.9

Wace, H., 295n.4, 308n.4
Waddington, W. H., 314n.8
Wagner, H., 268n.5
Waldmann, H., 312n.3
Walker, D., 277n.62
Walker, R. J., 298n.2
Walkley, A. B., 296n.3
Warren, T. H., 291n.1,n.10
Way, A. S., 131
Weber, R., 308n.4
Wegener, E. P., 303n.6
Weil, J., 154
Weiser, A., 268n.5
Wellhausen, J., 52, 332n.4
Wendland, H. D., 268n.5
Wendland, P., 295n.2
West, S., 122
Westcott, B. F., 223
Weymouth, R. F., 299n.1
Whiston, J., 143–44
White, G. E., 288n.10
White, H. G. Evelyn, 319n.15
Whiting, H. J., 268n.5
Whitlam, G., 296n.4
Wichern, J. H., 9–10
Wiedemann, T., 293n.1
Wilcken, U., 127, 182, 320n.2,n.3
William of Moerbeke, 137
Williamson, G. A, 303n.2
Willis, W. L., 310n.3, 311n.2, 312n.6
Winninger, P., 42, 267n.17
Wolff, S., 296n.4
Woodham-Smith, C., 266n.3
Woozley, A. D., 78
Wrege, H. T., 57
Wright, F. A., 297n.16, 300n.2, 301n.1

Youtie, H. C., 180, 187, 299n.4, 311n.3, 321n.5

Zahn, T., 242
Ziebarth, E., 314n.5
Zorell, F., 293n.6